ROADSIDE GEOLOGY of Pennsylvania

Bradford B. Van Diver

ECHO POINT BOOKS & MEDIA, LLC

Published by Echo Point Books & Media
Brattleboro, Vermont
www.EchoPointBooks.com

All rights reserved.
Neither this work nor any portions thereof may be reproduced, stored in a retrieval system, or transmitted in any capacity without written permission from the publisher.

Copyright © 1990, 2014 Bradford B. Van Diver

ISBN: 978-1-62654-841-1

Cover image courtesy of the Commonwealth of Pennsylvania

Cover design by Rachel Boothby Gualco,
Echo Point Books & Media

Editorial and proofreading assistance by Christine Schultz,
Echo Point Books & Media

Printed and bound in the United States of America

Preface

A glance at the 1980 edition of the Geologic Map of Pennsylvania spread upon the wall gives reason enough for a book. The map is truly a work of art, with a marvelous combination of patterns and colors that dazzle as they reveal in a moment the incredible variety and intricacy of the state's geologic foundation. The colors, of course, are artificial, but the patterns they so beautifully outline, are the real erosional expression of hundreds of millions of years of geologic history.

On the ground, that spectacular Pennsylvania story is told piecemeal, but in much greater detail in its roadcuts, natural outcrops, landscapes, rivers, springs, mines, oil and gas wells. It is also revealed in its human history of migration, settlement, exploitation of the land and resources, and conflict.

That's what this book is about. It is more than just a description and explanation of roadcuts and other rocks visible from a passing car. It is equally a guide to many other facets of the state, including both geologic and human history, on and off the road. It is not comprehensive; that would require much too much space.

The roadguides are embellished with a number of "focal points," detailed discussions of geologic phenomena or off-the-road sites. You will recognize each of these departures by a boldface heading, and you will know you are back on the route description when you see a black dot at the head of the following paragraph.

This book is designed primarily for people with a burning curiosity about the geological world around them, but with little or no formal training in geology. Therefore, I have kept geological jargon to an absolute minimum and basic explanations to a maximum. Certain terms are unavoidable, and those are explained both in the text where first used and in the Glossary in the back of the book. For background, the book begins with extensive introductory material on geology in general and Pennsylvania geology in particular. I urge you to read this first before trying to follow the road guides; it will help you to read the rocks. For example, when you stand before, or drive past, a roadcut in Ordovician limestone, you will see not just gray stratified rock, but in your mind's eye will be able to envision a vast inland sea or a submerged continental shelf some 475 million years ago. Or folded and faulted sedimentary beds of Mississippian and Pennsylvanian age will bring forth a scene of continental collision between ancestral Africa and North America some 300 million years ago. Or when you view Precambrian gneisses and schists of the Pennsylvania Piedmont, you will see the core, all that's left of a billion-year-old mountain range. This is the essence of geology, bringing history and Earth dynamics to life.

Naturally I didn't obtain the basic information for this book all by myself. A book of this type leans heavily on the publications of many others. Fortunately my task was greatly facilitated by the Pennsylvania Geological Survey, the oldest (1836) and one of the best state surveys in the United States, which has a prodigious output of geological publications, both for the professional and layman. Most of my material is borrowed from those publications, including the 1980 edition of the Geologic Map of Pennsylvania, and also from Guidebooks of the Annual Field Conferences of Pennsylvania Geologists, from U.S. Geological Survey publications, and from the Centennial Field Guide, Volume 5, of the Northeastern Section of the Geological Society of America.

I am grateful to William D. Sevon of the Pennsylvania Geological Survey, Harrisburg Office, for helpful discussions and the loan of two photographs of the Conewago Falls potholes, and to Helen L. Delano of the Pittsburgh Office for some vital information concerning a couple of outstanding large roadcuts on I-79. Thanks also to Dr. C. Gil Wiswall of West Chester University for his help with the plate tectonics aspect of Pennsylvania geology.

My wife, Bev, and I had a lot of fun traveling the roads of this beautiful and fascinating state and visiting all the sites I describe. Bev is my indispensable helpmate. When we collect field data, I do all the driving and nearly all the observing while she records. She often sees and notes on her own, things she is curious about and that you, the reader, might also wonder about, and thus she makes substantial contributions to content. Her entries are all the better because she has no formal training in geology. She is also an English teacher, and when she takes my handwritten copy and puts it on our little computer, she is quick to spot difficult passages for me to rewrite. With her considerable help, we generally come out with a pretty readable product, and for this and all of her many contributions, I am deeply appreciative.

All of the photographs, except for the Conewago Falls potholes, are my own. So are the drawings, except for two depicting the ridges and Susquehanna Gaps north of Harrisburg. These are the superb artwork of my friend and former student, Theresa (Terry) Jancek. Many of my drawings, however, are less than original; they are adaptations, as noted, from illustrations that appear in the body of the literature.

In following this guide, let me urge you to stop your car, where you can safely and legally do so, to get closer to the geology of Pennsylvania and to ponder. You will find this a fascinating state, and the more you learn about it, the more you will be able to appreciate. Have fun!

Table of Contents

Preface iii

Getting Started 1
   Rocks — Fossils of Time 1
   The Pennsylvania Landscape 11
   Pennsylvania and the Ice Age 16
   Plate Tectonics 23
   Oil and Gas 35
   Coal 39

I. Allegheny Plateau 45
   I-70: West Virginia Border—New Stanton 49
   I-76: Ohio Border—Somerset 53
   I-79: West Virginia Border—Zelienople 59
   I-79: Zelienople—Erie 69
   I-80: Ohio Border—Brookville 77
   I-80: Brookville—Bellefonte 81
   I-81: Wilkes-Barre—New York Border 89
   I-84: Matamoras—I-380 Interchange 95
   I-90: Ohio Border—New York Border 99
   US 6: Scranton—Mansfield 105
   US 6: Mansfield—Lantz Corners 113
   US 6: Lantz Corners—Meadville 119
   US 15: Williamsport—New York Border 127
   US 62: Franklin—Junction US 6 131
   US 219: Bradford—McGees Mills 139
   US 219: McGees Mills—Maryland Border 145
   US 220: South Waverly—Williamsport 149

II. Valley and Ridge Province **159**
    I-70: Breezewood—Warfordsburg 163
    I-76: Somerset—Carlisle 167
    I-80: Bellefonte—Bloomsburg 179
    I-80: Bloomsburg—Delaware Water Gap 183
    I-81: Junction I-78—Wilkes-Barre 197
    I-380: Scranton—Crescent Lake 209
    US 15: Amity Hall—Williamsport 215
    US 209: Matamoras—Delaware Water Gap 221
    US 209: Stroudsburg—Jim Thorpe 227
    US 220: Williamsport—Port Matilda 233
    US 220: Port Matilda—Maryland Border 239
    US 322: Amity Hall—Martha Furnace 247
    PA 61: Sunbury—Frackville 253

III. Southeastern Pennsylvania **259**
    I-76: Carlisle—Philadelphia 261
    I-78: Junction I-81—Allentown 275
    I-81: Maryland Border—Junction I-78 279
    I-83: Maryland Border—Junction I-76 291
    US 15: Maryland Border—Amity Hall 295
    US 30: Breezewood—Gettysburg 303
    US 30: Gettysburg—Lancaster 311
    US 222: Lancaster—Allentown 317
    PA 9: Philadelphia—Junction I-80 320

Glossary **330**

References **337**

Index **340**

CENOZOIC

- **Q** — Quaternary, unconsolidated sediments
- **T** — Tertiary, unconsolidated sediments

MESOZOIC (Triassic and Jurassic (?))

- **TRd** — diabase dikes
- **TRs** — diabase sills and sheets
- **TRr** — sedimentary rocks, many redbeds, shales, sandstones, conglomerates

PALEOZOIC

- **P** — Permian Dunkard group, some coal
- **uP** — upper Pennsylvanian, contains most of west PA soft coal
- **lmP** — lower and middle Pennsylvanian, includes Llewelyn group in east PA which extends to uP and contains most anthracite coal
- **M** — Mississippian, undivided, Burgoon Pocono ridge formers, redbeds
- **D** — Devonian, undivided, abundant Catskill formation, redbeds
- **uS** — upper Silurian: shale, sandstone, some carbonate
- **lS** — lower Silurian: mainly Tuscarora and Clinton, or Shawangunk, ridge formers
- **Oh** — Hamburg Klippe: mainly shale and graywacke, low grade metamorphism, highly deformed
- **O** — Ordovician: undivided, abundant carbonates in lower and middle, sandstone and shale in upper
- **C** — Cambrian: undivided, many carbonates and quartzites

LOWER PALEOZOIC or PRECAMBRIAN

- **Xg** — mainly gneiss formed by metamorphosis of igneous intrusives
- **Xw** — Wissabricken schist and other, mainly metasedimentary rocks

PRECAMBRIAN

- **PC** — South Mt. metavolcanics, many varieties of gneiss in east PA

- bedrock formation contacts
- joints or faults, including thrust faults
- thrust fault enclosing Hamburg Klippe; barbs on side of overthrust slice
- syncline
- anticline
- syncline

approximate axes of rocks only shown on Allegheny Plateau, with outcrop patterns

Getting Started

ROCKS — FOSSILS OF TIME

The geologic story of Pennsylvania is written in its rocks. Each rock is like a page in time, filled with fascinating details about Earth, as surely as if they were printed. The story is there, but the language is different from the printed word. It is sometimes easy and sometimes difficult to decipher. Reconstructing Earth history is a matter of reading the rock records and placing them in their proper places and order of occurrence.

Reading the rocks is what geologists do, but it is not their world alone. With a little background, anyone can learn to do it. That's what this chapter, and, in fact, this book, are all about.

Theoretically, a rock record exists somewhere in the world for every step of the journey through geologic time; rocks were always forming somewhere. Conversely, no single place on Earth preserves a complete record of its geologic history. In our restless Earth, repeated mountain uplifts have placed rocks above sea level where some were destroyed by weathering and erosion. Their destruction produces gaps in the record, unconformities. The major task of geologists is to piece together, or correlate, the fragmentary record so as to develop a continuous flow of Earth history.

Correlation is the determination of time-equivalence between parts of the record preserved in different places. The sedimentary rock record, for example, may contain a limestone formation with distinctive fossils that can be identified in two different localities. If the older sedimentary rocks under the correlative limestone are best preserved in one place, and the overlying younger rock record is more complete in the other, the two records can be combined on the basis of the limestone equivalence. This produces a more complete geologic history and it is only one of a number of ways to accomplish correlation. This is relative dating: the two limestone exposures are the same age, but we don't know how old; the rocks below are relatively older, and those above are relatively younger. These methods work best with stratified sedimentary rocks.

Since the early part of this century, a variety of radiometric clocks has made possible absolute dating of rocks in terms of actual years of age. One process that sets the clock ticking is the crystallization of a mineral that incorporates within its atomic framework a small amount of radioactive uranium. The uranium will decay to non-radioactive lead at a steady rate. Knowing the decay rate allows us to measure the amount of remaining uranium relative to lead to determine the age of the mineral, in years.

Numerous radiometric clocks are now in use. Some are useful for very old rocks, others for relatively young ones. They are especially helpful in dating igneous activity or metamorphism. Radioactive carbon measures very recent events in sediments, sedimentary rocks, and volcanic rocks that contain plant or animal remains.

The combined data derived from relative and absolute dating, and correlation of rock records from all over the world, yield the modern Geologic Time Scale, a kind of calendar of Earth history. Its basic units, called periods, are divided according to the appearance or extinction of certain life forms. Periods are lumped into larger units called eras, and subdivided into smaller ones called epochs.

The rock record of Pennsylvania spans an unusually broad range of geologic history. In guiding you through its corridors of time, I refer repeatedly to ages of the rocks in terms of the periods, eras, and epochs of the Geologic Time Scale. Don't try to memorize all these strange names, but do refer to the chart from time to time. This is your map to navigate the vast seas of geologic time.

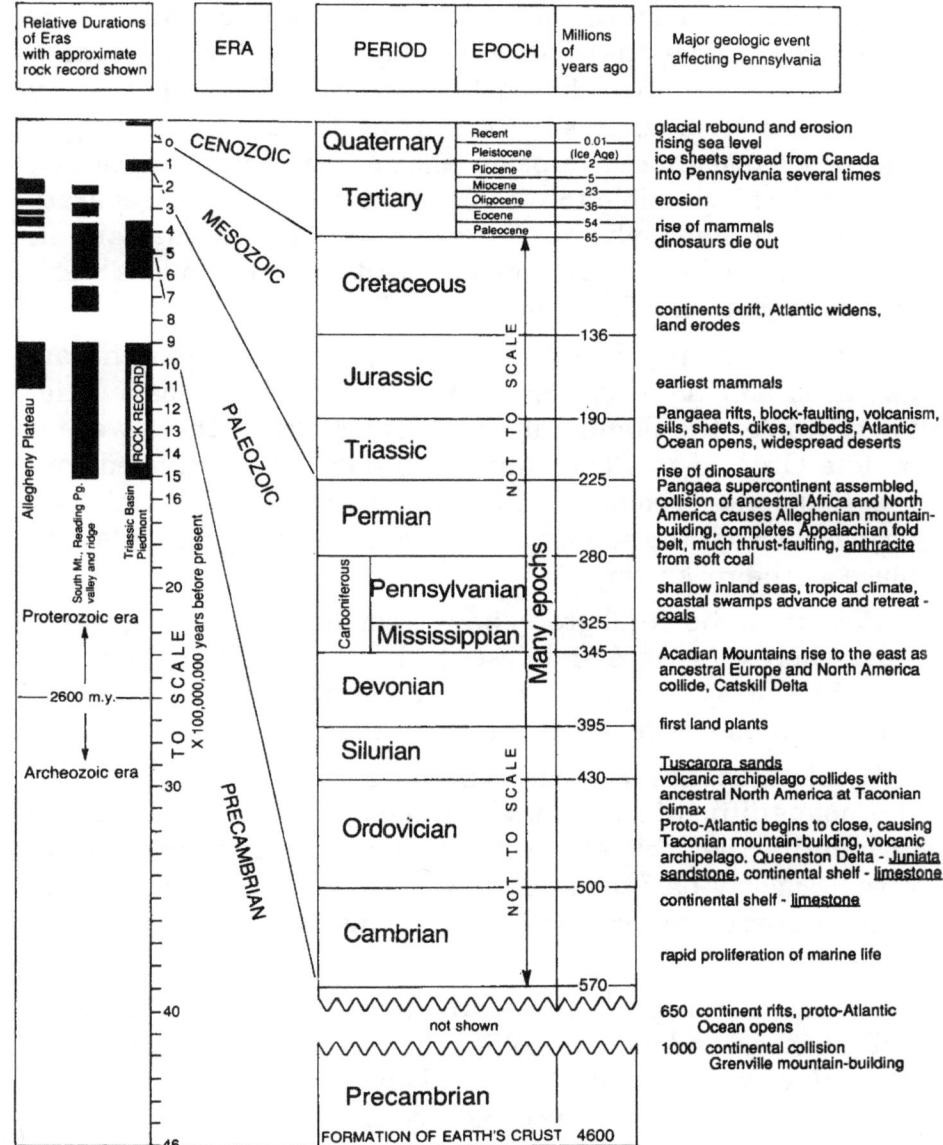

Geologic Time Scale

You will see the terms formation and group in many places in the roadguides that follow. Geologists use formation as a basic unit to classify rocks, particularly stratified sedimentary rocks. The term implies a certain uniformity of rock type, character, resistance to erosion, age and stratigraphic position that makes it possible to recognize the formation in separate outcrops. In Pennsylvania, for example, whitish quartzite in rather thin beds that jut from the crests of many ridges mark the lower Silurian Tuscarora formation.

A group is a larger unit that may have more diverse character, an assemblage of two or more formations. In Pennsylvania, for example, the dominant carbonate rocks of the lower-to-middle Ordovician Beekmantown group are subdivided into the Bellefonte, Axemann, Nittany, Larke, and Stonehenge formations; but the subdivisions and names vary from place to place as the rocks vary.

Both formation and group names come from the geographic sites where they were first described. The term formation may be replaced by the name of the dominant rock type, as in Tuscarora quartzite or Martinsburg shale.

What about the rocks? How can you read the stories in them? It is not as difficult as you might suspect; in fact, just being able to identify the rock and knowing how different rocks form reveals nearly half the story. So let's begin by looking at the different rock types you will see in Pennsylvania and how they formed.

Geologists recognize three broad classes of rocks: igneous, sedimentary, and metamorphic. The rock cycle illustrates, in a general way, how they form and interrelate. Despite their connotation of permanence, all rocks are temporary in the context of geologic time, subject to weathering, erosion, metamorphism, melting — recycling.

Sedimentary Rocks

Sedimentary rocks are most important in Pennsylvania for they underlie all but a small percentage of its land area. They formed from sediments, the unconsolidated debris produced by weathering and erosion of rocks of all kinds, mainly above sea

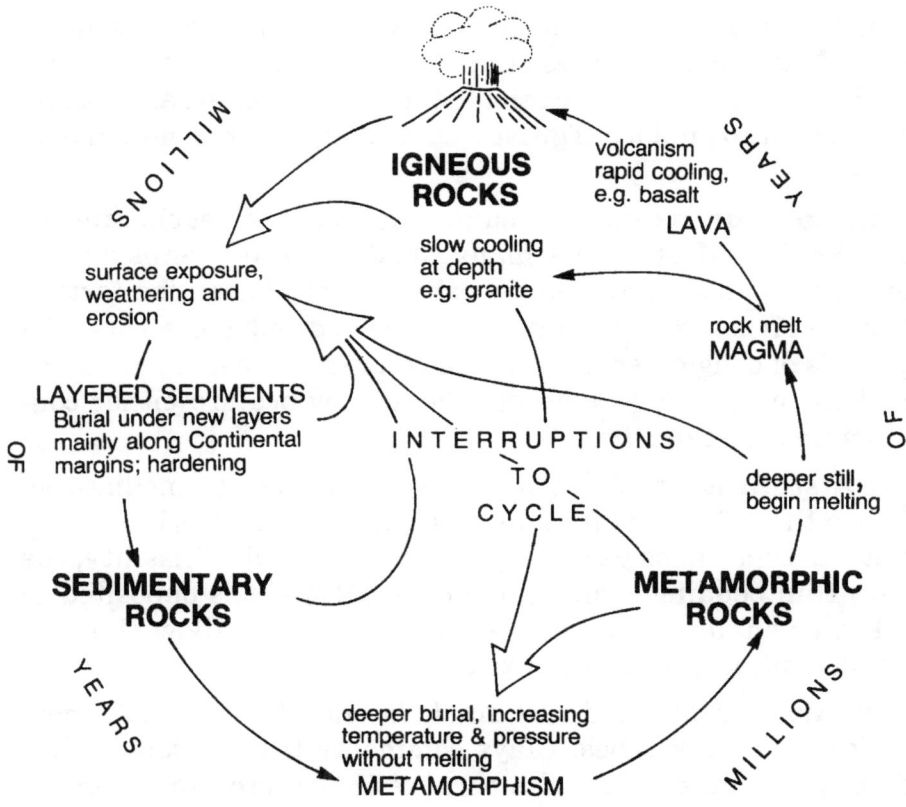

The rock cycle, showing the three major classes of rocks, how they form, and how they interrelate.

level. Rivers transported most of the sediments to the shallow seas that covered the region during much of the past and deposited them in flat layers. The strata piled up over many millions of years, and the buried sediments were cemented to rock by new minerals precipitated between the grains. The character of sedimentary deposits reflects the source rocks, environment, and modes of transportation and deposition. If the rock contains fossils, the remains of plants and animals, so much the better; they tell us of past life and evolution.

Conglomerate is consolidated gravel, generally with a wide range of grain size. Well-rounded coarse rock fragments record rolling and abrasion by streams or wave action. They represent deposition relatively close to their source because long transportation would grind the material down to sand or mud. Most conglomerates accumulated on or near land. Those in Pennsylvania resist erosion, stand up as ridges, and locally break up into jumbles of great blocks, that some people call rock cities.

Sandstone is cemented sand, like that of a beach, dune or river delta. Most of the many sandstones of Pennsylvania originated as shallow sea deposits, materials carried farther from their source, ground up and sorted out more than the gravels of conglomerates. The backbones of most of the ridges in Pennsylvania's Valley and Ridge province are either conglomerate or sandstone.

Siltstone is consolidated silt of a grain-size intermediate between that of a very fine sand or shale mud. It looks like shale, but can sometimes be distinguished by its lack of fissility, the very thin bedding common in shale. Most silt is deposited in lakes or shallow seas where calm conditions allow the suspended silt grains to settle to the bottom.

Shale, consolidated fine mud, is one of the state's most widespread rock types. Clays commonly travel even farther from their source than silts because they more easily remain suspended in the water; thus, they commonly represent somewhat deeper marine deposition. Shales are soft and tend to weather in recess beneath ledges of more resistant sandstone, conglomerate, limestone and dolostone caprocks.

Turbidite is a sediment or rock deposited from a turbidity current. These are muddy mixtures of sediment and water that are denser than the surrounding clear water; they flow under their own weight. Many turbidites have graded bedding in which grain size decreases gradually from the bottom to top of individual beds.

Conglomerate, sandstone, siltstone, shale, and turbidite are all products of mechanical erosion, transportation, and deposition and are called detrital rocks.

Limestone is made of the mineral calcite, a calcium carbonate dissolved from pre-existing rocks and then precipitated from solution, either directly or to form the hard parts of aquatic organisms, mostly in marine settings.

Dolostone is similar to limestone but is made of dolomite, a mineral like calcite, except in containing more magnesium. Apparently, most dolostone results from chemical reaction between seawater and limestone.

Many pure limestones and dolostones probably represent deposition farther from land than shale. Few, however, are pure. Limestones and dolostones are soluble in acidic water — rain water is generally acidic enough to dissolve them.

Coal is also a sedimentary rock. It forms from peat, a deposit of partially decayed plant matter.

Rock salt is a chemical sedimentary rock that was precipitated in natural evaporating basins like Great Salt Lake or the Dead Sea, so it is called an evaporite. Pennsylvania has large reserves of salt that are of great potential economic value, but all of it is deeply buried; you won't see it.

The other major classes of rocks, igneous and metamorphic, record Earth history too, but in different ways.

Igneous Rocks

Igneous rocks form when molten rock, magma, crystallizes. Crystallization of different magmas produces a wide range of intrusive, or plutonic, igneous rocks, of which granite is the most common. Intrusive rocks tend to be coarsely granular with interlocking textures. They typically consist of several different minerals.

If magma erupts at the surface, it becomes lava or volcanic ash. Crystallization of different lavas produces a wide range of volcanic igneous rocks. Basalt is the commonest volcanic rock. Volcanic, or extrusive, igneous rocks tend to consist of mineral grains too small to distinguish without a microscope. Geologists generally use the colors of volcanic rocks as a guide to their chemical composition. Most black volcanic rocks are basalt; rhyolite comes in light colors that suggest a bulk composition similar to granite. Each tells a different story.

Here are some of the most important igneous rock types:

Basalt is a fine-grained, dark gray to black rock composed principally of plagioclase and pyroxene, but you'll need magnification to identify these minerals. The dark color comes mostly from the black pyroxene, which contains abundant iron and magnesium. Basalt typically forms lava flows. Diabase, a rock of the same composition but composed of larger grains of plagioclase and pyroxene, is fairly common in eastern Pennsylvania. Nearly all of Pennsylvania's diabase formed during Triassic and Jurassic time. It occurs in the Gettysburg and Newark basins as dikes and sills. **Dikes** form as magma intrudes the country rock along fractures. **Sills** are sheets of igneous rock sandwiched between sedimentary beds. The many dikes in and near the Gettysburg and Newark basins fill joints that opened as the supercontinent of Pangaea began to split up, stretching the crust, the same process that produced the basins.

Gabbro is a dark, coarse-grained intrusive rock with essentially the same mineral composition as basalt and diabase. Gabbros occur in the Wilmington complex and Baltimore mafic complex in southeastern Pennsylvania.

Granite is a generally pale rock composed mostly of fairly large crystals of quartz and feldspar, with variable, but small, amounts of dark minerals, such as black biotite mica or hornblende. **Rhyolite** is a pale volcanic rock that forms when molten granite magma erupts. **Pegmatite** is a granitic rock of very coarse grain-size that occurs sparingly in short dikes cutting the Precambrian to early Paleozoic rocks of the Piedmont province. Some pegmatites contain exotic minerals in large well-formed crystals of great interest to collectors.

Metamorphic Rocks

Nearly all of the exposed metamorphic rocks of Pennsylvania are in the South Mountain, Reading Prong, and Piedmont provinces, and the Anthracite region in northeastern Pennsylvania. Elsewhere they are restricted to narrow contact zones adjacent to Triassic and Jurassic dikes.

Metamorphic rocks form through recrystallization of older igneous, sedimentary, or metamorphic rocks. The transformation, called metamorphism, may be roughly compared with changes in the interior of a glacier. As a glacier moves, its millions of interlocking crystals of ice are repeatedly broken, distorted, and recrystallized. Glacier ice is generally hard and brittle; yet the net result of all these slow movements is plastic deformation of the whole mass, as though it were taffy. Similar changes occur in hard rock, but under much higher pressure and temperature, generally requiring deep burial.

Practically all of the metamorphism in Pennsylvania is regional, that is, it affects a large area. Regional metamorphism takes place in deeply buried rocks at the edges of continents and is often associated with igneous intrusion and volcanism. In general, the deeper the burial the more intense the metamorphism. Regional metamorphism typically involves pervasive microscopic shearing and granulation of the rock, while new minerals crystallize along the shear planes. Many of the new mineral grains grow parallel to each other, which is the principal cause of the strong layering that is the trademark of these rocks. The minerals in regional metamorphic rocks are keys both to the original rock types — thus the pre-metamorphic history — and to the temperatures and pressures of the metamorphism.

Metamorphic temperatures may also exist near igneous intrusions. Their effect on the intruded rocks, called contact metamorphism, generally does not extend more than a few feet from the intrusion. These rocks rarely show much evidence that deformation accompanied their metamorphism. Some show drastic changes in composition, the result of mineral matter moving from the magma into the intruded rock. The metamorphic rocks of eastern Pennsylvania all recrystallized during Precambrian or early Paleozoic time.

Here are some of the major rock types you will see:

Schist is the general term for a group of rocks that tend to split, or cleave, along the foliation surfaces established as the mineral grains grew in parallel alignment. Mica schist, the most common type, often sparkles on cleavage surfaces because of the parallel arrangement of shiny flakes of mica, like matted leaves on the forest floor. Most schists are products of medium-to-high grade metamorphism of shales. Many contain knots of red garnet or crystals of distinctive minerals that help define the temperatures and pressures of metamorphism.

Schists full of dark-green or black hornblende are usually called amphibolites. Most of those recrystallized from basalt.

Gneiss also contains a streaky foliation, but generally consists of larger mineral grains than those in most schists. The coarse texture of gneiss may be inherited from original igneous intrusive rocks, or produced by intense metamorphism of original sedimentary rocks. Most gneiss contains light and dark bands of different mineral composition. These bands are an important key to distinguish gneiss from igneous intrusive rocks of similar mineral composition.

Quartzite is the hard and brittle metamorphic equivalent of quartz sandstone, distinguished from it by window-glass-like curving fracture and lack of sandy feel.

Marble is metamorphosed limestone or dolostone. Most of it in Pennsylvania is massive, coarsely granular and almost white. The Franklin marble contains scattered flecks of finger-smudging, silvery graphite. Outcrops or hand specimens often sparkle as sunlight reflects from shiny surfaces of calcite or dolomite crystals.

Serpentinite is a dense, dark green to black rock formed from iron- and magnesium-rich rocks like those of the modern ocean floor. Elongate, lenticular slivers of serpentinite of the Pennsylvania Piedmont may well be pieces of ancient ocean floor that moved upward along deep-rooted faults and metamorphosed. Verde antique is the name for polished serpentinite used as ornamental stone.

Slate, the product of low-grade metamorphism of shale, cleaves into thin flat sheets used for flooring and roofing. It is fairly abundant in Lancaster, York, Lehigh, and Northampton counties.

Phyllite is a very fine-grained schist principally formed from shale by metamorphism beyond the grade of slate. It has a somewhat irregular cleavage and a pearly appearance produced by minute crystals of metamorphic mica. Most of the phyllite in Pennsylvania is interlayered with marble in the Cambrian Elbrook formation on the western side of South Mountain in Franklin and Cumberland counties. The Octoraro phyllite that occupies much of the Piedmont south of Lancaster and Chester is pale green.

THE PENNSYLVANIA LANDSCAPE

The colored geologic map of Pennsylvania, published by the Pennsylvania Geologic Survey shows bedrock divisions that are also physiographically distinctive provinces.

The landscapes of most of the state are products of differential erosion of hard and soft sedimentary rocks. Where these rocks lie flat, or nearly so, as on the Allegheny Plateau, the topographic contours follow closely the outcrop patterns of individual strata. Outcrops follow the trends of the streams which cut into and expose the lower tiers of the layercake. In the Valley and Ridge province where the rocks are folded, differential erosion leaves the more resistant layers standing, while the weaker rocks are reduced to a low level. The result is an outcrop pattern and a landscape that beautifully reflects this folding.

These two sharply-defined physiographic provinces make up about five-sixths of the land area of the state. The remaining

Physiographic provinces of Pennsylvania. Note how closely province boundaries follow geologic boundaries shown on the following bedrock map. —Adapted from Pennsylvania Geological Survey (1982).

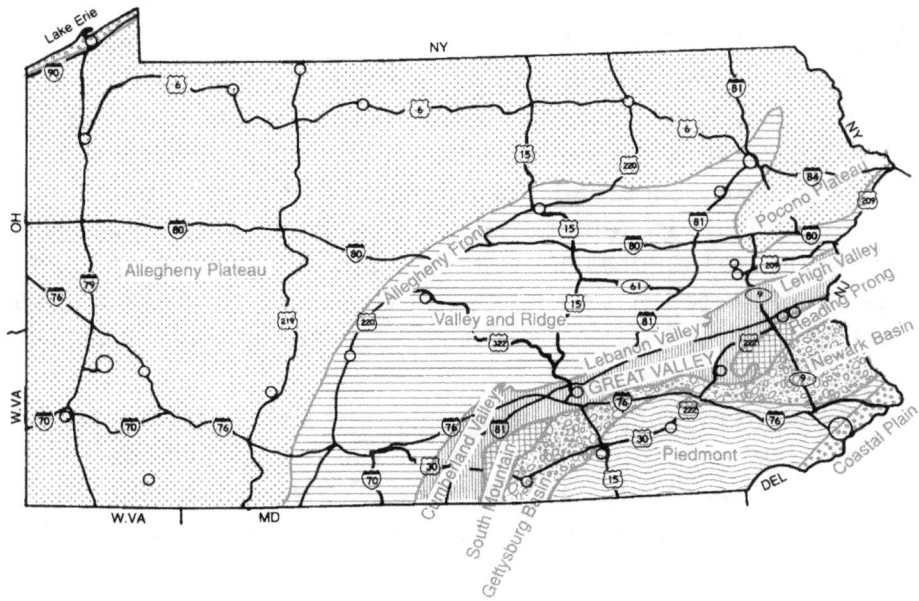

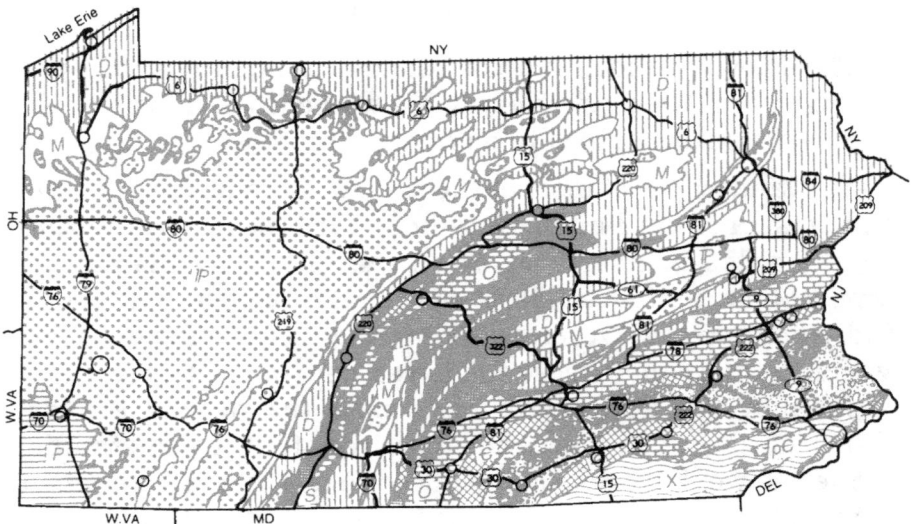

General geologic map of Pennsylvania. Symbols indicate ages in terms of geologic periods, as shown inside the front cover. —Adapted from Pennsylvania Geological Survey (1982).

bedrock-controlled provinces, all in southeastern Pennsylvania are the Great Valley, Triassic Lowlands, Piedmont, Reading Prong, and South Mountain. In addition, there is a narrow strip of Atlantic Coastal Plain in the extreme southeastern edge of the state and a similar strip of Erie Lowlands in the extreme northwestern part.

The Allegheny Plateau is part of the much larger Appalachian Uplands province that extends from western New York in a south-southwestern direction all the way to central Alabama and forms a broad western rampart to the Appalachian mountain belt. In Pennsylvania, as in New York, the topography of this region does not suggest a plateau, for streams have deeply dissected it and glacial erosion and deposition have further modified part of it. In general, the surface is rather mountainous.

The plateau label does accurately reflect the condition of the bedrock, which in outcrop consists of flat to slightly folded sedimentary beds of Silurian, Devonian, Mississippian, Pennsylvanian, and Permian ages. The folding is too broad and gentle to be visible in most outcrops. The anticlines are economically important, for they are the sites of the oil and gas fields of western Pennsylvania. The folds trend parallel to the much more severe folds of the Valley and Ridge province and have the same origin.

The eastern border of the Allegheny Plateau is marked by a prominent scarp called the Allegheny Front that faces the Valley and Ridge province and is carved from upturned beds. The same scarp continues south and north over almost the entire length of the Appalachians. In New York, for example, it is the 3000-foot eastern front of the Catskills.

The Valley and Ridge province proper is the backbone of the Appalachian Mountains that stretches from the Hudson Valley of New York to central Alabama. The term refers to that part of the mountain chain that is dominated by folding and thrust faulting. In Pennsylvania, it appears as a series of long, parallel, sharp-crested ridges that repeatedly double back on themselves and are separated by broad, rather flat valleys. In satellite imagery, the ridges appear to wrap gracefully around the southeastern corner of the state and then curve smoothly

Drainage basins of Pennsylvania. Note the strong alignment of streams in the Valley and Ridge province controlled by geologic structures, and the randomness of flow directions in the west, where the bedrock is only weakly deformed.

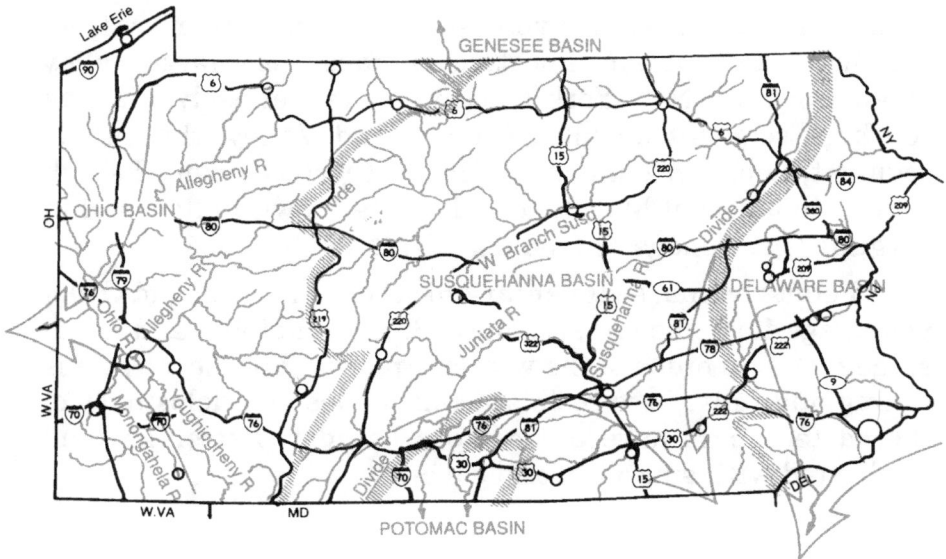

back toward north. It is a region of many breathtaking overlooks where the average height of the mountains is 800 to 1000 feet, with a maximum of 1800 feet between ridge-top and valley floor. The ridges are held up mostly by beds of resistant sandstone, quartzite, and conglomerate, separated by less resistant shales, siltstones, and in the older Paleozoic strata, limestone and dolostone. Erosion has truncated the upturned hard beds and scooped out the soft materials in between. Just how that may have happened deserves further discussion.

The erosional history of Pennsylvania landscapes has been the subject of heated controversy for more than 100 years. The debate has focused on the peculiar drainage patterns and the even summit elevations of the Valley and Ridge province.

The drainage is unusual in that several major streams cut right through the resistant ridges, creating water gaps. Pennsylvania is, in fact, a showcase of water gaps and also wind gaps. Presumably, the latter originated as water gaps but later were abandoned by the streams that cut them.

How did the streams manage to breach the hard rock barriers? Streams aren't supposed to do that. They're supposed to flow between the hard ridges, hollowing out the softer rocks between while tributaries drain the flanking slopes. These questions will be discussed in the text.

The Great Valley is a topographically distinct part of the Valley and Ridge province that is also more intensely folded, highly cleaved, and extensively faulted. The name derives from the fact that it is an almost continuous, very large valley extending from New York to Georgia. In Pennsylvania, the southern segment is called Cumberland Valley, the middle segment, Lebanon Valley, and the northern segment, Lehigh Valley. In adjacent Maryland, it is Hagerstown Valley, and in Virginia, the famous Shenandoah Valley.

In Pennsylvania, sedimentary rocks of Cambrian and Ordovician age floor the Great Valley. Shale, siltstone, and sandstone make up much of the western part where the surface is rolling and hilly. Less resistant limestones and dolostones of the eastern part have eroded to a broader, flatter lowland perforated with sinkholes and caves.

South Mountain and Reading Prong provinces both expose Precambrian rocks. South Mountain is the northern tip of the Blue Ridge province that continues to Georgia and ranges from 5 to 50 miles wide. It locally consists of several ridges, and contains the highest mountain in the East, Mt. Mitchell, 6684 feet, in North Carolina. The Reading Prong is similar, but less prominent. It is on the southwestern end of a Precambrian mass that continues across New Jersey as the Ramapo Mountains and southeastern New York as the Hudson Highlands. It also shows up in scattered outcrops in the Berkshires of Massachusetts and the Green Mountains of Vermont with an intermittent cover of Paleozoic rocks. Lower Paleozoic rocks drape over the flanks of both South Mountain and Reading Prong.

Only a small part of the Piedmont province proper crosses Pennsylvania; it extends from southern New York to Alabama and attains a maximum width of 125 miles. It is like a gigantic stair tread on the eastern side of the folded Appalachians. The Pennsylvania Piedmont consists almost entirely of Precambrian and lower Paleozoic rocks. Most of the rocks are metamorphic, with complex deformation that generally is truncated by the erosion surface. Gneisses, schists, and quartzites underlie hillier parts, while less resistant marbles form valleys, some of which are water reservoir sites.

In Pennsylvania, as elsewhere, the Piedmont surface slopes gently toward the coast. Its eastern limit, called the Fall Line, is where it dives under the gently-dipping Cretaceous strata of the Coastal Plain province, that lie unconformably on top of it. The Fall Line passes through Philadelphia in a northeastern direction so that only a narrow strip of the Coastal Plain lies within the state.

The Triassic Lowlands are a downfaulted segment of the Piedmont filled with predominantly red Triassic sedimentary rocks shot through with diabase sills, sheets, and dikes. The northeastern half in Pennsylvania is part of the Newark Basin which continues across northern New Jersey and terminates in Rockland County, New York. The southwestern half, called the Gettysburg Basin, terminates in adjacent Maryland. Other similar basins are the Hartford Basin in Connecticut and

Massachusetts, the Culpepper Basin in Virginia, and the Deep River Basin of North Carolina. All are roughly parallel to the modern east coast and all formed as the supercontinent of Pangaea split up, beginning about 200 million years ago. Crustal stretching caused block faulting, and the down-dropped block basins filled with sediments. The faulting also initiated pressure-relief melting in the asthenosphere, and basaltic magma intruded the sediments. In Pennsylvania, the Triassic beds dip toward the north and northwest to where they are broken along steeply-inclined faults.

The Triassic Lowland surface is a rolling plain 400 to 600 feet above sea level, interrupted by ridges composed of the more resistant diabase that locally rise to 1200 feet.

The Lake Erie Plain is a strip two to five miles wide along the lakeshore in northwestern Pennsylvania. It is a miniscule segment of the Central Lowlands province that includes the St. Lawrence, Champlain, and Hudson-Mohawk lowlands and the Ontario, Erie, Huron, and Michigan Lake basins; it extends westward to the Great Plains, southward to Texas, and northward to Saskatchewan. To be sure, there are differences among these several regions, but they all consist of low, fairly even topography, built upon nearly flat-lying sedimentary strata. In Pennsylvania, the surface gradually rises from 572 feet at the lake to about 800 feet at the foot of the Erie scarp. The scarp borders the Allegheny Plateau, broken only by beach ridges of glacial lakes; Wittlesey is the higher and Warren is the lower.

PENNSYLVANIA AND THE ICE AGE

During Pleistocene time, vast ice sheets repeatedly spread across Canada, the northern United States, and northern Europe. In so doing they greatly modified the landscape. No one knows why the climate fluctuated between glacial and interglacial conditions.

Ice reached Pennsylvania during the two most recent glaciations, covering the northwest and northeast corners of the state. Other states, particularly in the mid-west, contain evidence of earlier advances. If ice also reached Pennsylvania during those glacial episodes, the evidence has been destroyed.

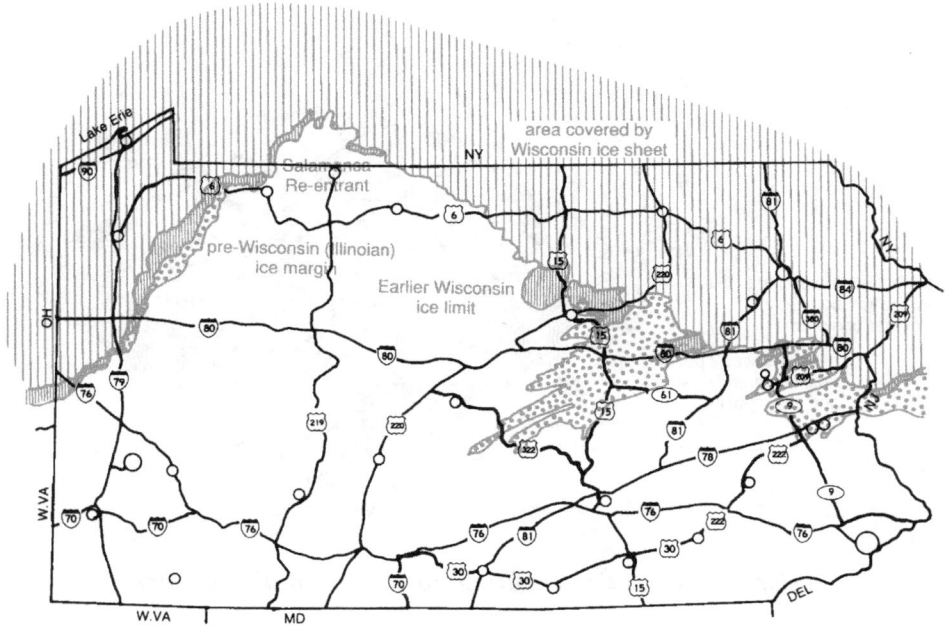

Glacial map of Pennsylvania, as indicated by drift deposits. Drift left by the earliest glaciation is preserved only where later glaciers did not override it.
—Adapted from Pennsylvania Geological Survey (1982)

Pennsylvania's glacial landscapes are in places distinctly, and in other places only subtly different from those of the rest of the state. To understand why, let's make an imaginary summer visit to the lower margin of a modern valley glacier in Alaska, one that moves like a ribbon down a mountain valley and issues onto a plain. What do we see? Lots of dirty ice, so loaded with rock waste as to almost completely conceal the ice itself. In front of the glacier are chaotic mounds of debris, called till or moraine, containing rocks of every size, most of them angular, many with scraped and polished surfaces, in a matrix of sand, silt, and clay. There are hummocks and hollows, some ponds, and some half-buried blocks of ice beyond the glacier terminus. Meltwater everywhere drains solution-widened crevasses, gullies the glacier and till, forms ponds and issues from tunnels at the base of the glacier. The water is milky because it is loaded with rock flour formed by milling of ice-bound rocks against each other and bedrock. Multiple braided

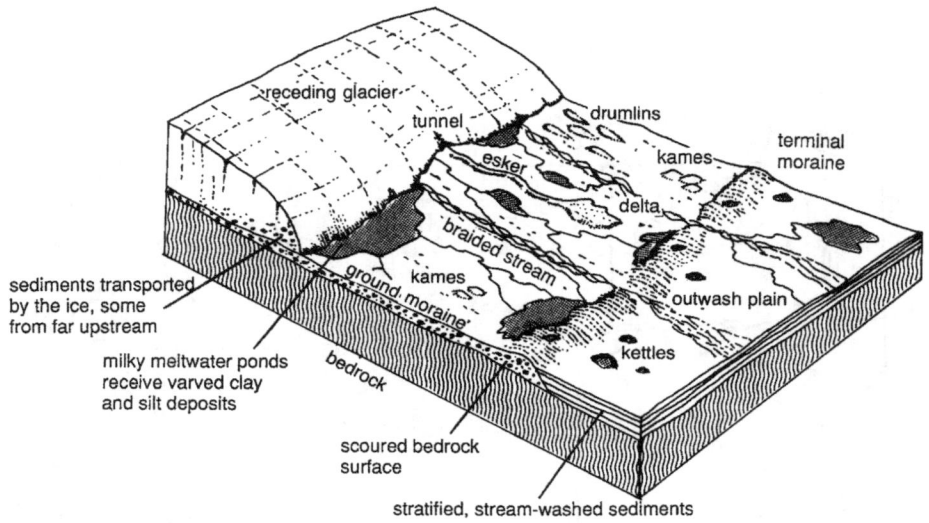

Erosion and deposition resulting from continental glaciation.

channels carry the sediment-choked melt away from the moraine across a gently sloping outwash plain from which a few mostly buried icebergs project. Farther downslope is a large milky-green lake dammed by another terminal moraine that must have formed long ago at an earlier stand of the glacier.

Now let's take a time leap forward and imagine the glacier gone, rapidly melted after its long still-stand. What do we see? The terminal moraine now is a ridge marking the former ice boundary. It is even more hummocky than before because the buried ice has melted creating new hollows called kettles. Some are filled with water, thus kettle lakes, contained by sediments with so much flour, or clay and silt that water can't drain through them. The surface is stony and messy and little vegetation has managed to take hold in it. Viewed from atop the moraine the land upstream is flatter and lower, but also littered with stony ground moraine that was either smeared under the moving ice or dumped in place when the glacier melted. Many of the stones, called erratics, are foreign to this place, transported long distances by the ice and unlike the surrounding bedrock. Here and there are clusters of mounds, kames, formed where streams entered temporal ponds in, alongside, or near the ice, and dumped their sediments in deltas. Digging into one of these, we see sediments different from the moraine; they're relatively clean, sandy, stratified, cross-bedded, better size-sorted, and the gravels are better-rounded. In fact, they look just like stream sediments, which is what they are.

In place of streams that flowed under the ice are long, narrow, meandering ridges of sand and gravel, eskers, marking the former channels. They are now partly covered with a stand of white pine, which likes the sandy soil. From the air the eskers look like rivers of sand, an apt expression, since they are river deposits. We also see a number of streamlined hills, drumlins, elongated in the former direction of ice flow. Digging into them we find till again, and we reason that these were mounds that were reshaped under the ice.

Bedrock outcrops on the valley floor and sides show surfaces polished and scratched by the rock waste in the passing glacier. The valley itself was gouged in the process, making it wider and deeper with steep walls and flat floor. Glacial erratics are everywhere.

Looking the other way, downstream from the terminal moraine, the hummocks grade quickly to an outwash plain pockmarked with kettle lakes. Farther down, there is an extremely flat, dusty surface in place of the former, moraine-dammed lake. Digging there, we find varved clays, composed mostly of glacial rock flour deposited from meltwater. In the flowing, turbulent streams, the finely pulverized rock flour remained in suspension, but it settled in the quiet water of the lake. Looking more closely, we see that the varves are like flat tree rings, each consisting of light gray clay that grades to a thin dark band at the top. Each represents one year of accumulation; the thicker, lighter part settled out in summer, and the thin, dark part in winter, when relatively little new sediment entered the lake. The exceptionally fine materials that did settle in winter were heavily charged with organic matter, which makes them dark. The varves vary in thickness because the climate changed from year to year; the thicker ones correspond to warmer years when more ice melted and more fine sediment reached the lake.

The example I just described is a valley glacier, not an ice sheet like those that spread across vast expanses of land during the ice ages. Valley glaciers are moving rivers of ice that form in the mountains over long periods during which only part of the winter snows melt each summer. The snow converts to ice as it is buried. The ice piles up to the point where it cannot support its own weight and begins to flow downslope; a glacier is born. Rocks fall into it, the ice plucks more from the valley floor and sides, and all become part of the moving mass. Like a giant grist mill, the glacier grinds the rocks together and against bedrock of the valley floor and sides, generating copious amounts of rock flour. The glacier continues snaking downvalley until it reaches a level where the ice melts as fast as it comes in from upstream. If the climate stabilizes for long enough, the rock waste carried to the terminus will melt out in heaps forming a moraine. If the climate cools over a long time, the glacier will expand downslope, overriding the moraine, possibly remolding some of it to drumlins. If the climate warms, the ice margin will recede upvalley, seeking a new balance between supply and melting.

Ice sheets form in high latitudes where there may be few, if any, mountains. The ice simply piles up until it begins to spread radially. To use a time-worn analogy, it creeps outward like pancake batter poured onto a hot griddle. Otherwise, the mechanisms and erosional and depositional features are basically similar to those of valley glaciers. However, the outer limits of the ice are controlled less by altitude than by latitude.

At the height of the ice ages, ice overran the mountains of the northeastern United States, grinding them somewhat and rounding them off. Many valleys that were more or less aligned with the ice flow were also gouged by tongues of ice that moved independently at the base of the glacier, producing erosional and depositional features similar to those of valley glaciers. Such valleys are preserved in the glaciated regions of Pennsylvania, but none as striking as the Finger Lakes of New York.

The basins of Lakes Erie and Ontario served as spreading centers. Apparently, the ice piled up in these depressions and then moved radially from them as if from separate ice sheets. Perhaps the best evidence is in New York, where the long axes

Pleistocene glaciation of North America, showing ice flow lines derived from drift deposit forms and glacial scour features.

of thousands of drumlins radiate from Lake Ontario. The effects in Pennsylvania are less evident.

In the waning stages of the last ice age, the ice margin eventually melted back to the Lake Erie basin while the St. Lawrence outlet for the Great Lakes was still blocked by ice. Meltwater, banked against the glacier on the north and high ground on the south, filled the basin to a level much higher than the present Lake Erie, forming glacial Lake Wittlesey. At a later stage, the water level dropped and stabilized again, forming Lake Warren. Today, inland from Lake Erie in Erie County, remarkably well-preserved beach ridges record these two ancient lakes.

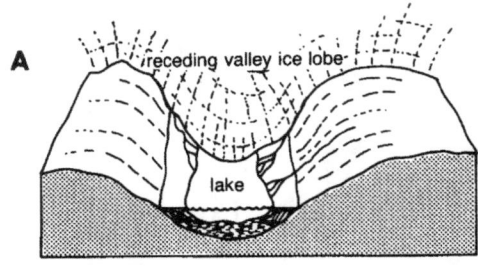

A. Continuous strips of delta sediments form from meltwater streams issuing from the sides of a valley lobe into a lake as the ice recedes upvalley. Such deposits are called kame terraces.

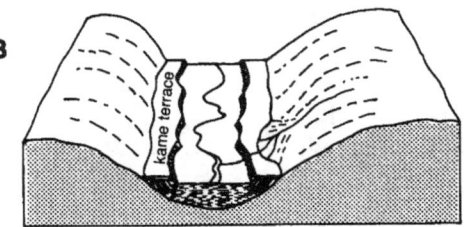

B. After the ice is gone, lack of meltwater greatly diminishes stream flow leaving kame terraces standing above the meltwater lake-and-outwash sediments plain.

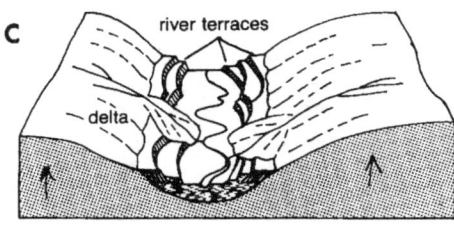

C. Glacial rebound causes the river to cut down, producing ever lower floodplains. Partial removal of floodplains and underlying deposits leaves a multiply-terraced valley.

Proposed origin of terracing on Pennsylvania's major rivers.
—Adapted from P.J. Fleischer, 1977

All glacial sediments, including those deposited directly from the ice and those reworked by streams, are referred to as drift. Northeastern Pennsylvania contains one important type of drift not mentioned above that forms striking terraces along the Susquehanna, Delaware and other major rivers. This is material deposited by the streams when they were major drainageways for meltwater from the last receding ice sheet. The valleys filled to considerable depths, and the terracing probably began long after the glacier disappeared, as a result of uplift of the land. At the height of the last ice age, the mass of the ice depressed the land. When the ice melted, the land gradually rebounded. The rivers responded by cutting downward into the drift and moving it downstream to Chesapeake Bay and the continental shelf. The removal has not been complete. Instead the rivers snaked from side to side in everchanging meanders as they cut into the fill, creating new floodplains at ever lower levels and incompletely carving up the old ones. So patches of the old levels survive as flat terraces on the valley sides. Much of Philadelphia is on these terraces.

PLATE TECTONICS
Pennsylvania and the Great Compression

Pennsylvania is only a small part of a very large collision margin mountain belt - the Appalachians. This great linear range extends continuously for nearly 2000 miles from Newfoundland to Alabama and ranges from 70 to 390 miles wide. It is part of a much larger mountain system that split up into pieces that dispersed with the moving continents during the last 200 million years. The Caledonian mountains of Great Britain, eastern Greenland and northeastern Scandinavia were originally contiguous with the Appalachians. Other mountains of the same age in western Europe, northwestern Africa, and the Canadian arctic may have been separate branches of the system. The Ouachita Mountains of Arkansas and Oklahoma and the Marathon Mountains of Texas may also have been continuous with the southern Appalachians. That great mountain belt formed by collision between large tectonic plates with both oceanic and continental components, a process that generally requires many millions of years.

The crust responds to such collisions in various ways. Layered rocks corrugate into alternating ridges and troughs called anticlines and synclines. Gently dipping thrust faults may slice the rocks and pile the slices high on top of each other. These two mechanisms, in particular, thicken the crust so it presses deeper into the underlying mantle while mountains rise above. In a sense, the mountains resemble icebergs with deep roots floating in the Earth's mantle. In the mountain core, where temperatures and pressures are high, the rocks become metamorphosed. Some may even melt, generating magma that subsequently rises into the higher rocks and may even break through the surface in volcanic eruptions. Thus, rock deformation, metamorphism, igneous intrusion, and volcanism are all closely related in compressive mountain-building. The collective term geologists use for all these processes is orogeny.

To fully understand Appalachian mountain-building, let us first consider the internal structure of Earth and second, an area of geology called plate tectonics. We have long known from the study of earthquake waves that our planet has a solid, more or less spherical, core about 1320 miles in diameter, most likely composed of iron. Around this are concentric shells of, first, a liquid outer core approximately 1440 miles thick, then the mostly solid mantle, about 1740 miles thick, that comprises the bulk of Earth's mass. The section of mantle between 60 and 150 miles below the surface, asthenosphere, apparently is partly liquid and mobile. By contrast, the thin outer shell of Earth, the lithosphere, is rigid, and moves on the weak rocks of the asthenosphere. The uppermost lithosphere is crust, a thin crust of dense basalt under the oceans, a much thicker crust of lighter granite and related rocks under the continents. The behavior of the lithosphere and asthenosphere are of most interest in explaining Appalachian mountain-building, and for that matter, plate tectonics in general.

Now, plate tectonics. The whole lithosphere is broken into a small number of enormous slabs, tectonic plates, that fit together like a global jigsaw puzzle. Most plates contain areas of both oceanic and continental crust.

The North American plate encompasses North America, Greenland, the western half of Iceland, and the half of the north Atlantic Ocean west of the mid-Atlantic ridge. The South American plate includes all of that continent and the half of the south Atlantic west of the mid-Atlantic ridge. The African plate includes that continent and large portions of the ocean basins on all sides of it.

The lithospheric plates are mobile and have been moving slowly during much of geologic time, constantly changing the face of the Earth. Plate boundaries are ruptures, along which the plates move away from each other, toward each other, or horizontally past each other.

A prime example of a pull-apart plate boundary is the great submerged mid-Atlantic ridge that traverses the full length of the Atlantic ocean. The mid Atlantic ridge has a deep rift valley along its crest that opens as the plates on opposite sides pull away from each other. As the plates separate, basalt magma

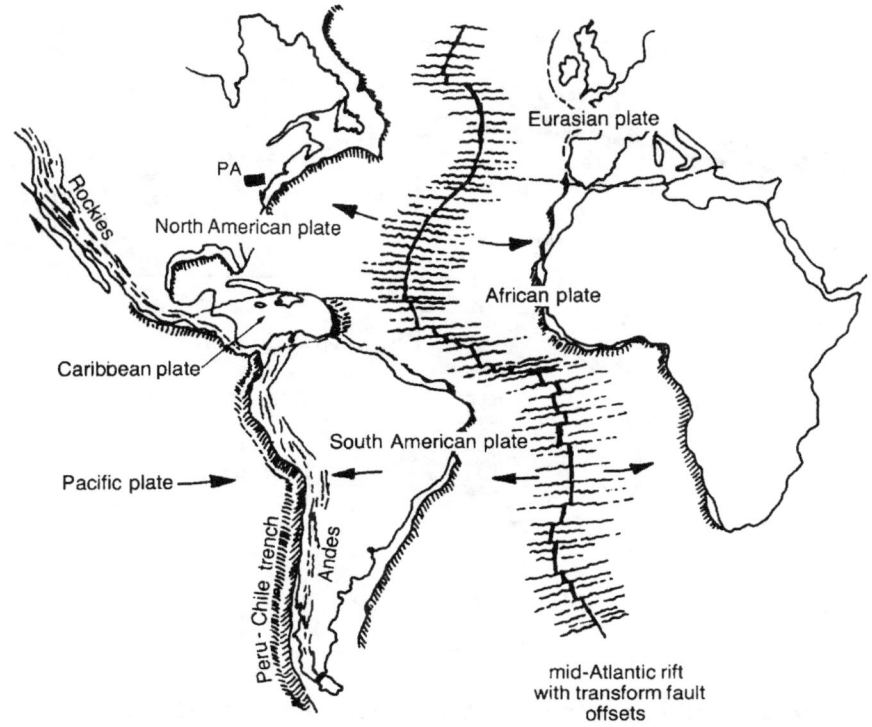

Modern plate tectonic geography of the Atlantic Ocean basin and bordering continents.

rises from the asthenosphere along deep fractures, and erupts onto the floor of the rift valley to become new oceanic crust. We can see that process at work in Iceland, the only place where rift lava erupts above sea level. Elsewhere, the oceanic ridge system is submerged, as is most of the oceanic crust.

Mount St. Helens and the other Cascade volcanoes of the Pacific Northwest coast lie parallel to a converging plate boundary where the dense oceanic lithosphere of the Juan de Fuca plate descends beneath the edge of lighter granitic rock of the North American continent. The descending slab of oceanic lithosphere heats as it sinks, and finally begins to melt, forming new magma for igneous intrusions and volcanic eruptions. Called subduction, this is the common process at convergent boundaries, where two plates collide. Subduction destroys old oceanic crust.

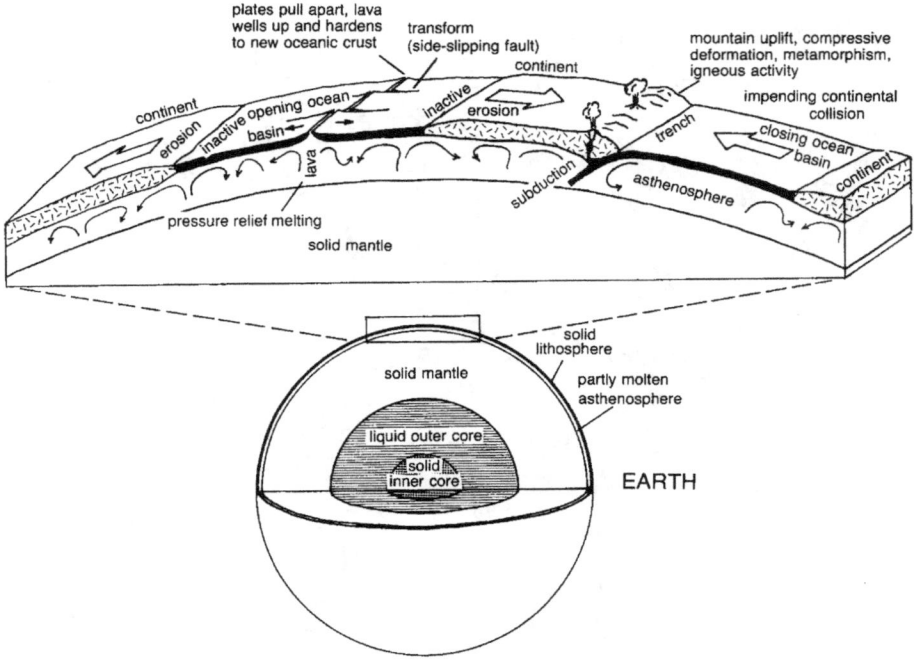

These are some of the major elements produced by movements of tectonic plates over the partially molten and mobile upper mantle, called asthenosphere.

The San Andreas fault is an outstanding example of the type of boundary where plates slide sideways past each other. A series of jerks and pauses accompany the movement of western California, part of the Pacific plate, as it slowly drifts north past the rest of the state and the continent. That movement provides some basis in fact to the fears of Californians that a slice of their earthquake-wracked state may one day push off to sea. Millions of years from now the long slice of California west of the San Andreas fault may actually detach from the rest of the state and move north towards Alaska with the rest of the Pacific Ocean floor.

Movements of the plates with respect to each other have changed through time. Ocean basins have opened and closed as the riding continents alternately rifted and collided. Each rifting produced land masses shaped differently from those that assembled earlier because each rupture occurred along a different line. Most of today's continents, therefore, are assemblages of fragments of earlier continents as well as new igneous and metamorphic crust generated by mountain-building at convergent margins.

The rocks and landscapes of Pennsylvania's mountains are products of four mountain building orogenies, each followed by a lengthy period of crustal stability and erosion.

We begin with the earliest event, the Grenville orogeny of about one billion years ago, late Precambrian time. This probably resulted from collision between a North America much smaller than the continent we know, and another continent. It produced a range of high mountains, the ancestral Adirondacks, now so deeply eroded that only their roots remain.

Those roots are intensely deformed igneous and metamorphic rocks. Vast exposures are in the Grenville province of Canada along the eastern side of the Canadian Shield, a broad, 180 to 300 mile-wide strip of land that stretches 1300 miles from the modern Adirondack Mountains of New York and Lake Huron, to the northeast coast of Labrador. Grenvillean rocks also surface in the Green Mountains of Vermont, Berkshires of Massachusetts, Hudson Highlands of New York, and Ramapo Mountains of New Jersey. In Pennsylvania they appear in the Reading Prong at the southern tip of the Ramapo Mountains, South Mountain, at the northern tip of the Blue Ridge, and the Piedmont province.

The Grenville continent remained intact until about 650 million years ago, when new driving forces began to tear it apart along rifts that opened to become Proto-Atlantic Ocean, an early version of the modern Atlantic. The rift system gradually widened that ocean basin during the next 200 million years.

The next three mountain-building orogenies occurred during the Paleozoic era, each one the result of another plate collision. By the end of Paleozoic time, nearly all the Earth's continental crust had assembled into a single great supercontinent, Pangaea.

The Taconian orogeny occurred in middle and late Ordovician time about 445 to 435 million years ago as North America and Europe converged. In the early stages of that convergence,

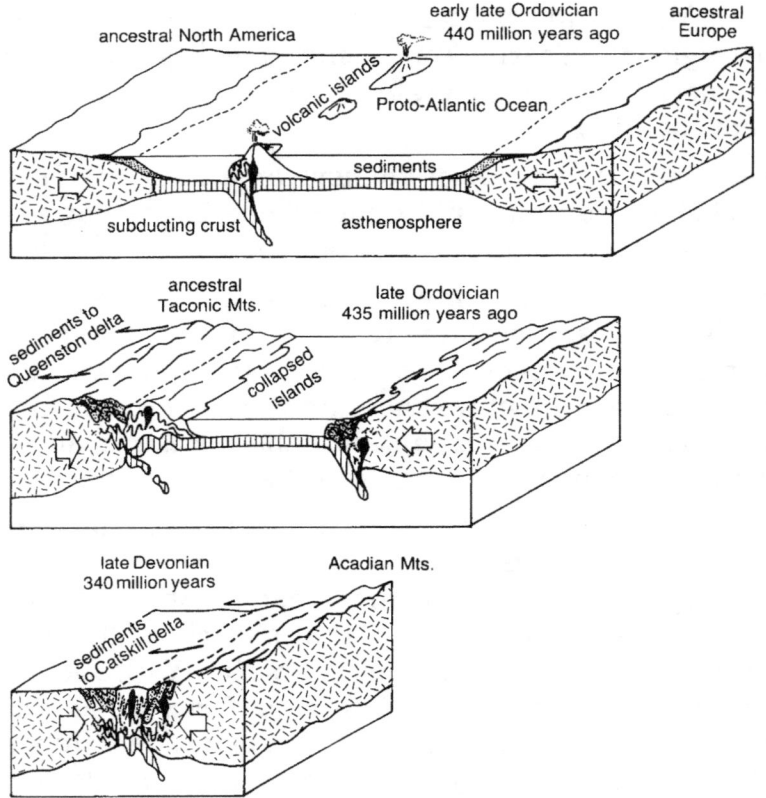

This is basically how geologists think the northern Proto-Atlantic Ocean closed, producing the Taconian and Acadian orogenies, and the Queenston and Catskill deltas, respectively. The final closure joined North America and Eurasia into a northern supercontinent often referred to as Laurentia.

the oceanic lithosphere ruptured somewhere east of North America, and the western segment sank beneath the eastern. Magma from the melting slab rose and erupted to form a volcanic island arc. Meanwhile, the deformation associated with the compression progressed landward as subduction consumed the oceanic slab that separated the island arc from the continent. The Taconian orogeny climaxed with intense folding, thrust-faulting, metamorphism, and igneous activity, as the chain of volcanoes collided with the continent. These effects are abundantly represented in New England, eastern New York, and the Maritime Provinces of Canada, where the compression appears to have been concentrated. This suture between the island arc and North America also appears in the Piedmont rocks of eastern Pennsylvania.

In the rest of Pennsylvania, however, the Taconian orogeny is recorded less by deformation and metamorphism than by the nature and order of deposition of the sediments eroded from the rising mountains, which were far to the east. The Taconian sedimentary package contains three distinct units. The oldest is a very thick section of latest Precambrian to middle Ordovician sandstones and carbonate rocks deposited on a continental shelf during a long period of tectonic quiet that preceded the plate collision and the mountain building it caused. The rocks contain many shallow water features, like mudcracks, that must have formed during periods of emergence above sea level.

Higher in the sedimentary section in younger rocks, the carbonates give way to the middle unit that signals a drastic change in sedimentation and the beginning of uplift in the east. The volcanic island arc was approaching the continental margin. Older sedimentary rocks that had accumulated along the seaward edge of the continental shelf were compressed to form a ridge, while the inner part of the continental shelf buckled down to form a basin, the kind geologists call a foredeep. That basin rapidly filled with black muds and dark turbidite sediments, muddy sandstones dumped from underwater landslides into the deep water of the foredeep. Some geologists call such assemblages of sediment flysch.

The third sedimentary unit accumulated during the final stage of orogeny, when sediments were shed in such great quantity from the rising mountains that they gradually displaced the marine waters. The resulting deposits form a gigantic sedimentary wedge, the Queenston Delta. It consists mostly of sandstones, many of them red. They were laid down in shallow sea water or on dry land. The Queenston Delta is thickest along a northeast-trending line from the Catskill Mountains of New York through central Pennsylvania where it reaches 4000 feet. It thins northwestward across the state, to about 1500 feet at Lake Erie. Two of its main units are the Juniata and Tuscarora sandstones which, because of their resistance to erosion, make prominent outcrops along many ridge crests.

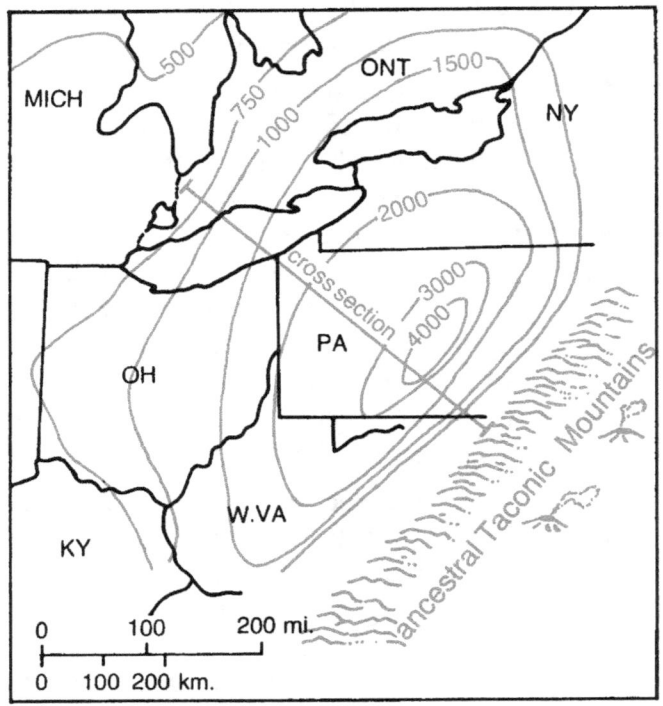

This map shows the outline and thicknesses in feet of the late Ordovician Queenston Delta, formed during continental convergence of Europe and North America. The delta is made up of sediments resulting from the erosion of the ancestral Taconic Mountains that moved northwestward and displaced an inland sea. —Adapted from Eicher, D.L. and McAlester, A.L., 1980, *History of Earth*, Prentice-Hall, Inc., Englewood Cliffs, New Jersey

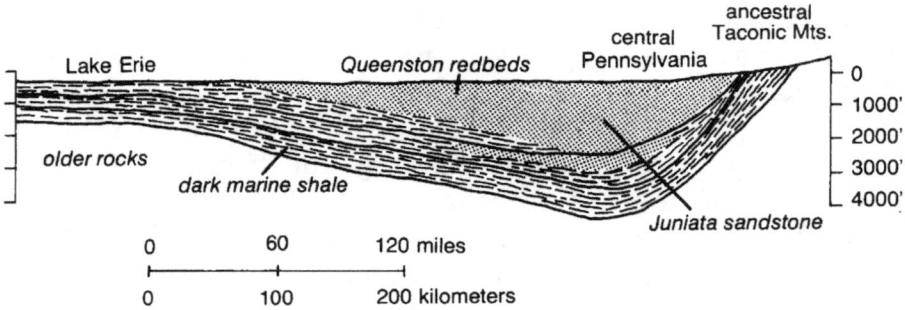

Reconstructed cross-section along the line shown on the map, developed from total formation thicknesses, with deformation eliminated.

The lower Silurian Tuscarora sandstone, and its equivalent in eastern Pennsylvania, the Shawangunk conglomerate, mark the end of the Taconian orogeny, and the beginning of 30 million years of tectonic quiet that preceded the Acadian orogeny. In the interval, the continental margin was eroded to a low level, a new continental shelf developed, and accumulated a new sequence of sedimentary rocks. The setting was similar to that which preceded the Taconian orogeny, but did not last as long. Carbonates and other shallow-water shelf deposits are not as thick or nearly as extensive as those that formed earlier.

Why the tectonic quiet? After the collision between the volcanic island chain and continent that climaxed Taconian mountain building, much ocean still remained between North America and Europe, and it continued to close as the plates converged. The closure must have been accommodated by subduction, but it could not have been along the earlier line.

The Acadian orogeny occurred in middle and late Devonian time, about 375 - 335 million years ago. It produced yet another range of mountains east of the Taconian Mountains, which were already deeply eroded. Large expanses of metamorphic rocks and numerous granite intrusions in New England and the Canadian maritime provinces attest to the intensity of this event. No such rocks exist in Pennsylvania because it was too far west of the Acadian Mountains. However, sedimentary rocks associated with formation of the Acadian Mountains do exist in Pennsylvania. They closely resemble the sedimentary rocks associated with the Taconic Mountains.

As before, the dominantly carbonate based formation is followed by a section of flysch sediments that consist mostly of dark shales and muddy sandstones. A new delta, the Catskill Delta caps the sequence.

The Catskill Delta is a gigantic sedimentary wedge that resembles the older Queenston Delta and covers approximately the same area. It is thicker than the Queenston Delta because it formed during the erosional destruction of a much larger range of mountains. As before, the deposits thin out northwestward as the sediments also get finer-textured. Redbeds are common.

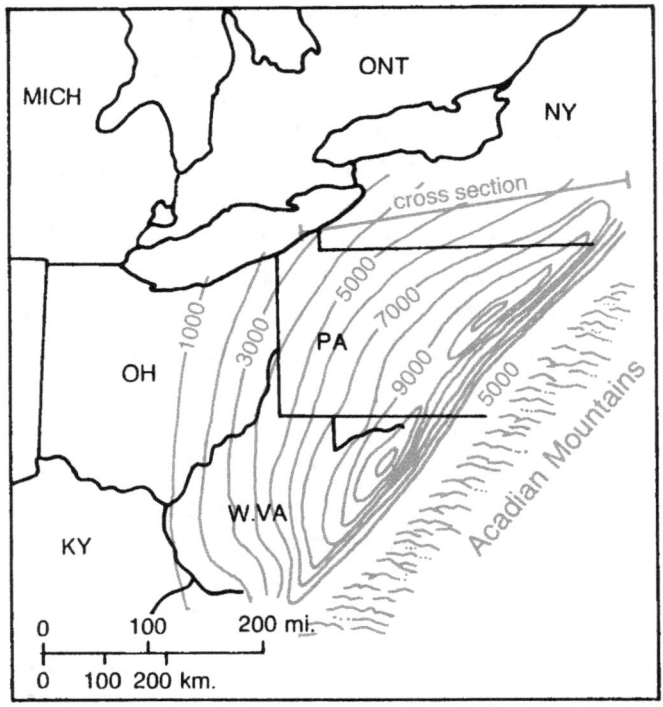

This map shows the outline and thicknesses in feet of the Catskill Delta. The delta formed in middle to late Devonian time as ancestral Europe collided with North America, causing Acadian mountain-building. It is made up of sediments derived by erosion of the mountains. —Adapted from Eicher, D.L. and McAlester, A.L., 1980, *History of Earth*, Prentice-Hall, Inc., Englewood Cliffs, New Jersey

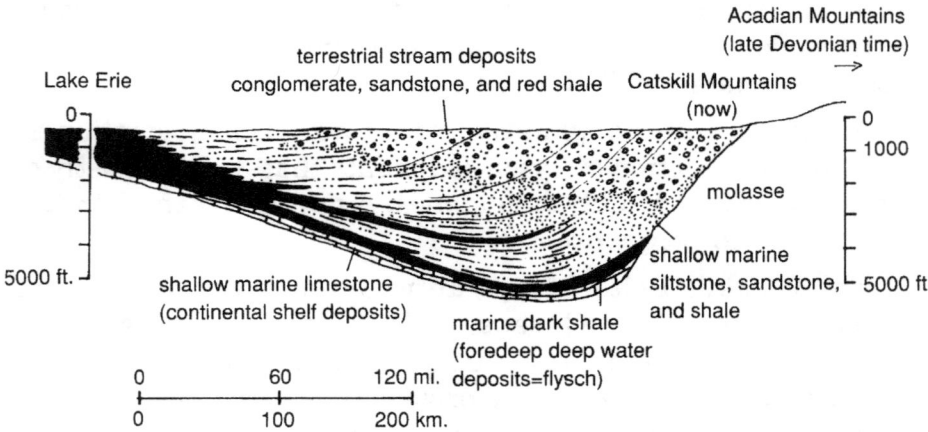

Reconstructed cross-section along the line shown on the map, developed from total formation thicknesses, with deformation eliminated.

The upper unit in this delta, the late Devonian Catskill formation, may often be recognized by its distinctively reddish to greenish to gray conglomerates, sandstones, and shales. It is widespread in the state, forming the caprock of the eastern half of the Allegheny Plateau and cropping out along the length of the Allegheny Front and on the sides of many ridges in the Valley and Ridge province. It records the rise and fall — erosion — of the Acadian Mountains.

The Acadian orogeny resulted from the closing of the Proto-Atlantic Ocean basin and collision between North America and Europe to form a continent some geologists call Laurentia. In the process, a large continental fragment, Avalonia, which had been separated during the opening of the Proto-Atlantic, was caught in the vice; it is now part of the Appalachians, but it has not been recognized in Pennsylvania.

The Alleghenian orogeny occurred during the Pennsylvanian and Permian periods, about 300 - 220 million years ago. It followed so closely on the heels of the Acadian orogeny and the mountains rose so quickly that delta followed delta with no carbonate or flysch sedimentary deposits in between. The red shales, sandstones, and conglomerates of the upper Mississippian Mauch Chunk formation mark the beginning of the uplift. These are followed by the coal-bearing deposits of Pennsylvanian and Permian age that record shallow marine deposition with a widely fluctuating sea level that caused great swings in the position of the coastline.

The Alleghenian orogeny also produced the spectacular folding, thrust-faulting, and other deformation of all rocks of Pennsylvanian and older age in the Valley and Ridge province, including the anthracite coal region. On the Allegheny Plateau, Permian and older rocks were folded into broad open anticlines and synclines floored by flat-lying thrust faults at depth. Many of the anticlines now contain oil and gas reservoirs. The lesser intensity of Allegheny Plateau deformation stems from the fact that the basal thrust faults are much shallower there than they are under the Valley and Ridge province, and also the Allegheny Plateau was farther away from the uplift. Involvement of Permian rocks indicates that deformation continued into the Permian period.

The Alleghenian orogeny resulted from collision between North America and Africa, which was then part of a southern continent, Gondwana, along with South America, Antarctica, Arabia, India, and Australia. This completed the assembly of an enormous supercontinent, Pangaea, surrounded by a world-girdling ocean. The coals and other features of the rocks formed during the final assemblage of Pangaea indicate that Pennsylvania's climate then was tropical. Paleomagnetic data show that the place was almost precisely on the equator.

Pangaea did not last long by geologic standards. It began to break up during the Triassic period about 200 million years ago as the modern Atlantic Ocean began to open and the outlines of our present continents were established. In the early stages of the opening of the Atlantic Ocean, the crustal stretching produced numerous block-fault basins, including the Newark and Gettysburg basins of Pennsylvania. As they subsided, the

Triassic rift basins of eastern North America.

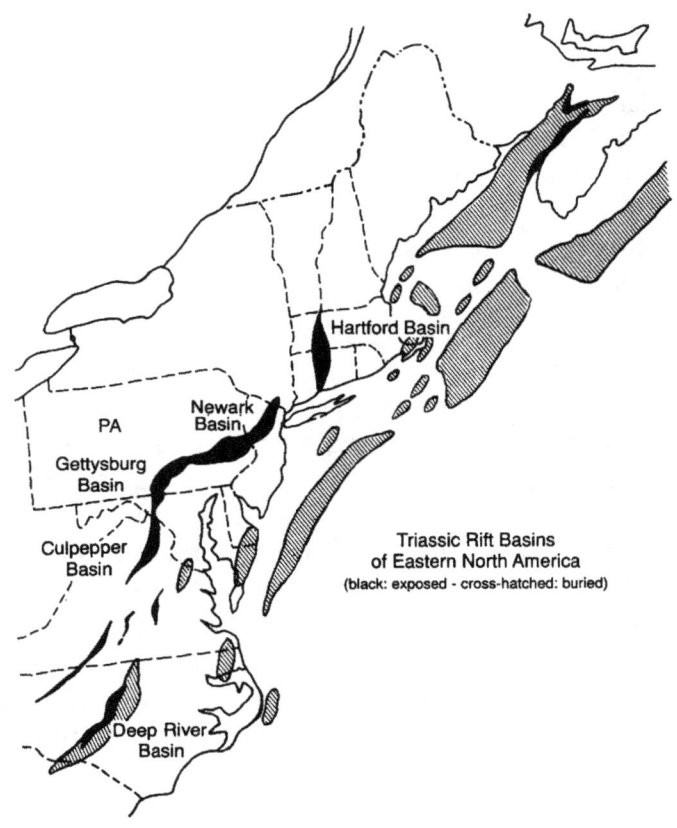

basins filled with red sediments and basaltic lava flows and shallow diabase intrusions. By mid-Jurassic time, some of the fractures connected to form a dominant rift zone that split the originally broader Appalachian Mountains down the middle and a linear sea filled the gap — the juvenile Atlantic Ocean.

Today, 150 million years later, the Atlantic Ocean basin is still widening, continually splitting apart along the mid-Atlantic rift system at an average rate of about 2 inches per year. Meanwhile, North America and the adjacent half of the Atlantic crust remain part of the same lithospheric plate, and the Appalachians are slowly wearing down.

OIL AND GAS

Petroleum is the general name for oil, gas, asphalt, tar, and other solid forms of hydrocarbons, compounds of hydrogen and carbon, that are widespread in sedimentary rocks.

Petroleum, like coal, is a biological product, the organic debris of former life — buried, transformed, and preserved in sediments — therefore, fossil fuels. Unlike coal, oil, and gas generally do not concentrate in the source rocks where they formed, but rather in porous, permeable reservoir rocks to which they migrated. Apparently, they form best in the continental shelf or in rocks deposited on the floor of an inland sea where all manner of marine life, both animal and plant, accumulates in quantity on the seafloor and mud buries it before oxidation can destroy it. Temperature rises with deepening burial, causing chemical reactions that convert the organic matter to hydrocarbons. Rising pressure beneath the accumulating pile of sediments squeezes the oil, gas, and water out; and they migrate generally upward. If, in their ascent, they meet with a shale, rock salt, or other caprock through which they cannot pass, they puddle to form a reservoir; the oil, in time, will float to the top of the water and the gas to the top of the oil. Most good reservoir rocks are permeable sandstones or limestones through which the oil and gas can migrate freely.

Oil is widely distributed in Devonian rocks of western Pennsylvania. Natural gas is even more widespread, occurring in rocks of Cambrian, Silurian, Devonian, Mississippian, and

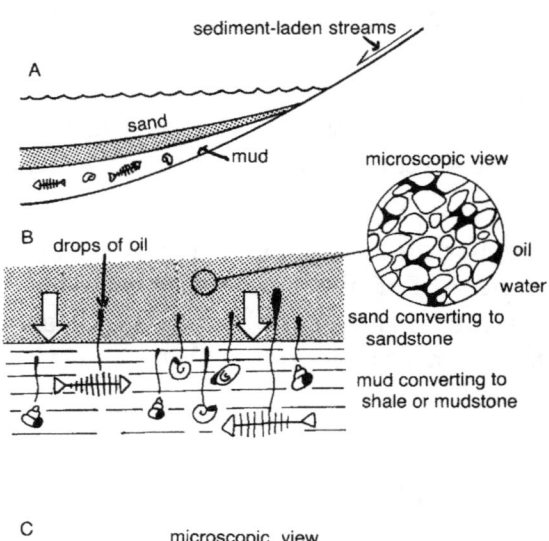

Oil and gas form from organic matter that accumulates on the seafloor, as in A above. Deep burial and heating converts this matter to hydrocarbons that are squeezed out by compaction, as in B. Oil, gas, and water continue migrating upward through porous rocks like sandstone until they encounter an impermeable layer like shale, as in C, where they accumulate and separate, the hydrocarbons floating to the top.

Pennsylvanian ages. Very little gas occurs in Cambrian rocks, and none is known in Ordovician rocks. Cambrian and Ordovician strata also lie at great depth below the surface where the gas would be expensive to extract. On the other hand, Permian rocks are non-productive because they lie at such shallow depths that oil and gas are easily flushed out by the natural movement of near-surface groundwaters.

Virtually no producing fields exist in Pennsylvania east of the Allegheny Front. The oil and gas potential of that region is not well-known, in part because of relatively little exploration. It is possible, however, that deformation of the Appalachian Mountains province was so intense that any petroleum that might have been present was either destroyed by overheating or lost to the surface. Most petroleum is in only gently deformed rocks as, for example, in the Allegheny Plateau, where it generally collects in the crests of broad open anticlinal folds, or in slightly uptilted beds truncated by faults and juxtaposed with shales.

Oil and gas, like coal, have contributed immensely to the economic development of Pennsylvania. The petroleum industry began here in 1859 with the completion of the world's first commercial oil well, and the state soon after became the world's first center for oil production.

Titusville by Oil Creek, where oil seeps had been known for many years, was the location of the first well. Before 1850, only small quantities of oil were collected for use as a natural medicine. Crude oil was rarely burned for lighting because it made too much smoke. The introduction of distillation in 1850 changed the whole picture. The refined product, called carbon oil, burned far more cleanly, and a large market for it developed virtually overnight, making natural seeps inadequate to meet the demand. With this impetus, the search began for new ways to extract oil from the ground in large quantities, but with limited success. In 1859, "Colonel" Edwin L. Drake was dispatched to Titusville by the newly formed Seneca Oil Company of New Haven, Connecticut, to drill a well. Drake built an engine house and derrick on the site, installed an engine and boiler to supply power, and began drilling in mid-August. The drillers first drove an iron pipe 32 feet through soil and soft

Regions of shallow and deep gas and/or oil in western Pennsylvania.
—Adapted from Pennsylvania Geological Survey, Map 10

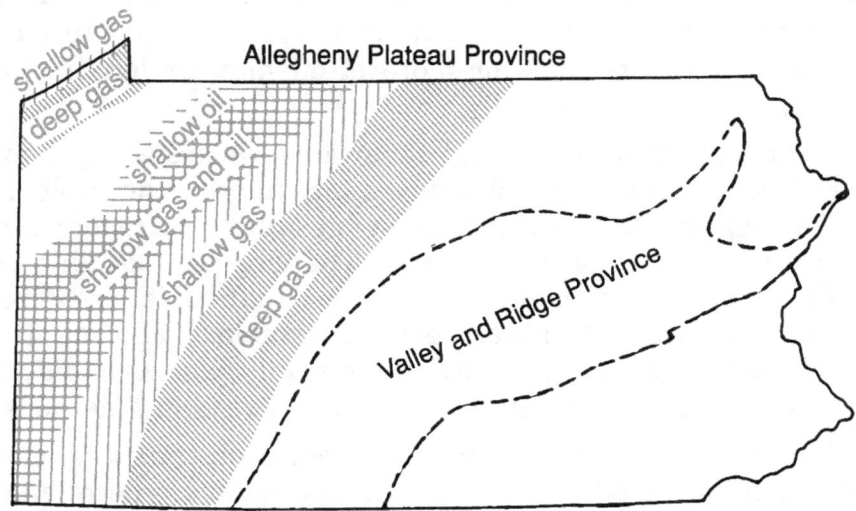

sediments. They then ran the drill stem through the pipe and commenced to drill bedrock, making about three feet a day. They struck oil almost by accident on August 28 at a depth of only 69 feet!

The news of the Drake discovery spread like wildfire, prompting a mad rush to obtain land at any price and drill a well. In a short time, the entire Oil Creek valley was leased or purchased; and scores of wells were being drilled. By the end of 1861, 74 producing wells were yielding a total of 1200 barrels of oil a day. By 1864, this figure had jumped to 6000 barrels a day!

The commercial development of natural gas came after the oil boom of the 1860s to early 1880s. During these years, the gas from the oil wells was used only to drive the oil well machinery and to heat buildings. Rapid expansion of the gas industry began in 1884 when gas was piped from the Murrys-ville gas field to Pittsburgh. One of the main factors in the development of the steel industry in Pittsburgh was piped gas, which for many years sold at such a low price that the cost of manufacturing was greatly reduced.

From 1859-95, Pennsylvania led the nation in crude oil production. Since then, production has generally declined and, were it not for secondary recovery methods developed principally within the state, the industry here would now be dead. Survival hinged more on finding efficient ways to get oil out of the ground than on finding new reservoirs to tap. Without secondary recovery, 60% or more of oil may be left in the ground.

Oil and gas deposits are non-renewable; we are using them faster than nature can build new ones. In fact, the geologic record suggests that it takes at least one million years for oil to form from original organic matter. The long-term decline in production unquestionably indicates gradual depletion of existing reservoirs. To be sure, much recoverable oil and gas still remain in the ground, but they will certainly run out in only a few tens of years, even with new and better recovery methods.

What is secondary recovery? The world oil industry uses several methods, all aimed at forcing the reluctant petroleum out of the reservoir rocks and into the well. Pennsylvania has used three extensively: vacuum, water flooding, and air-gas drive. Basically the vacuum method involves the use of negative pressure at the well head to suck the oil and gas out. Both of the other methods require drilling new nearby wells for injecting water or air-gas under pressure, which forces the oil and gas toward the recovery well. The advantage of water flooding was discovered about 1876-77 when water accidentally entered an abandoned well on Oil Creek, increasing the output in nearby producing wells. This type of unpressurized flooding was practiced sporadically until the introduction of pressure flooding about 1927-28.

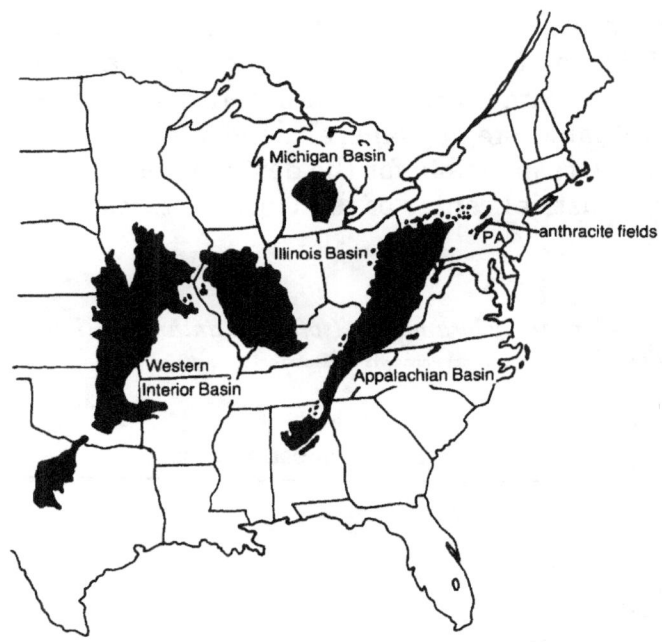

Major coal fields of the eastern United States.

COAL

It is no accident that in many parts of the world, rocks of Pennsylvanian geologic age are known as "Coal Measures." Although coal formed locally during every period since the spread of terrestrial vegetation in Devonian time, Pennsylvanian rocks contain most of the high-rank coal. In these formations lie the great coal fields of the British Isles, France, Germany, and the smaller fields of Belgium, Silesia, and the Donetz basin of Russia, as well as the major coal fields of eastern North America. These fields together produce more than 80% of the world's coal. It is fitting that this great coal-bearing system, as the rocks of each geologic period are called, should be named for the chief coal-producing state, the one where it was first comprehensively studied.

Coal is by far Pennsylvania's most important mineral resource; about one third of the state is underlain by coal deposits. In the past, one-fourth of the total coal output in North America came from just the anthracite region of eastern Pennsylvania, which is small compared to the great bituminous coal fields of the western half of the state. The fabulous Pittsburgh seam of bituminous coal has been called the world's most valuable single mineral deposit. More than any other factor, it is responsible for the development of Pittsburgh as a major industrial center of the United States.

Regions underlain by bituminous (soft) and anthracite (hard) coal in Pennsylvania. —Adapted from Pennsylvania Geological Survey, Map 11

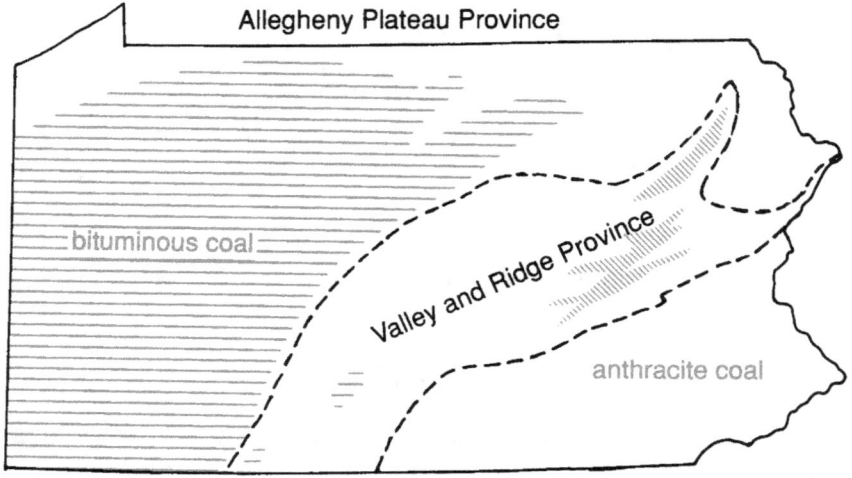

The earliest record of mining the Pittsburgh seam is on a "Plan of Fort Pitt and Parts Adjacent," dated 1761. The mine was named Ward's Pit for Major Edward Ward, who opened it. British troops first worked it to obtain fuel for Fort Pitt, which it overlooked. Growth of the coal industry proceeded slowly until the mid-1800s, depending largely on its use by glass and brine industries. The discovery of the coking process then sparked an industrial revolution that turned Pittsburgh into Steel City, USA.

Coke forms when bituminous coal is heated to a high temperature in an oxygen-deficient environment. This drives off the volatiles and leaves a spongy residue that is ideal for the smelting process in blast furnaces. The fixed carbon and ash of the coal are fused together, forming a material strong enough to support the burden of the ore and limestone flux in the furnace, while permeable enough to permit free passage of the air blast through the burning mass during smelting. This generates the high temperature and carbon monoxide needed to draw oxygen from the ore, leaving molten iron that is tapped off at the bottom of the furnace.

Today, of course, coal has many other uses besides smelting. It is burned for heat or to generate electricity. It is refined to make medicines, plastics, synthetic rubber, fertilizer, cosmetics, food products, paint, dyes, and synthetic fibers.

Coal forms from peat that accumulates in swamps. Under ideal conditions, inorganic sediments cover the peat, and compression and elevated temperatures gradually convert it to lignite and then bituminous coal. The process drives off moisture and increases the carbon content from a low of about 25% in lignite to a high of about 90% in high-rank bituminous coal. The rank of the coal is a measure of its carbon content — the higher the carbon, the higher the heat value. Anthracite is metamorphosed coal found in deformed rocks of mountainous regions. It is a clean-burning, hard coal with up to 94% carbon. The anthracite fields of eastern Pennsylvania were a small part of the bituminous field until they were caught up in Valley and Ridge folding and metamorphism.

The bituminous field of Pennsylvania is just the northern tip of the enormous Appalachian field that extends south to

Alabama, and covers thousands of square miles. Other large Pennsylvanian-age coal fields are the Michigan, Illinois and Midcontinent fields. Altogether, these cover more than 250,000 square miles!

Why are Pennsylvanian coals so abundant? First, many of the large swamp-loving plants had evolved only a short time before the Pennsylvanian period and simply were not available to make coal before then. Even more important were the special climatic and lowlands conditions that combined to produce widespread coastal swamps. Rock magnetism data indicate that eastern North America was approximately on the equator during the Pennsylvanian period and the climate was tropical. The strongest clue lies in the cyclicity of coal deposits. The coal seams seldom occur alone; instead, a sequence of sedimentary beds in a particular coal region may contain as many as 30 coal seams separated by sandstones, shales, and limestones.

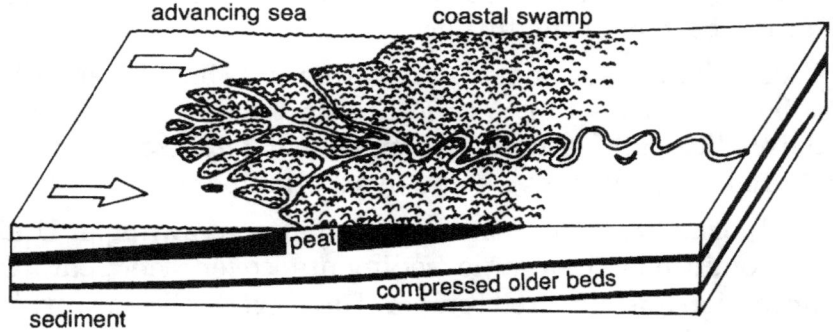

Coal seams like those of Pennsylvania formed from peat that accumulated in swamps along the coast of a shallow inland sea. When the sea advanced inland, new sediments covered the peat, setting the stage for its preservation and conversion to coal. Alternating advances and retreats produced many coal seams. —Adapted from Edmunds, W.E. and Koppe, E.F., 1968, Coal in Pennsylvania, Pennsylvania Geological Survey Ed. Series 7

Each seam is interpreted as part of a cycle of sedimentation that involved advance and retreat of a shallow inland sea, with coal swamp conditions following the changing shoreline. The enormous lateral extent of individual coal seams, therefore, indicates an equally great shift of the shoreline. The inland basins must have been very large and shallow for the coastline to migrate so far.

What caused the many sea level changes that produced multiple coal seams? The most inviting answer to that question comes from much farther afield. Elsewhere in the world, the rock record indicates that late Paleozoic time brought continental glaciaton similar to that of the much more recent Pleistocene ice age. The great southern continent of Gondwana then straddling the south pole was the platform for ice sheets that apparently expanded and contracted in response to climatic fluctuations.

It is well established that at the climax of the last ice age, sea level dropped more than 300 feet because so much water was locked up in the ice; the Atlantic coast was far out on the continental shelf — more than 100 miles out in places. Furthermore, there is a depositional record in the northern United States of three preceding glacial advances, all of roughly similar areal extent, all separated by lengthy periods of retreat. It is possible that there were additional advances and retreats for which no land-based glacial deposits remain because they were destroyed by later advances. Sea level dropped during each advance, rose during each retreat, and coastlines changed accordingly.

Continental glaciation during Pennsylvanian time may have caused similar sea level fluctuations. Laurentia, the northern supercontinent, with North America attached, and Gondwana, the southern supercontinent, were separated by the Tethys Sea, but were slowly drifting together. Their convergence initiated, and eventually climaxed, the Alleghenian uplift. As the mountains rose along the North American margin, the crust northwest of them warped gently down into a foreland basin, and ocean water poured in to form a shallow inland sea. This was the receiving basin for sediments spread northwest from the rising mountains, and for the coal swamps. Its shallow depth and gently sloping coastal plain made it extremely sensitive to even slight sea level changes as the southern glaciers expanded and contracted.

Many geologists envision a situation in which coal-forming peat was preserved only during periods of glacial recession when sea level rose as the land ice melted. A drop in sea level would have subjected the swampy debris to decay and erosion. Where the cover sediments achieved sufficient thickness during sea level rise, the forming coal seams were protected from erosion during the next coastline retreat.

I
Allegheny Plateau

On the satellite image of Pennsylvania, the Allegheny Plateau is sharply defined, distinctly different from the rest of the state. It looks like a plateau, or tableland and it covers about three-fifths of Pennsylvania, spanning the northern and western parts of the state. Its apparently uniform, stream-dissected texture contrasts strikingly with the finely-etched pattern of sharp-crested ridges in the adjacent Valley and Ridge province. The rocks of both regions bear indelible imprints of the continental collision between North America and Africa that produced the Alleghenian orogeny during the Pennsylvanian and Permian periods.

In western Pennsylvania, the major streams—the Allegheny, Monongahela, Youghiogheny—trace random paths across the Allegheny Plateau that reflect flat, or nearly flat, beds. Nearer the Allegheny Front, the upper branches of the Susquehanna system follow the trends of the more pronounced folds that occur there. Nevertheless, the overall appearance is strongly suggestive of a plateau, especially by comparison to the bordering valleys and ridges, which obviously reflect severe deformation.

On the ground, you will find the Allegheny Plateau does not look like a plateau at all; there is considerable relief. The greatest relief, and also the highest elevations in Pennsylvania, are near the Allegheny Front, where resistant rocks have been uptilted to high levels. Elsewhere the topography results largely from stream dissection over the last 200 million years. In

Landsat satellite mosaic of Pennsylvania. Note the distinctive topographic textures of the Allegheny Plateau, Valley and Ridge province, and Great Valley. Compare these and other provinces with the bedrock geology shown on the general geologic map.

northeastern and northwestern Pennsylvania, ice age glaciers also left their imprint on Allegheny Plateau topography, scouring in places and dumping sediments in others.

In roadcuts and outcrops, you will seldom see a whole anticline or syncline; the folds are much too large and open. Instead you see slices of the sedimentary layercake, flat or slightly inclined, with steeper inclines near the Allegheny Front because the folding is more pronounced. Elsewhere the folds are nothing more than gentle waves several miles wide.

The gentle Allegheny Plateau folds are economically important for they are the sites of many oil and gas reservoirs that have contributed greatly to the historic development of Pennsylvania as a major industrial state. The oil is now largely depleted, but there appears to be considerable natural gas left. Look for the many old wells that dot the landscape. The gas wells are not much more than pipes and pressure gauges leading from the ground to storage tanks. Oil wells may be recognized by their old, rusting pump-jacks; occasionally you may see one of these bobbing up and down on a still-active well.

The Allegheny Plateau is also soft coal country. Coal has been, and is, an even greater source of the state's wealth. The plateau landscape, in many places, is scarred with unreclaimed strip mines. The more recent mine sites are, by law, reclaimed. The principal coals occur in Pennsylvanian rocks that cap the Allegheny Plateau in almost all the western part of the state. The same coal-bearing formations also surface in the cores of synclinal folds in north central Pennsylvania near the Allegheny Front. Eastern Pennsylvania coal is similarly preserved in synclines in the Valley and Ridge province; there, however, deformation and mild metamorphism converted it to hard anthracite.

There is also some coal in the Permian rocks, found only in the southwestern corner of the state, but most is not commercially recoverable.

Outcrops of coal-bearing formations, or coal measures, often appear thinly bedded and varicolored. Coal seams are black and weather recessively. They are sandwiched mainly among shales, sandstones, and limestones. Colors and textures vary from layer to layer, giving a banded appearance. Outcrops are often crumbly because the rocks, in sum, offer only weak resistance to erosion.

Upper Devonian rocks of the Catskill Delta complex dominate the northern edge and eastern half of the Allegheny Plateau. Red to greenish gray, commonly cross-bedded sandstones, shales, and locally, conglomerates distinguish the Catskill formation. They derive from sediments shed from mountains that rose in the region of the present coastal plain during the Acadian orogeny.

The Mississippian Pocono, or Burgoon, sandstone is a prominent ridge-former in the central and eastern Allegheny Plateau. It is separated from the Catskill formation by transitional strata of the Huntley Mountain formation. At the top of the Mississippian rock section are the distinctive redbeds of the Mauch Chunk group directly under the Pennsylvanian strata.

The northwestern margin of the Allegheny Plateau in Pennsylvania, called the Erie scarp, faces a narrow strip of Erie Lowlands. The scarp cuts through and exposes Mississippian and Devonian beds.

I-70
West Virginia Border—New Stanton
58 mi./93.5 km.

Between the West Virginia border, and mile 23, I-70 crosses part of the Pittsburgh Plateau. Rocks that appear in numerous cuts belong to the lower Permian Greene (younger) and Washington (older) formations. Both are coal-bearing sequences like those of the underlying Pennsylvanian rocks, but the Greene formation contains no workable coal seams; the Washington formation only two.

Roadcuts are colorfully banded with flat-lying beds of buff to grayish sandstones and siltstones, gray to brown to black shales, and some coal, limestone, and even redbeds. As usual, most of the sandstone beds form ledges because they are more resistant to erosion than the enclosing shales, and the shales tend to break down to chips that shingle the lower parts of the cuts.

This is, of course, part of the Allegheny Plateau. The plateau label befits the structure of the bedrock strata, which are nearly flat-lying, but not the rollercoaster landscape, a product of long-term erosion.

Alleghenian deformation created broad open folds that involve Permian beds, the youngest Paleozoic formations in the state. They show that Alleghenian deformation continued at least into Permian time. All of the preserved Permian and underlying Pennsylvanian beds, as well as the upper Mississippian Mauch Chunk formation are part of a great wedge of sediments that spread northwestward from the eroding mountains into a shallow inland sea.

Between miles 23 and 40, just west of the Monongahela bridge, the road is built on Pennsylvanian-Permian Waynesburg formation and the underlying Pennsylvanian Monongahela and Conemaugh groups, all of which contain productive coal seams. The fabulous Pittsburgh coal marks the base of the Monongahela group. The Monongahela River cut through its upper layers into the Conemaugh group that crops out in a narrow strip for several miles along its banks. Piles of black waste rock identify a strip mine on the south side of the highway near mile 37.

Between miles 40 and 47, at the Youghiogheny River, the road is principally built on Monongahela strata. Both the Monongahela River and Youghiogheny, its tributary, meander widely over this section of the plateau, apparently unaffected by the structure of the bedrock.

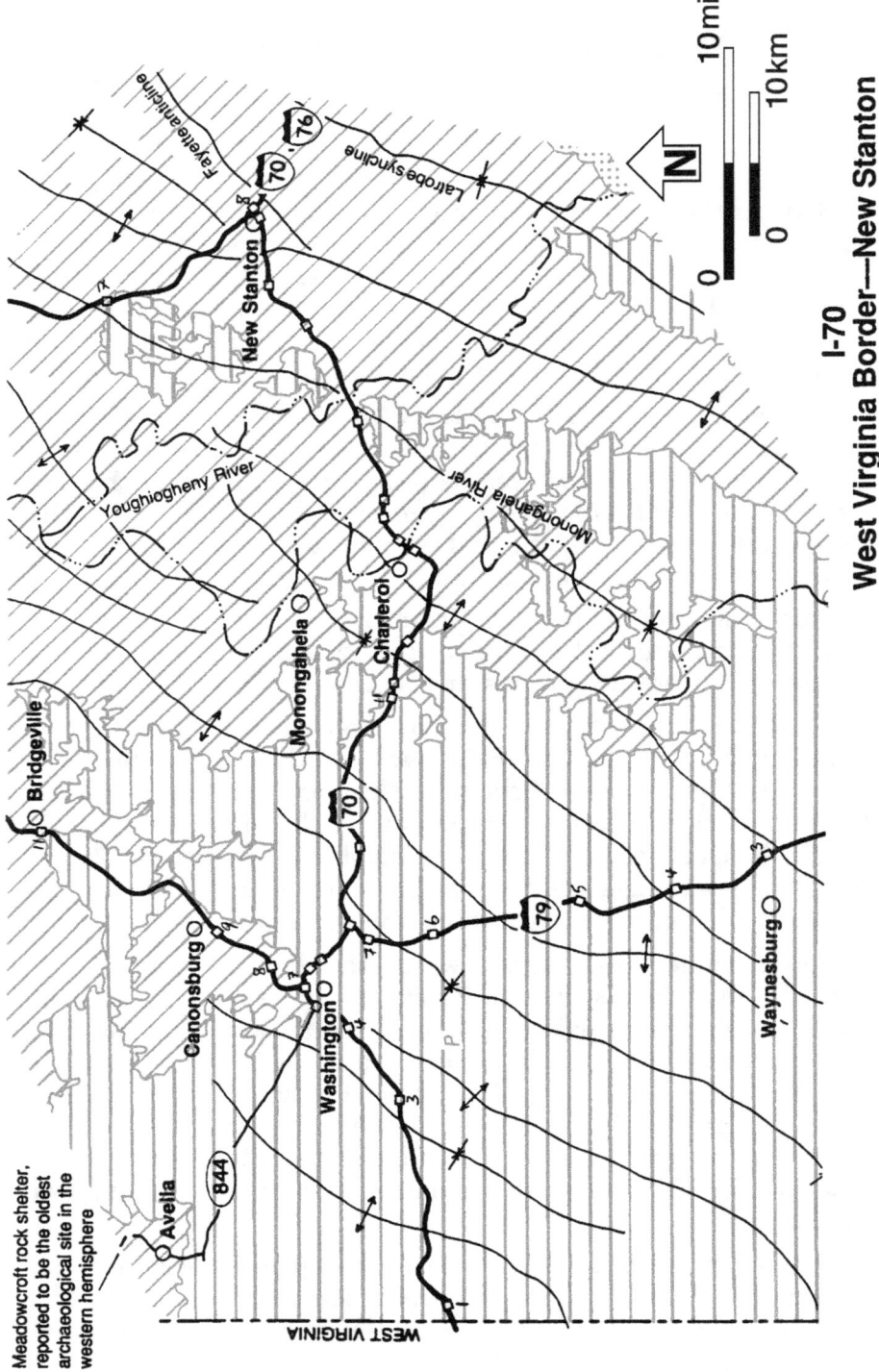

The beds are not quite flat, but gently corrugated into very large-scale, northeast-trending anticlines and synclines, parallel to the Allegheny Front and the more intense folds of the Valley and Ridge province. The folding in both regions has the same origin — collision of North America and Africa. The style and intensity of deformation differ.

Folded rocks of both regions are part of a stratified package that rides on the same basal thrust fault, a sole thrust. Bedding beneath the fault is virtually undeformed and lies nearly flat. The sole thrust fault consistently follows the weakest of these beds.

Under the Valley and Ridge province, the sole thrust slices through beds of the lower Cambrian Elbrook and Waynesboro formations at depths of 25,000 to 35,000 feet. Under the Allegheny Front, the fault breaks out of the Cambrian beds, and rises to the upper Silurian Wills Creek formation and Salina group. Then it levels off in these even weaker beds at a depth of less than 10,000 feet.

The extremely thick package of sedimentary rocks above the sole thrust in the Valley and Ridge province presented great resistance to movement. The compression associated with continental collision was accommodated by fault splays that branched from the sole thrust and steeply ascended. Some of these splays are exposed at the surface today, but most did not reach that high. Movement along them produced the intense folding that is a trademark of this region.

The much thinner rock package under the Allegheny Plateau offered less resistance to horizontal fault movement, so the rocks are only mildly, and very broadly, folded. The upturned beds exposed along the Allegheny Front are the eroded limb of an anticline formed where the rocks flexed from the ramp incline onto the flat thrust of the Allegheny Plateau.

From the I-70 bridge, the Youghiogheny River meanders north for 15 miles as the crow flies to join the Monongahela at McKeesport. A few miles beyond that, the enlarged Monongahela joins the Allegheny at the Golden Triangle of Pittsburgh to form the Ohio River.

Between miles 47 and 58 at the I-76 junction, the highway cuts diagonally across the northwest limb of a large anticline. Cuts expose Monongahela beds near mile 47. From there to mile 58, the road progresses deeper and deeper into the stratigraphic section through Conemaugh beds and finally into the underlying Allegheny group. The I-76 junction lies on the axis of the fold.

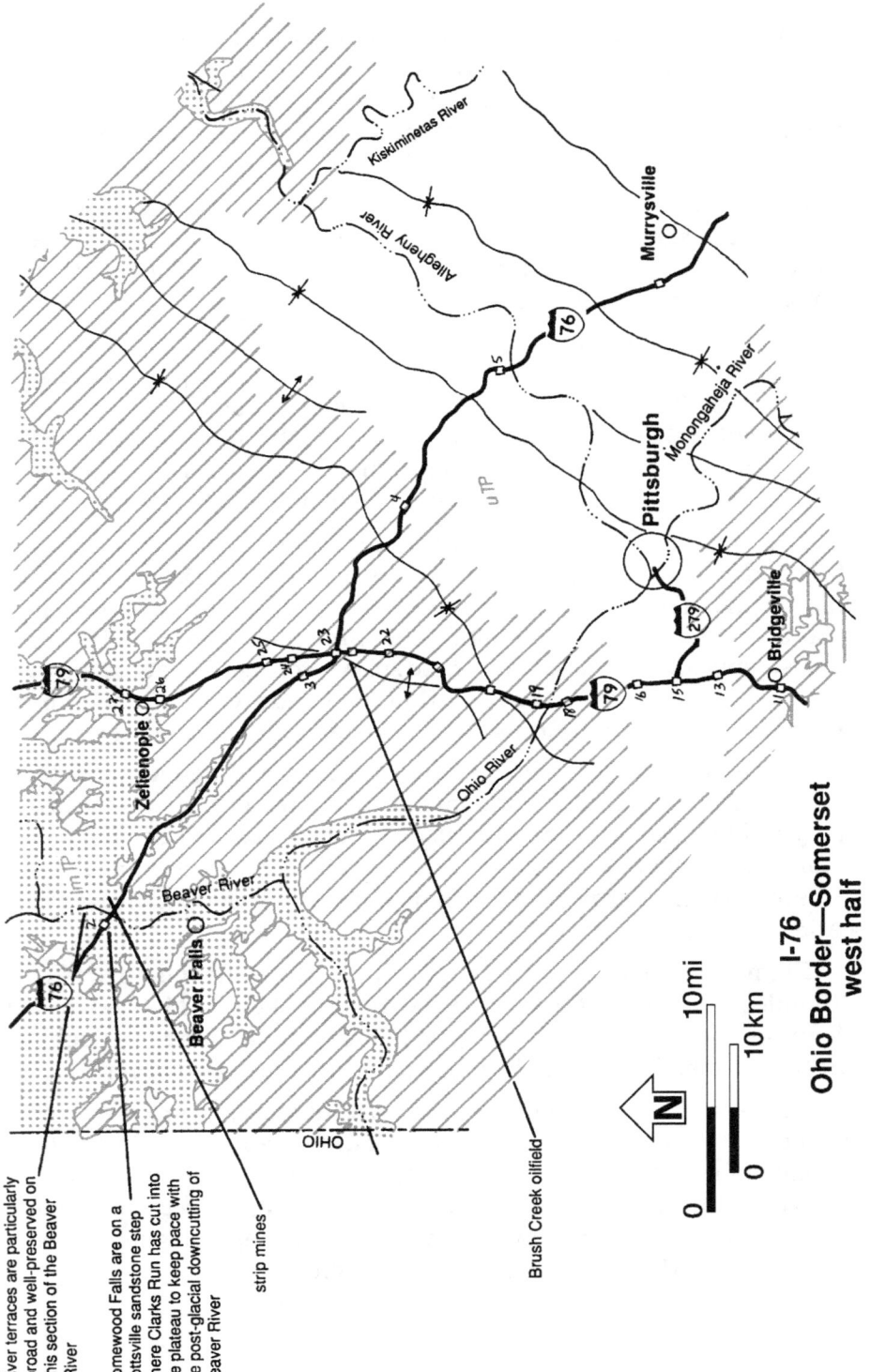

I-76
Ohio Border—Somerset
110 mi./177 km.

The Pennsylvania Turnpike

Between the Ohio border and mile 13, the turnpike crosses glaciated Allegheny Plateau. Relief is very low because invading ice scraped the hills, and glacial drift partly fills the valleys. A thin blanket of glacial moraine covers much of the land, both high and low. The more hummocky surface you see approximately between miles 10 and 13 is end moraine, till deposited at the southernmost reach of glacial ice about 20,000 years ago; Beaver Falls lies near its outer edge. The ice margin then lay several miles inside that of an earlier glaciation. The margin continues from here northeast to Warren County and the New York border.

Bedrock is all Pennsylvanian age, mainly the coal-bearing Allegheny group. The region has been extensively strip-mined, and a few of the mines are visible from the highway. In 1989, one mine lay north of the road near mile 10, another, with water-filled trenches, south near mile 11; you may be able to recognize several others — especially if they are unreclaimed.

Strip Mining

Strip mining, or surface mining, as opposed to underground mining, is the most economical and widely-used method for extracting coal from Pennsylvania's plateau country. It is used because almost all of the valuable coal seams are flat-lying and near the surface. The method involves blasting and removal of overburden, the rock that overlies the coal, followed by removal of the coal. The thicker the overburden relative to the thickness of the coal seam, the more costly the operation becomes, so the method has its limits.

Coal mines on the plateau are either area or contour strip mines. Area strip mining is employed in relatively flat areas where the open pit can be enlarged in all directions, and the overburden is of more or less constant thickness. Contour mining is done where relatively deep coal crops out on hillslopes in areas of higher relief. The cost of removing the coal increases as the open pit penetrates deeper into the side of the hill because more overburden must be removed; therefore the open pit follows the contour of the hill, penetrating only to an economical depth. The deeply carved valleys provide an ideal contour-mining environment for the deeper coal seams. Both types of mines

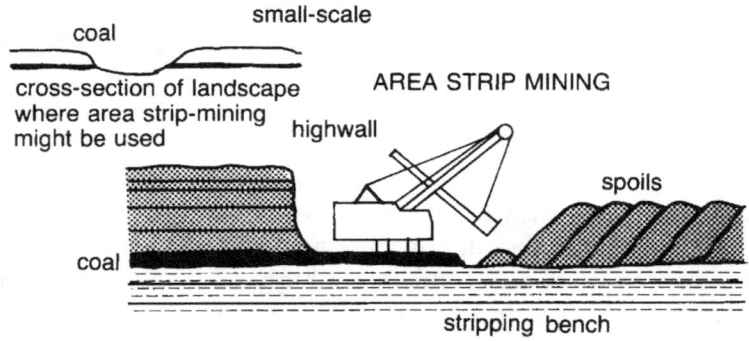

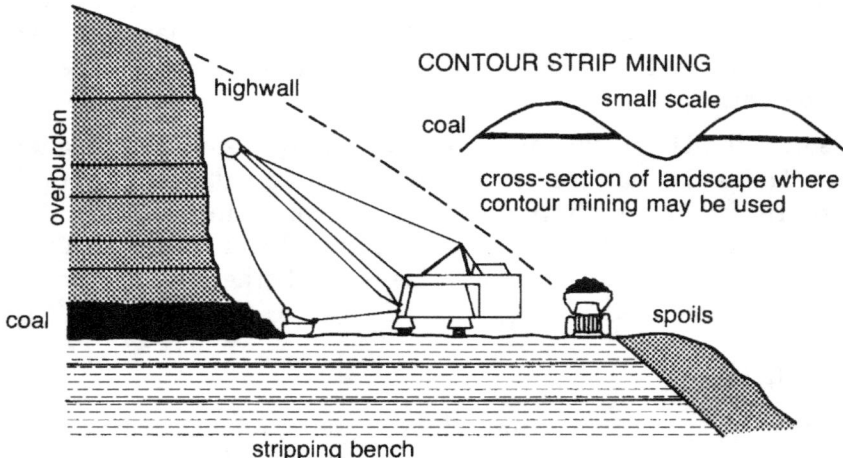

are visible from the Pennsylvania Turnpike. Many of the unreclaimed pits are full of water.

Removal of overburden in strip mining creates barren spoil piles next to the mined-out pits. Current reclamation law requires that the pits must be filled, recontoured, covered with soil, and revegetated. Most of the older mines have not been reclaimed.

• Exit 2 is on a broad river terrace on the west side of and about 160 feet above the Beaver River. Several similar terraces exist on the other side of the river to the north and south. The terraces are glacial outwash sediments deposited as the glaciers of the last ice age melted then carved into terraces as the river eroded those deposits.

Entrenchment of the main streams is largely responsible for the generally rugged topography of the Allegheny Plateau. Small tributaries also cut downward to keep pace with the main streams.

The gorge of one such tributary, called Clarks Run, is just south of Exit 2. Homewood Falls at its head cascades 30 feet over Pottsville sandstone beds. Waterfalls are common in the tributaries; their upstream recession creates and extends the gorges.

Watch for the unreclaimed strip mines across the river from Exit 2. Another is adjacent to the west-bound lane in Brush Creek valley at mile 19.

The road between miles 13 and 28 crosses Allegheny and Conemaugh shales and sandstones with some thin coal seams. The sandstone interbeds in some cuts project from the less resistant shales as ledges.

Near mile 31 by Warrendale, you cross the axis of one of the many northeast-trending anticlinal folds that gently corrugate the Allegheny Plateau bedrock. This one contains Brush Creek oilfield, one of many in a 70-mile-wide, northeast-trending zone that stretches from the southwest corner of the state to beyond the New York border. Most of the oil and gas in this region came from upper Devonian sandstones at shallow depths of between 1200 and 2800 feet below the surface. The oil is now largely depleted but some of the fields still produce gas.

All of the bedrock between miles 28 and 48 belongs to the Conemaugh group. Most cuts are in brownish shale and sandstone, some contain redbeds and thin limestones, and several have thin coal seams. The sequence represents alternating marine and non-marine deposition that may reflect pulses of uplift of the Alleghenian Mountains east of here. It may also reflect sea level changes independent of Alleghenian uplift. Watch for a strip mine near mile 46.

You cross the Allegheny River just south of Exit 5, with good views upstream and downstream of high cut-banks in flat-lying Conemaugh beds on the crest of a gentle anticline. Several abandoned oil and gas wells and reclaimed strip mines are nearby.

Between miles 48 and 68, the road crosses several plateau-style, large-scale open folds that are little more than gentle undulations in the bedrock. Cuts across the anticlines are mostly in Conemaugh beds and those in the synclines are mostly in the overlying Monongahela group. At the base of the latter unit is the fabulous Pittsburgh coal seam, mined since 1761.

Old strip and underground mines near miles 70, 71, and 80 are in the Pittsburgh coal seam, which varies from five to eight feet thick in Westmoreland County, really quite thick as coal seams go. Pittsburgh coal also underlies the area near mile 83 at a depth of more than 150 feet. Underground mine subsidence caused the turnpike pavement to settle, requiring costly repairs.

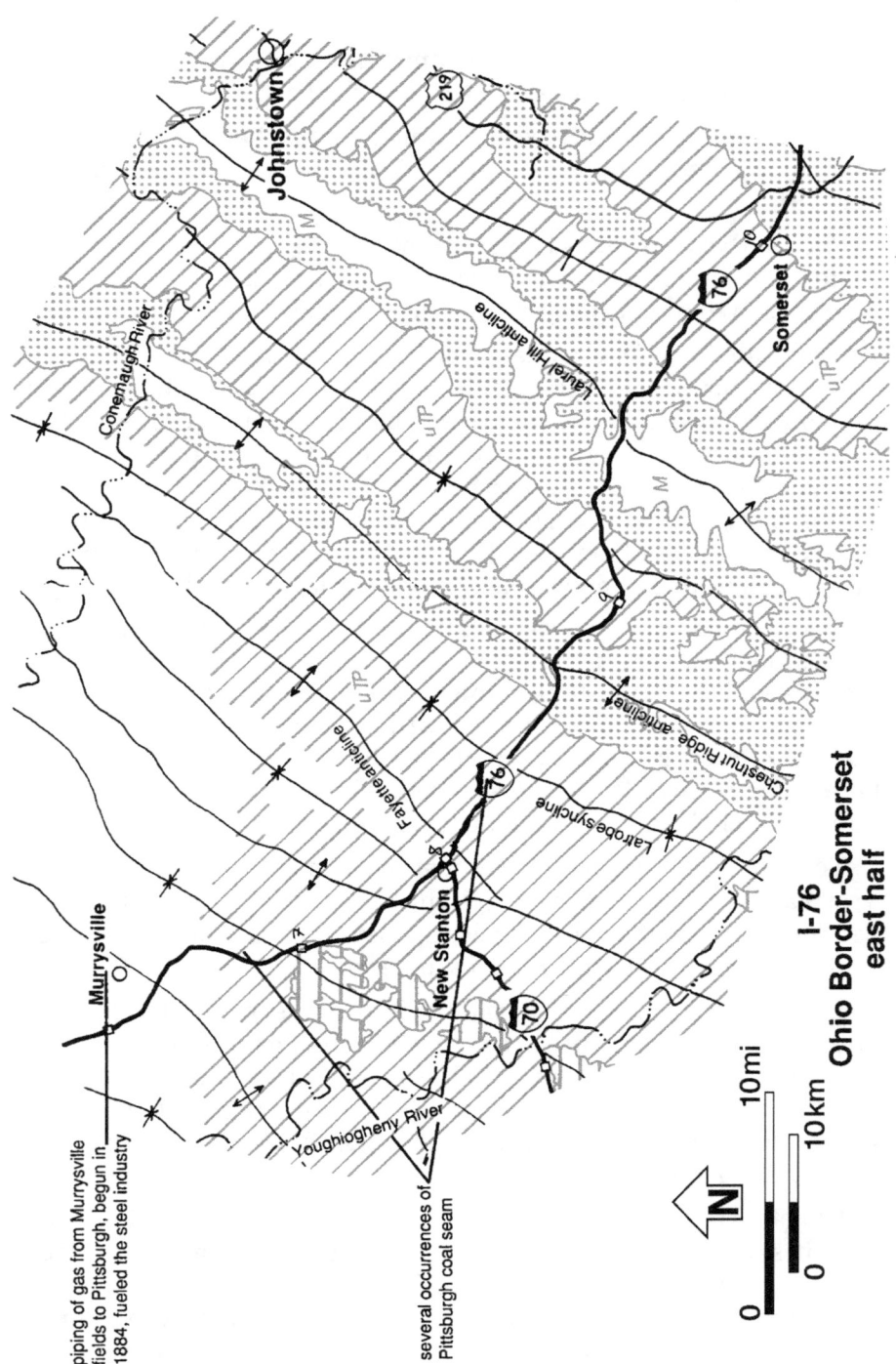

None of the folds of the eastern part of the Allegheny Plateau is obvious in single roadcuts; the bedding seldom dips more than a few degrees, and the folds are several miles wide from limb to limb. You never see a complete fold in cross-section. Nevertheless, some folds have been traced in a northeast-southwest direction for many tens of miles. The elongate outcrop patterns, shown on the general geologic map at the front of the book, make them very conspicuous. The easternmost anticlines form prominent northeast-trending ridges with intervening synclinal valleys, a subdued version of the Valley and Ridge landscape. Deformation of the two physiographic regions varies in intensity but has the same origin — collision between North America and Africa about 300 to 220 million years ago.

Plateau folding is less intense mainly because it involves a thinner slice of rock sliding over a basal thrust fault. The same thrust fault to the east apparently slices through nearly flat-lying Cambrian strata at great depth beneath the Valley and Ridge province. The much thicker pile of rock strata there offers greater resistance to compression and responds with more complex deformation. Numerous lesser faults branch upward from the basal thrust, creating smaller slices stacked westward against each other. The rocks within each slice are crumpled into tight folds.

At the Allegheny Front, however, the basal thrust ascends a ramp through more than half of the Paleozoic strata, then levels off in Silurian rocks. The thinner slice there apparently moves with relative ease, and deformation is milder.

Contact between the lower and middle Pennsylvanian Pottsville and upper Mississippian Mauch Chunk beds is near mile 86 on the axis of topographically prominent Chestnut Ridge anticline. Between miles 82 and 87, the road descends the stratigraphic section, passing through increasingly older rocks as it approaches the core of the anticlinal arch. From west to east, one first crosses Monongahela, then Conemaugh, then Allegheny, then Pottsville, and finally Mauch Chunk beds. All are part of the sediments shed from the Alleghenian Mountains rising somewhere in the east. Note the distinctive red color of the Mauch Chunk beds at the top of the cut near mile 86.

Going down the eastern flank of Chestnut Ridge between miles 87 and 91, the road passes successively younger formations as it approaches the trough of a syncline. The youngest are the Conemaugh beds exposed in the large roadcut just west of Exit 9, approximately on the axis of a shallow syncline between the Chestnut Ridge and Laurel Hill anticlines. The axis of the Laurel Hill anticline is near the western portal of the now-unused Laurel Hill Tunnel about mile 99, where red Mauch Chunk beds crop out again, with Loyalhanna sandy limestone at its base overlying buff, cross-bedded, lower Mississippian Burgoon sandstone. In the southwestern corner of the state, oil well drillers log the Burgoon and Loyalhanna formations together as Big Injun, which has produced gas and some oil. Laurel Hill ridge rises in places more than 1200 feet above the general plateau surface.

The enormous, long cut on the Laurel Hill tunnel bypass, approximately between miles 100 and 101, exposes more Mauch Chunk beds in the core of the anticline, with Pottsville sandstone on top. The contact is at road level near where the bypass rejoins the old tunnel road, between miles 101 and 102. The contact between the Pottsville and the Allegheny beds above it is less than a half mile farther east; that between the Allegheny and Conemaugh formations is near mile 103. Between miles 103 and 110, you cross another shallow syncline with Conemaugh beds in the trough and several old coal mine workings nearby.

The large cut near mile 104 exposes the 16-inch thick Harlem coal seam and the overlying Ames shaley limestone, which is full of fossils. Somerset is at the eastern limit of the region of deep gas wells that have produced from depths of 6000 to 14000 feet in Devonian, Silurian, and Cambrian sandstones.

I-79
West Virginia Border—Zelienople
86 mi./139 km.

From the West Virginia border northward almost to Exit 8, 39 miles, the I-79 road crosses Permian bedrock. These youngest Paleozoic rocks exist only in southwestern Pennsylvania, in the part of the Allegheny Plateau known as the Pittsburg Plateau. Most of the cuts expose colorfully banded beds of the Greene formation, containing sequences of grayish sandstones and siltstones, gray to brown to black shales and some coal, light gray limestone, and even redbeds. The underlying Washington formation, exposed near Exit 8, is similar, but contains the only workable Permian coals. Both formations are part of the Dunkard group, which also includes the underlying Pennsylvanian and Permian Waynesburg formation. All were laid down in a shallow inland sea that covered this region 290-250 million years ago, at a time when all the world's continents were finally coming together like a giant jigsaw puzzle to form the supercontinent of Pangaea.

Of course, you can't tell all of this from Pennsylvania's Permian rocks alone. What you can tell from the rocks in Pennsylvania is that conditions in Permian time were a continuation of those of the Pennsylvanian period. Shallow seas came and went, leaving behind the coal measures. Then came the crunch when North America and Africa collided and the Alleghenian Mountains began to build. That ended the Permian rock record in this part of the world.

Cross-section of Leidy gas field, showing entrapment of gas in folded and faulted anticline. Note that faults in Oriskany and Tully formations disappear before reaching surface. —Adapted from Wagner and Lytle (1969, p. 15)

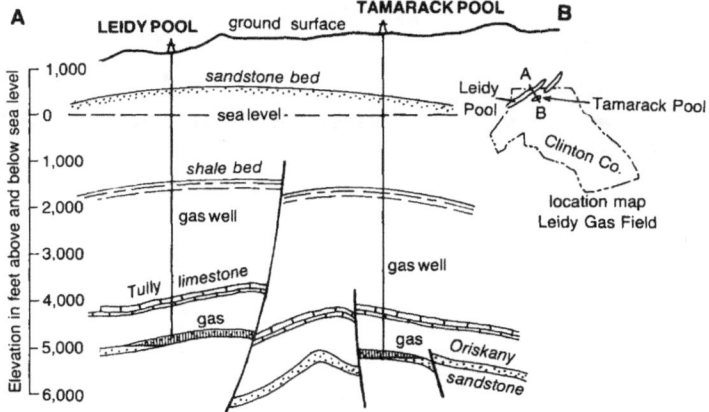

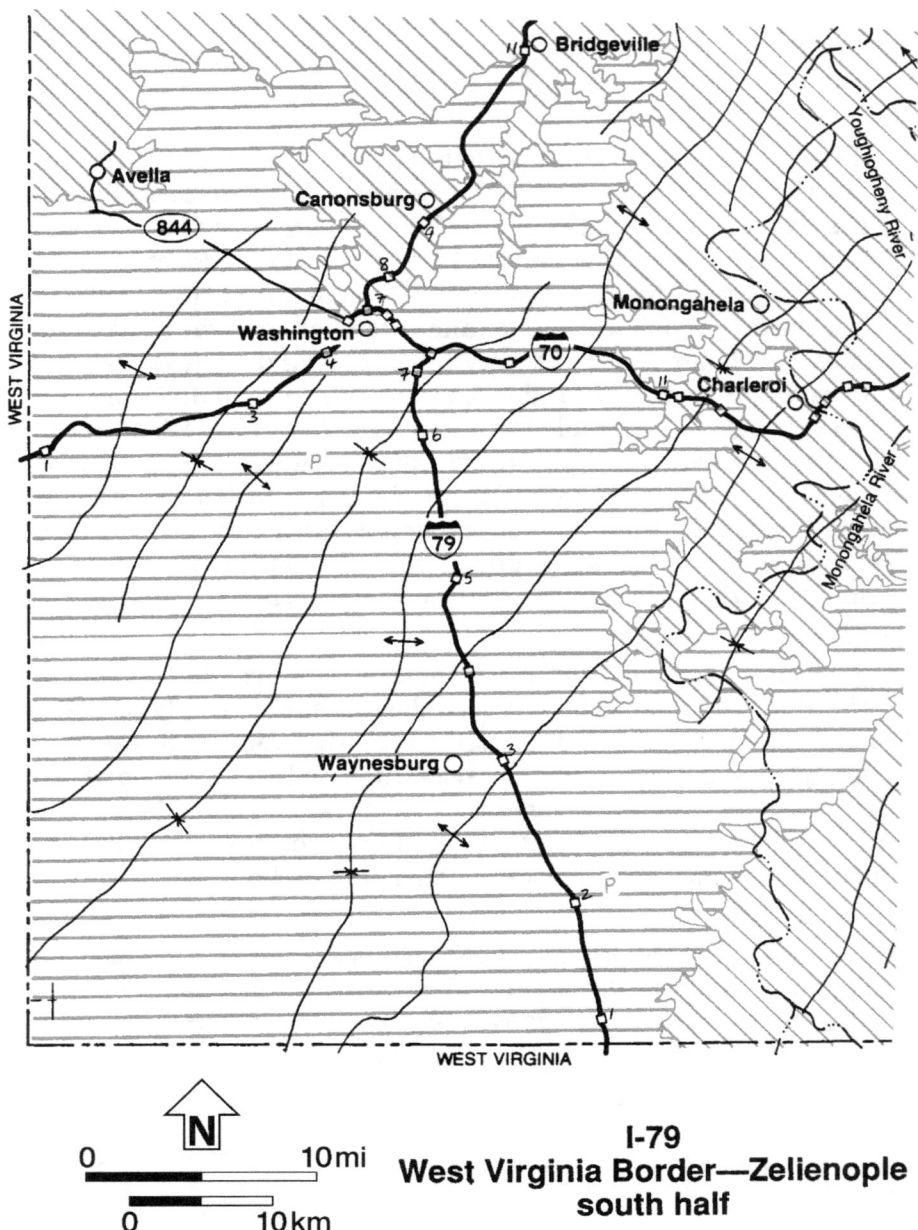

I-79
West Virginia Border—Zelienople
south half

Practically all of the coal mined in this region comes from Pennsylvanian beds. The sedimentary beds everywhere look flat, although they all dip gently southeast and are broadly folded into a very large system of anticlines and synclines that trend north.

The anticlinal arches contain the principal gas reservoirs of the region, in Mississippian and late Devonian beds at shallow depths of 1200 to 2800 feet. Some of the reservoirs also contain oil, but production has almost ceased. Gas has long been most important to the region, particularly to Pittsburgh. The discovery of gas in the Murrysville anticline east of the city in 1878 led to construction of a pipeline to Pittsburgh in 1883. Since then, natural gas has been the region's principal fuel. It was the most important influence on the industrial

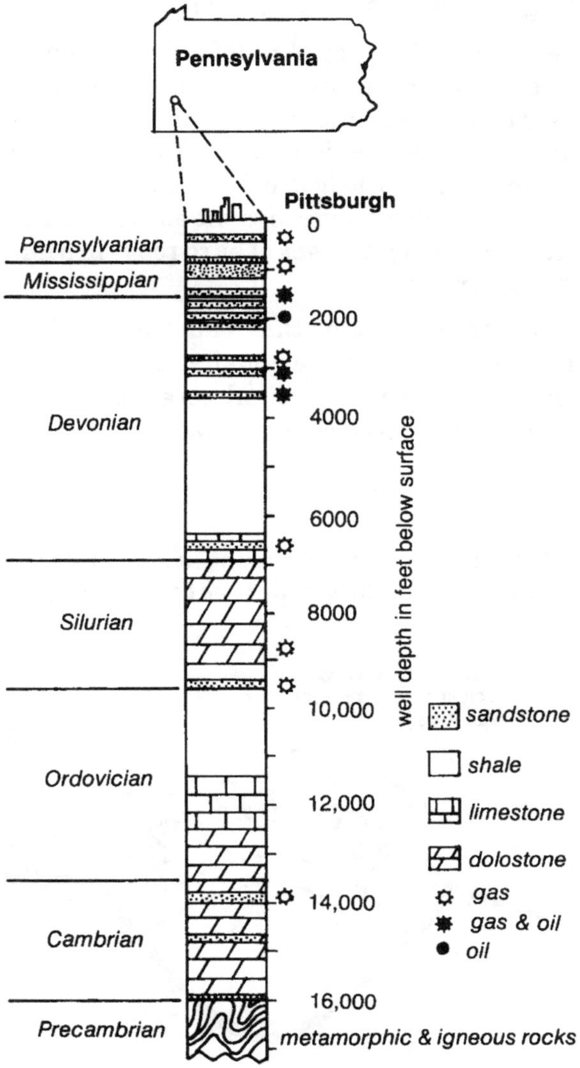

Stratigraphic section of the subsurface rocks at Pittsburgh, showing levels from which oil and gas have been produced. — Adapted from Wagner and Lytle (1969, p. 6)

61

revolution that made Pittsburgh into Steel City, USA. Since 1921, Pennsylvania has consumed more gas than it produces. Today, more than 85% of its gas comes from other states. Imported gas is pumped into old depleted gas reservoirs for storage during the summer, then pumped back out during the winter.

The moderate relief of this part of the Pittsburgh Plateau results from erosion of many small streams tributary to the Monongahela River, which flows northward east of the highway. Incision of the trunk stream is much deeper, and the tributary streams are attempting to keep pace, creating an attractive rollercoaster landscape.

Plateau Relief and Valley and Ridge Relief Compared

Most of the landscape relief of the plateau, including the glaciated portions, formed as streams eroded a sequence of flat layers of more or less equal resistance to erosion. What you see in stream banks and roadcuts is cross-sections of beds stacked up like books. The landscape does not express the bedrock. Stream dissection of folded rocks in the Valley and Ridge province to the east preferentially excavated the soft rocks, leaving high ridges of hard rocks in a landscape that does reflect bedrock structure.

It is not always true that Valley and Ridge anticlines are high and synclines are low. Occasionally, erosion inversion occurs, in which valleys develop in anticlinal cores, or ridges in synclinal cores, as shown in the cross-section below.

This schematic cross-section shows why ridges tend to develop over synclines and valleys tend to be carved out of anticlines.

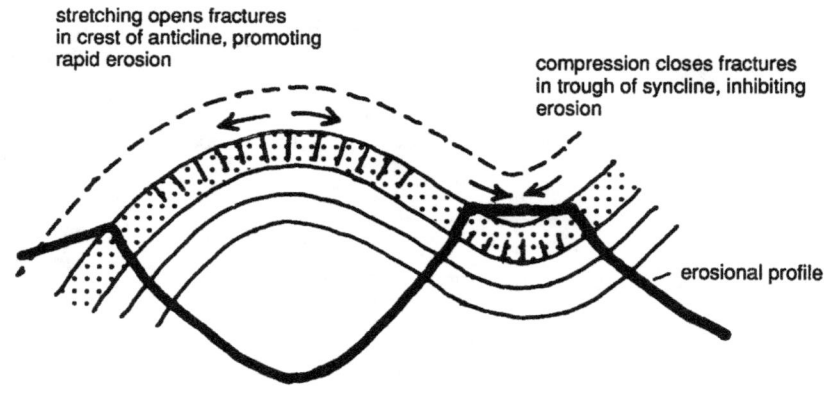

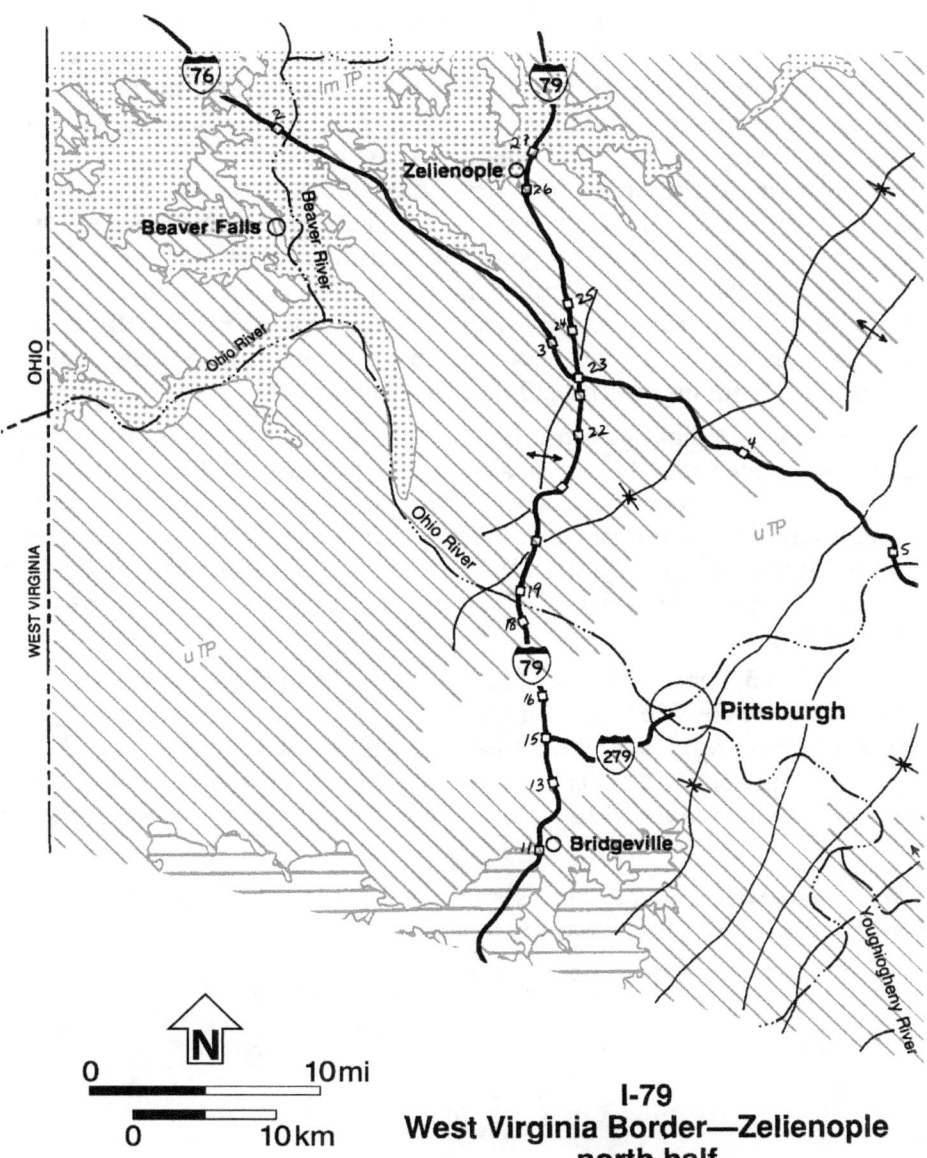

I-79
West Virginia Border—Zelienople
north half

63

- Between Exits 8 and 15, 18 miles, the road lies mostly on rocks of the Monongahela group in the upper part of the Pennsylvanian stratigraphic section. The basal three-quarters of this group is the Pittsburgh formation, which contains in its base the fabulous Pittsburgh coal seam, a continuous bed, several feet thick, that underlies almost all of Washington County, southern Allegheny County, southern Armstrong County, and western Westmoreland County. It has been mined continuously for more than 200 years, and geologists recently estimated that the reserves still in the ground in Washington County will last nearly 150 years at current production rates. The seam is completely mined out in Beaver County and virtually so in Allegheny, Armstrong, and Westmoreland counties. Practically all of it is mined underground, leaving the greater Pittsburgh region virtually sitting on air and subject to ground subsidence.

The Monongahela group differs from the Permian formations in containing few redbeds and an abundance of dolomitic limestones, in addition to shales and siltstones. Thin sandstones replace the thick sandstones that dominate the underlying Conemaugh group. Roadcuts in the Pittsburgh formation at the Bridgeville exit expose rocks fairly typical of the lower part of the unit, with light gray carbonates dominating.

At Exit 15, the I-279 interchange, cuts in the Conemaugh strata that underlie the Monongahela formation, expose interbedded light brown to buff sandstone and siltstone, and dark brown shales. The roadbed between Exits 15 and 27, 29 miles, crosses Conemaugh bedrock almost exclusively. Many cuts, some quite large, reveal

This cut by the Bridgeville exit exposes dark shales in the Pittsburgh formation. The pale rocks above and below are limestone.

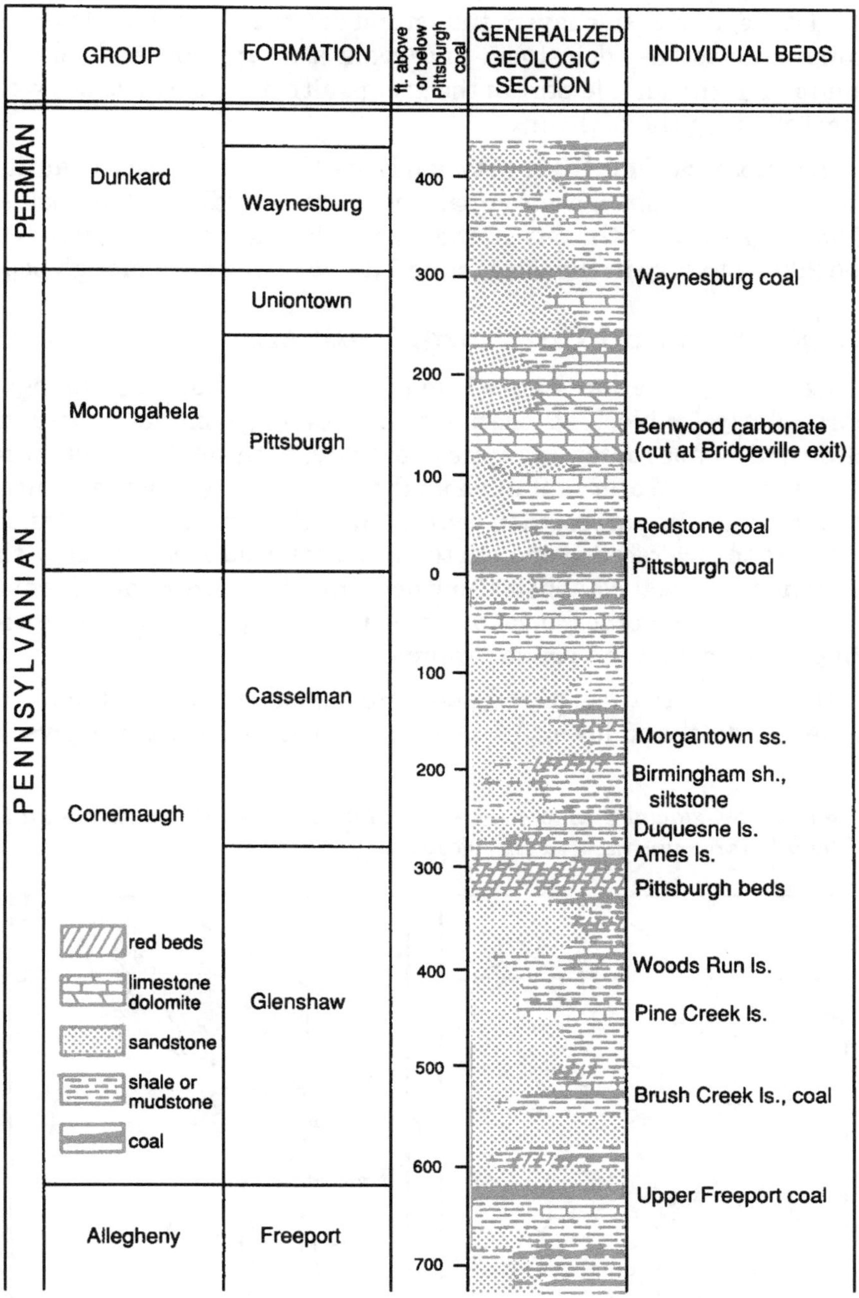

GROUP	FORMATION	ft. above or below Pittsburgh coal	GENERALIZED GEOLOGIC SECTION	INDIVIDUAL BEDS
PERMIAN				
Dunkard	Waynesburg	400		
PENNSYLVANIAN				
Monongahela	Uniontown	300		Waynesburg coal
		200		
	Pittsburgh			Benwood carbonate (cut at Bridgeville exit)
		100		
				Redstone coal
		0		Pittsburgh coal
Conemaugh	Casselman	100		
		200		Morgantown ss. Birmingham sh., siltstone Duquesne ls. Ames ls.
		300		Pittsburgh beds
	Glenshaw	400		Woods Run ls. Pine Creek ls.
		500		Brush Creek ls., coal
		600		Upper Freeport coal
Allegheny	Freeport	700		

Legend:
- red beds
- limestone dolomite
- sandstone
- shale or mudstone
- coal

Exposed coal measures in Allegheny County.
—Adapted from Wagner and others (1970, p. 46)

65

interbedded, rather dark gray to brown shale and light brownish sandstone, in massive layers that stand out as ledges. Thin layers of impure limestone, dolostone, and coals also appear. Redbeds are common in the middle part of the group, although they are not well-exposed along the highway.

You cross the Ohio River between Exits 17 and 19, with a beautiful view toward Pittsburgh encompassing the tall buildings of the Golden Triangle section of the city in the distant background. The river cut the 200-foot scarp in the foreground into flat-lying Conemaugh beds.

Glaciation and Pittsburgh's Rivers

Ice age glaciation caused many stream drainage changes, particularly by blocking channels and forcing meltwater through alternate escape routes. Many new routes went through passes in pre-glacial divides. Water first filled up the valleys in front of the ice until it overtopped the passes, then poured through, cutting new gorges. By the time the ice receded enough to uncover the old escape routes, the new ones were well-established and permanent. Most of Pennsylvania's narrow bedrock gorges formed in this way. Slippery Rock gorge and Pine Creek gorge, are prime examples.

In pre-glacial time, three separate major stream systems, the upper, middle, and lower Allegheny, drained western Pennsylvania

Pre- and post-glacial drainage networks in western Pennsylvania are quite different. —Adapted from Wagner and others (1970 p. 86-87)

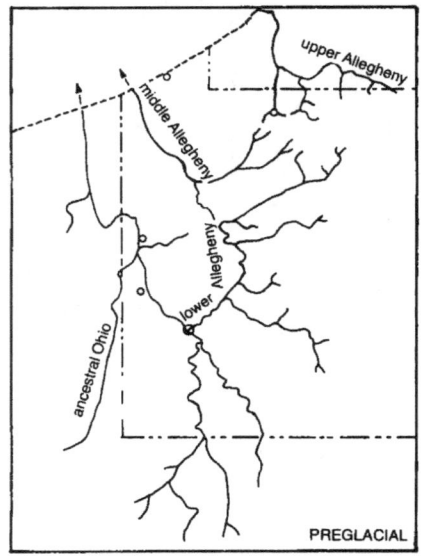
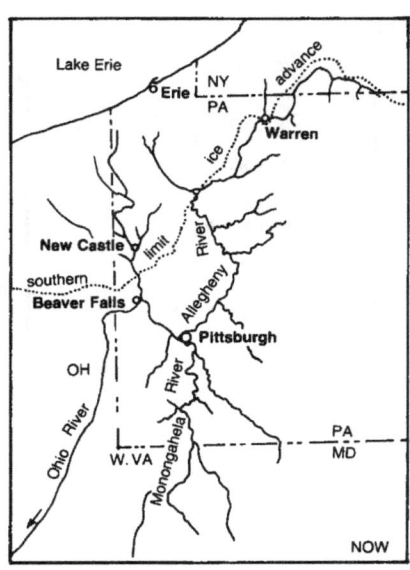

Cut on east side of I-79 just north of Exit 22 exposes interbedded sandstone and siltstone of the Conemaugh group, with a sharply-defined sediment-filled stream channel on the right side.

northward to a northeast-flowing master stream in the region of the present lakes Erie and Ontario, and emptied to the Atlantic Ocean via the St. Lawrence River. Even the ancestral Ohio River, shown on the "before" map, drained northward, tributary to the Monongahela of the lower Allegheny system. The Golden Triangle juncture at the future site of Pittsburgh was already there, but the Allegheny discharge must have been less than it is today because its drainage basin was much smaller. Drainage divides separated the three Alleghenies.

All the rivers today unite into a single Ohio River system that empties to the Mississippi and the Gulf of Mexico. No major streams now drain the plateau northward to Lake Erie. The Ohio, shown on the map, is no longer just a tributary, but is part of a much enlarged and extended master trunk stream in which flow direction is reversed. Its old headwaters drainage divide, by the West Virginia-Ohio border, has been destroyed, as have several other divides in Pennsylvania that permitted the joining of the systems.

• Zelienople is beside one of the tributaries of this new river network, Connoquenessing Creek. Several miles west at Ellwood City, the creek joins Beaver River which then flows south to join the Ohio at Rochester. Since the last ice age ended, the creek has cut deeper into the plateau, trying to keep pace with the downcutting of the master streams and leaving behind some lovely gorges and waterfalls.

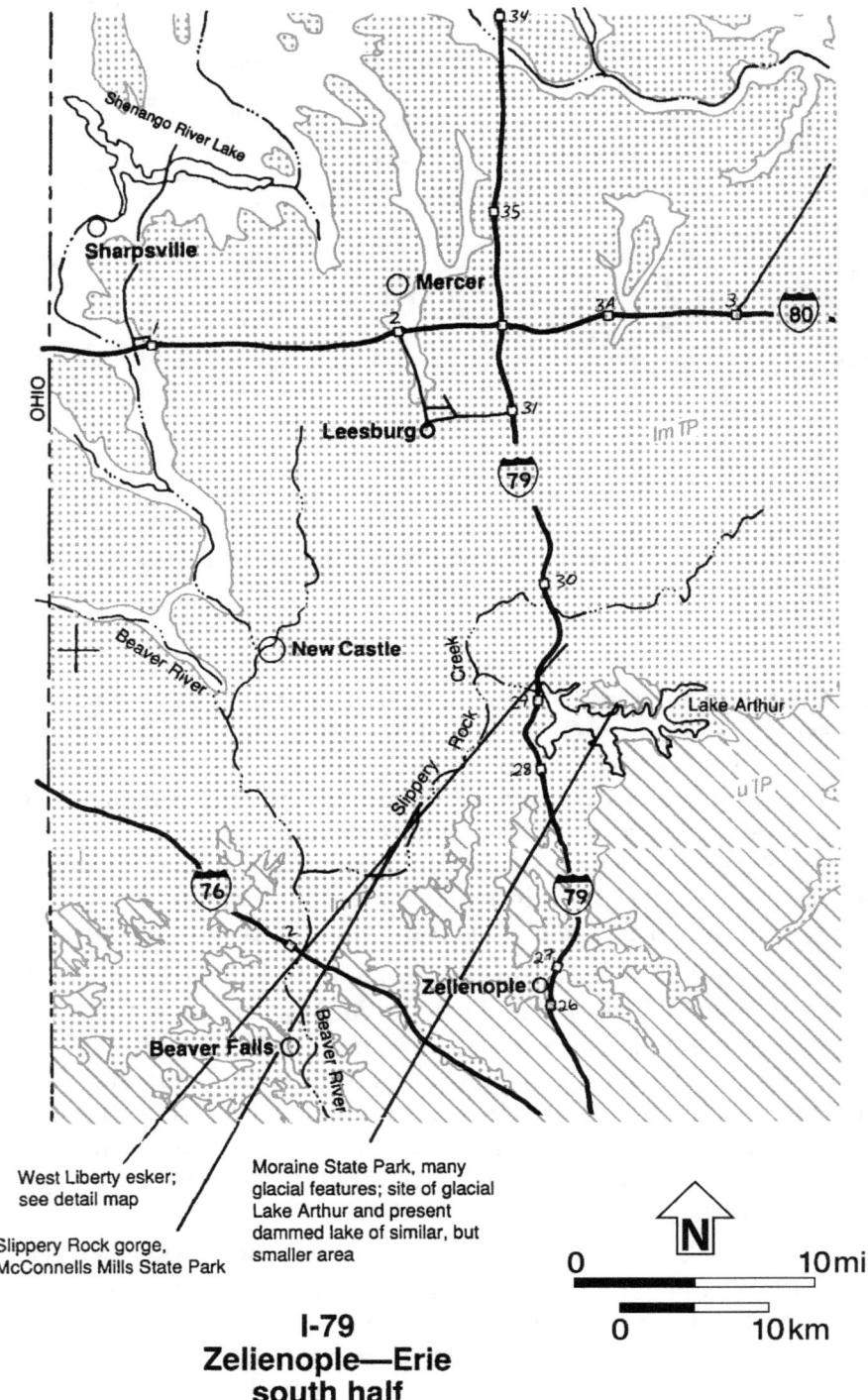

West Liberty esker; see detail map

Slippery Rock gorge, McConnells Mills State Park

Moraine State Park, many glacial features; site of glacial Lake Arthur and present dammed lake of similar, but smaller area

**I-79
Zelienople—Erie
south half**

I-79
Zelienople—Erie
95 mi./153 km.

Between Exits 27 and 29, 13 miles, I-79 traverses Pennsylvanian Allegheny group beds almost exclusively. You'll see layers of flat dark shale, some lighter sandstones and limestones, and a few coal seams. Moraine State Park lies west, and McConnells Mill State Park east of the highway between Exits 28 and 29. Exit 29 lies on the glacial border; the region west of it was covered by ice sheets; the region east of it was not.

Moraine State Park

This park, adjacent to the east side of I-79, is easily reached from Exits 28 or 29. The park is a showcase of glacial features, including the West Liberty esker. Its central feature, Lake Arthur, formed by damming of Muddy Creek near Exit 29, occupies the same basin as that of its predecessor, glacial Lake Watts, although it is much smaller. The original lake formed during the last ice age when ice blocked Muddy Creek approximately where the man-made dam is today. The lake then extended about six miles farther east than does the present one, and the water level was about 70 feet higher. The even larger and higher Lakes Edmond and Redmond formed contemporaneously in the ice-blocked Slippery Rock Creek drainage basin to the northeast, and water escaped from them to Lake Arthur via a spillway at West Sunbury. This is an area of extensive land disturbance by strip mining of coal from the Conemaugh and Allegheny groups. As a result, some of the glacial features have been destroyed. The lake also straddles the western Pennsylvania region of shallow oil fields. Numerous old wells are on its southern central side.

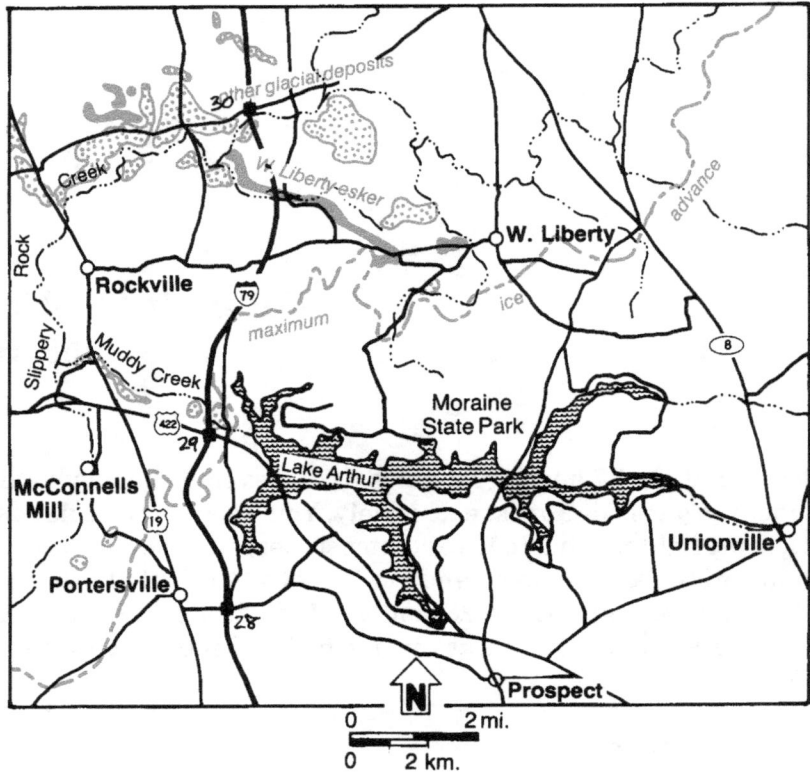

Sketch map of the Lake Arthur region showing glacial features. —Adapted from Lytle, W.S., Moraine State Park, Pennsylvania Geological Survey Park Guide #4

Slippery Rock Gorge

This gorge is the main geologic attraction of McConnells Mill State Park, a few minutes drive west of Exits 28 or 29. Torrents of meltwater excavated the gorge about 40,000 years ago, during the last ice age. The pre-glacial drainage system was quite different from the present one.

Cleeland Rock, about three miles west of Portersville served as a stream divide. One stream flowed north to join the pre-glacial Slippery Rock Creek near Princeton Station, which continued west from there. Another stream flowed south to join the pre-glacial Conoquenessing Creek at Wurtenburg. At one point, ice reached south almost to the rock, burying all but the upper reaches of the Slippery Rock drainage net. A system of interconnected lakes formed in front of the ice there, including Lake Watts, that spilled over the pass at Cleeland Rock and began to cut through it. The process continued during retreat of the ice front, eventually destroying the divide altogether, cutting Slippery Rock Gorge, and initiating some drainage reversals.

The bedrock floor of the gorge is covered with 20 to 40 feet of glacial and more recent rock debris. The lower gorge walls expose flat-lying sandstones and interbedded coaly shales of the Pottsville group; in the upper parts are coals, shales and limestones of the Allegheny group.

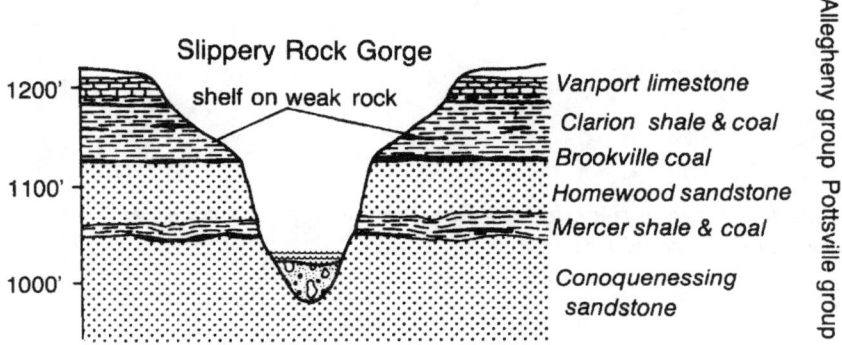

Schematic cross-section of Slippery Rock Gorge. —Adapted from Bushnell K,. McConnels Mill State Park, Pennsylvania Geological Survey Park Guide #9

Why Slippery Rock? The name apparently derives from a slab of Pottsvile sandstone near the mouth of Hell Run that a natural oil seep made slippery. Many such seeps prompted extensive drilling along the creek after completion of the first commercial oil well at Titusville in 1859. None was very successful; all are now abandoned.

The 1846 Lawrence Furnace is at the southern end of the park. This and other furnaces produced iron from ores of the Vanport limestone that crops out near the gorge rim.

• Between Exits 29 and 30, the road passes the contact between the Allegheny and underlying Pottsville groups. The rocks increase in age toward the north because the whole layer cake of beds, from the lowest, upper Devonian, to the highest, Permian layers, is tilted slightly to southeast and truncated at the surface by erosion. The simple pattern is, of course, complicated by modern stream dissection, which cuts through the uppermost beds and creates long, narrow, tree-like exposures of the older, underlying beds in the valley sides. These outcrop patterns, most conspicuous in northwestern Pennsylvania, are common wherever flat-lying or gently dipping, unfolded strata are intricately stream dissected, as they are here. Glaciation produced the attractive, smooth, rolling landscape by shaving the hilltops, rounding corners, and filling in some valleys.

The treeless ridge in the middle ground is the West Liberty esker as seen from Reichert Road.

West Libery Esker

West Libery esker, perhaps the best in Pennsylvania, is about 15 minutes from Exit 30. It is one of many glacial features in and around Moraine State Park. The esker is a low, rounded ridge with a surprisingly even crest; it resembles an abandoned railroad embankment. Three aligned segments of the esker once formed a continuous ridge about 6.5 miles long. The segment nearest West Liberty is by far the longest, about three miles.

The core of the esker is water-washed sand and gravel, stream deposits. In places it is covered with till dumped from the ice as it finally melted. The cover of till shows that those parts of the esker formed along a stream that tunneled through or beneath the ice. Other places have no till cover, suggesting that at least in its final version, the stream flowed in the ice open to the sky, perhaps following a system of crevasses. It is entirely possible for meltwater to flow for a way along a crevasse, then tunnel for some distance before breaking out again into an open channel. The esker records a period of glacial stagnation when sediment-charged meltwater filled every available hollow and flowed through every existing channel, finally issuing from the ice margin.

The esker's shape, as seen from the air, gives clues to its origin. Segments west of I-79 have sinuous trends, like a river. East of I-79, straight segments with angular bends recall a pattern of crevasses. The esker material is unconsolidated and easily eroded. The originally continuous ridge is now divided into three segments because Slippery Rock Creek and Taylor Run have cut through large parts of it over the past 20,000 or so years.

• Near Exit 34, by the Shenango River, you cross a major unconformity between lower Pennsylvanian Pottsville beds and the underlying lower Mississippian Shenango formation. Missing are approximately 30 million years of geologic record, either because no sediments were deposited during that period, or the region was eroded. In either case, the missing sedimentary section may represent an early pulse of the Alleghenian uplift.

Conneaut Lake

This, the largest natural lake in the state, is by US 6, a few minutes drive west of Exit 36. The Conneaut valley and surrounding landscape are replete with features formed as the ice of the last ice age melted. They resemble those of a typical valley glacier because the valley held a tongue-like extension of the ice sheet, which behaved much like an alpine glacier.

Conneaut Lake appears to be a kettle formed when the ice margin stood just to the north. Earlier, as the glacier melted, outwash sediments buried large blocks of ice. When the ice finally melted, Conneaut Lake filled the depression it left. It lies in a plain of outwash sediments confined within the valley that continues for several miles to the southeast along Conneaut Outlet. I-79 crosses Conneaut Outlet by Exit 35 where it is about 200 feet deep. Its wide, very flat floor contains Conneaut Marsh, one of the largest in Pennsylvania.

A large kame stands alongside Ellion Road between US 6 and Faust Road, one mile east of Conneaut Lake. It is an elongate hill about 100 feet high and a mile long, one of the largest of its kind in northwestern Pennsylvania. It trends north-south, beside the former ice margin, and probably formed in standing water as a small delta at the mouth of a meltwater stream issuing directly from the ice. This is apparent from the shape, position next to the ice, and pattern of internal layering revealed in quarries. The sediments are clay-free, water-washed sands and gravels with the fairly well-defined bedding typical of stream deposits.

Examination of the cobbles in glacial deposits is always an interesting exercise if you know enough of the regional bedrock geology to tell where the rocks came from, and how far the ice carried them. For example, this deposit contains granites and quartzites that most certainly were carried from Canada across the Lake Erie basin where the nearest exposures of these rock types exist on the far side

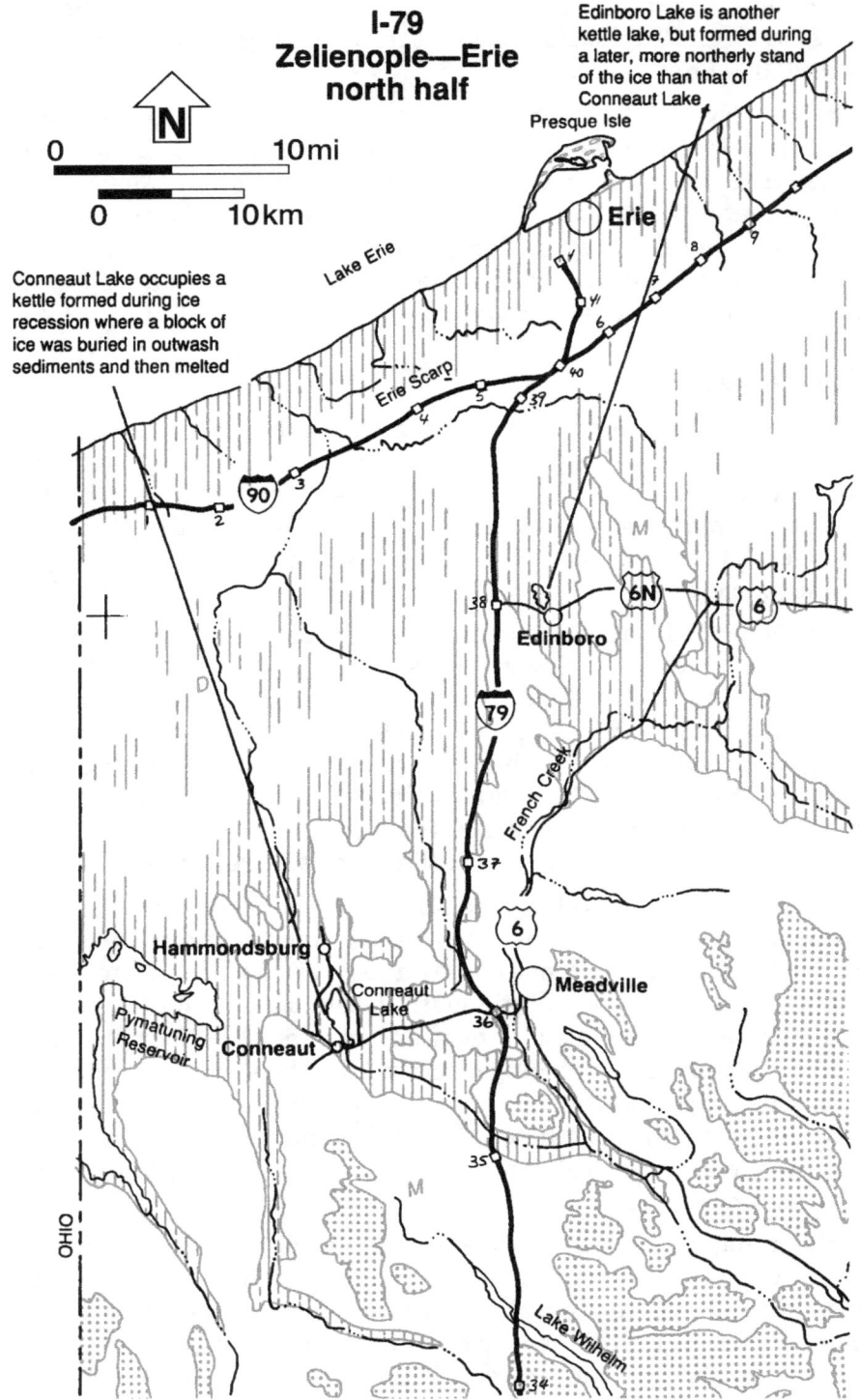

of the Lake Erie basin. The cobbles also include limestones from New York state, and siltstones and sandstones from closer sources in Pennsylvania. Boulders that are strangers to the bedrock of the regions where they are found, are called erratics.

More subtle glacial features are the hummocky terminal moraine that trends northwest and passes through Conneaut Lake village, a terrace on the eastern shore of the lake, outwash plain between the lake and the village of Harmonsburg to the north, and a thin ground moraine blanket in the regions bordering the valley.

• Between Exits 36 and 38, 21 miles, the highway crosses formations of the Mississippian Cuyahoga group. A very few roadcuts reveal horizontal layers of dark shales and a few thin beds of pale sandstones. This is plateau country with a glacially-smoothed rolling landscape of moderate relief. This section of plateau is not as deeply stream carved as the Pittsburgh Plateau, probably because it is in the headwaters of the Ohio River basin, where streams are smaller. A number of swamps alongside I-79 fill shallow, glacially-excavated depressions.

Edinboro Lake, three miles east of Exit 38, is a small kettle formed during a pause in ice recession from this region.

Between Exits 38 and 39, nine miles, you cross the moraine that delineates this later ice stand. You also pass the contact between Mississippian and Devonian bedrock. You can see dark shales at about mile 169 of the late Devonian Venango formation. Lighter shales, siltstones, and sandstones of the underlying Chadakoin formation crop out along a small creek by Exit 39.

The Erie scarp, at the northern edge of the Allegheny Plateau, lies between Exits 39 and 42. The scarp is about 200 feet high and is carved from the Chadakoin formation and beneath it, the Girard shale, neither of which appears in conspicuous outcrops.

At the US 20 interchange, you are on the glacial Lake Whittlesey beach ridge, at the break in slope marking the junction between the Erie scarp and the Erie plain. Lake Whittlesey was one of the earliest and highest versions of Lake Erie to fill the basin after the ice front backed off of the Erie scarp. A later and lower version, called Lake Warren, formed another beach ridge that you may cross a little more than a mile farther north. The northern shorelines of both lakes lapped against the ice front when it stood somewhere in the Erie and Ontario basins, and also blocked the St. Lawrence River.

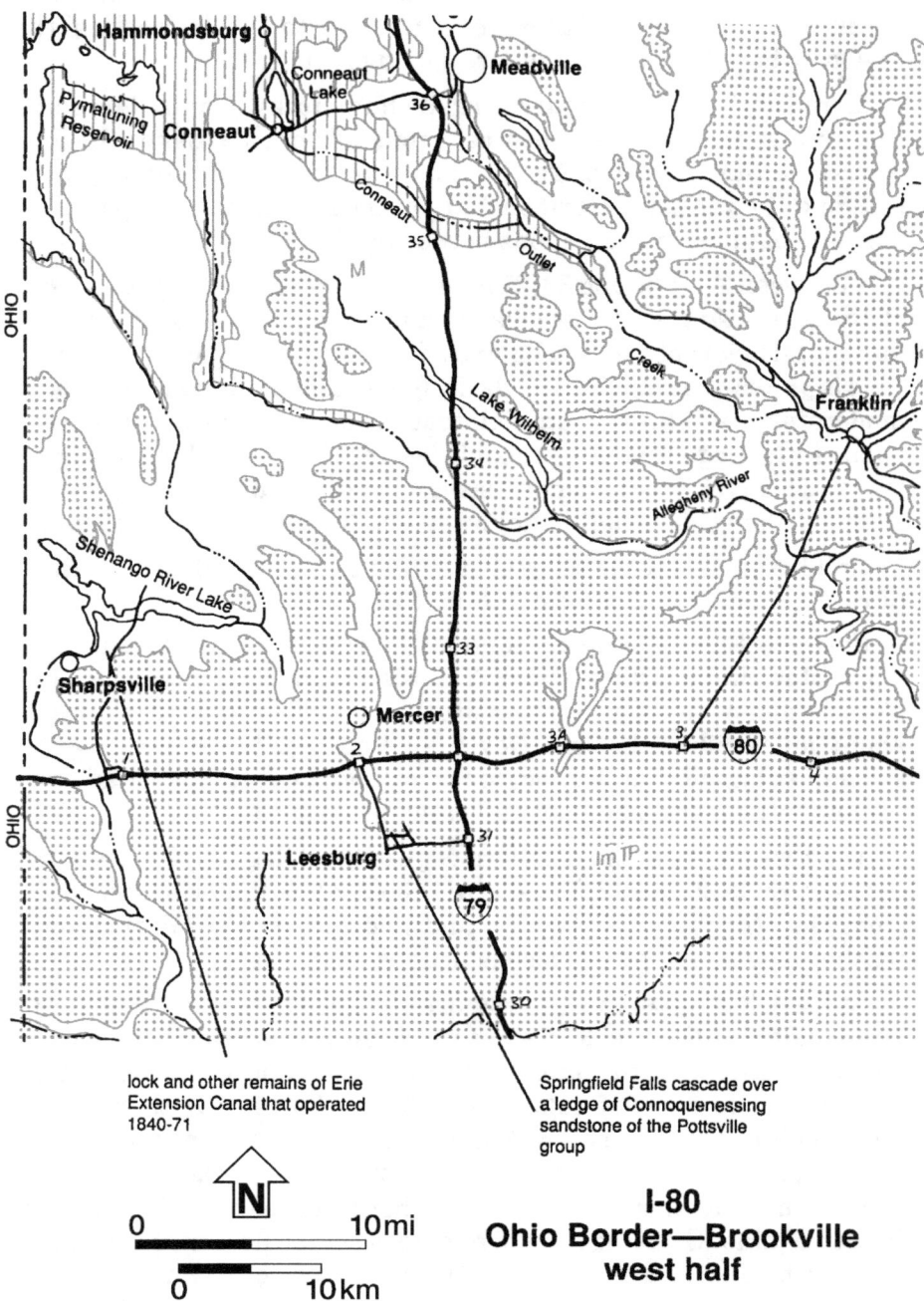

lock and other remains of Erie Extension Canal that operated 1840-71

Springfield Falls cascade over a ledge of Connoquenessing sandstone of the Pottsville group

**I-80
Ohio Border—Brookville
west half**

I-80
Ohio Border—Brookville
78 mi./126 km.

Nearly all the rocks exposed along this route belong to the Pennsylvanian Pottsville and Allegheny groups. Active and reclaimed strip mines are visible from the highway. The widespread bituminous coal of western Pennsylvania is part of the enormous Appalachian coal field that extends south to Alabama and covers thousands of square miles. Approximately 50 individual coal seams exist within the Pennsylvanian rock section.

This region also produces oil and gas from Devonian and younger rocks at depths of only 600-700 feet. In places you can see oil well pumps bobbing up and down, as well as gas storage tanks.

This highway crosses the northwestern Pennsylvania glaciated plateau between miles 0 and 25, a landscape that differs subtly but significantly from that farther east. This land lay beneath at least two ice age glaciers; each covered almost the same area. The glacial margin extends from Beaver Falls, about 30 miles south of I-80 near the Ohio border, northeast to northeastern Warren County. It is marked by terminal moraine of both glaciations; the hummocky topography near mile 25, for example, is the moraine of the last ice age.

Ground moraine covers much of the region west of the old ice margin. This material grossly resembles that of the end moraine, but the deposits either smeared under the active ice sheet or dropped from it as the ice melted downward. Ground moraine forms a thin blanket instead of a ridge.

Cleaner, water-worked outwash sediments also abound. All are sparingly exposed in roadbanks and in a few sand and gravel pits visible from the highway. Large boulders, glacial erratics, dot fields at mile 14. Swamps, visible from the highway at miles 11 and 14 are a telltale sign of glaciation. Some are hollows scooped out by the ice as it advanced over the land; others are kettles that fill basins left when lingering blocks of ice melted after the glacier retreated. In either case the result is poorly drained depressions isolated from streams.

Morainal deposits that line these depressions commonly contain much glacial flour, finely pulverized rock ground as ice-bound boulders milled against each other and the bedrock. Glacial flour is impermeable; it holds water in ponds and swamps. Some swamps are relics of

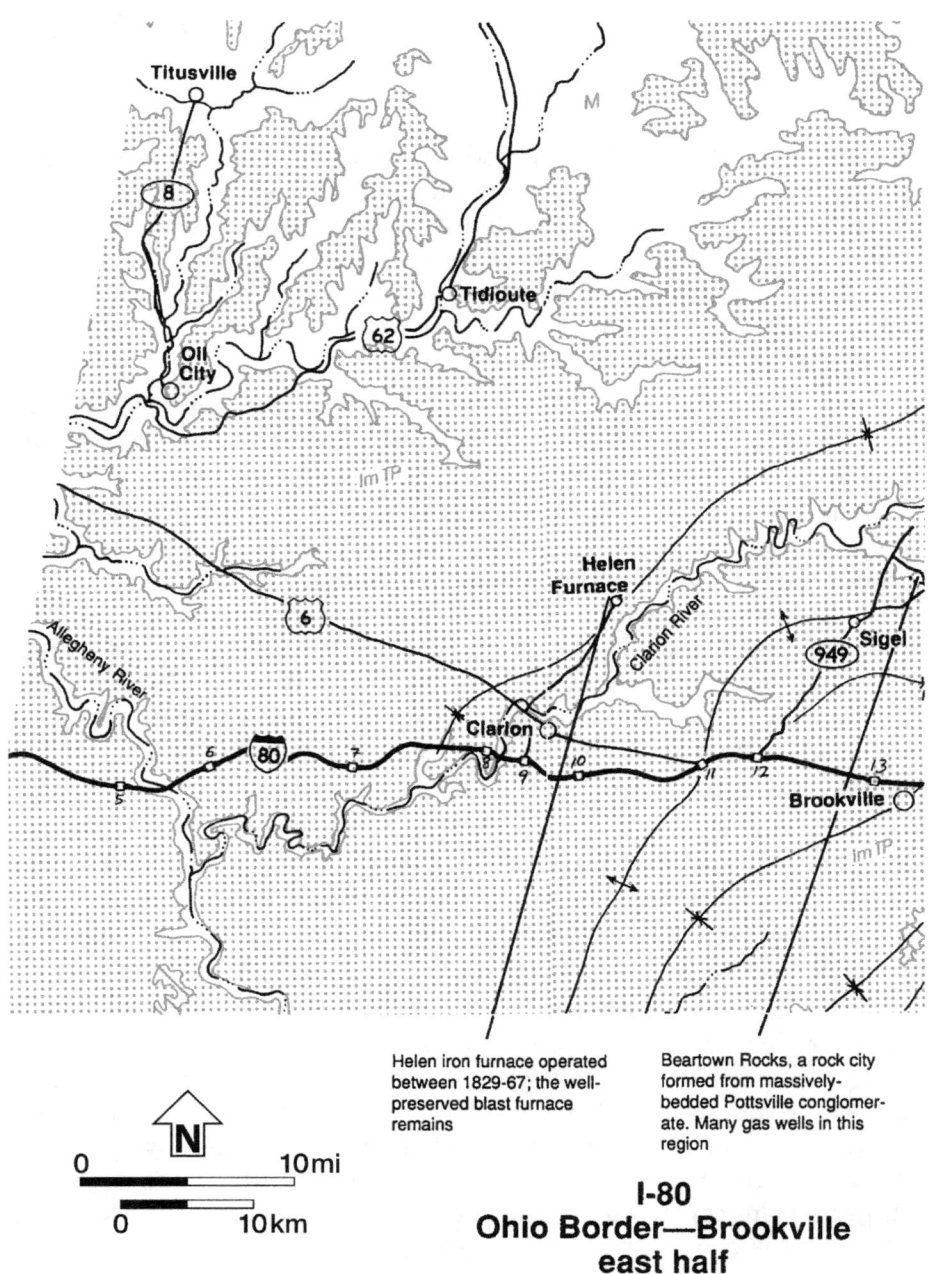

Helen iron furnace operated between 1829-67; the well-preserved blast furnace remains

Beartown Rocks, a rock city formed from massively-bedded Pottsville conglomerate. Many gas wells in this region

**I-80
Ohio Border—Brookville
east half**

meltwater lakes that originally covered much larger tracts of land, and were dammed against the ice during early stages of glacial retreat. Extremely flat sediment plains commonly mark old lake beds.

Glacial scour and sedimentation in this region left a subdued landscape that required very few roadcuts. One, at mile 12, shows light brownish sandstones of the Pottsville group.

Erie Extension Canal

Remains of the Erie Extension Canal, which operated from 1840 until 1871 and served the western Pennsylvania iron industry, lie along Pennsylvania 18 north of Exit 1. Near Sharpsville, for example, about five miles north, is the only surviving canal lock of this system. Another section of the canal is south of Greenville three tenths of a mile east of Pennsylvania 18 on the Wasser Bridge Road.

• Roadcuts are more numerous between miles 25 and 78, Exit 13, in part because this section has more relief, and in part because it is not covered with glacial sediments. You will see Pottsville and Allegheny beds, mainly light brownish sandstones, some massively bedded and cross-bedded, siltstones, and brownish to black shales, with a few coal beds.

You cross the Allegheny River at mile 45, and the Clarion River near mile 61. Both streams cut through the Pottsville formation, exposing narrow bands of the underlying lower Mississippian strata. The contact is a major unconformity representing a gap in the geologic record of about 30 million years. Remnants of the Mississippian rocks have been found in the base of the Pottsville formation indicating that the unconformity is an old erosion surface, not merely a period in which no sediments were deposited.

Beartown Rocks — A Rock City

Massively-bedded Pottsville conglomerates are particularly prone to form rock cities where they crop out on hillslopes. One fine example is Beartown Rocks in Kittanning State Forest about twelve miles northeast of Exit 12. Rock cities form because the thick beds break along widely separated joints into huge blocks that lean this way and that as the underlying softer, shaley beds erode. Frost heave and prying plant roots aid the initial opening of the joints.

The bedrock and topography of the Allegheny Plateau strongly favor formation of rock cities. Massively-bedded resistant conglomerates and sandstones are common in rocks of Pennsylvanian to upper Devonian age that crop out there, and many overlie weaker rocks. Stream erosion left these flat-lying beds projecting from gentle hillslopes, where they gradually break up into large blocks.

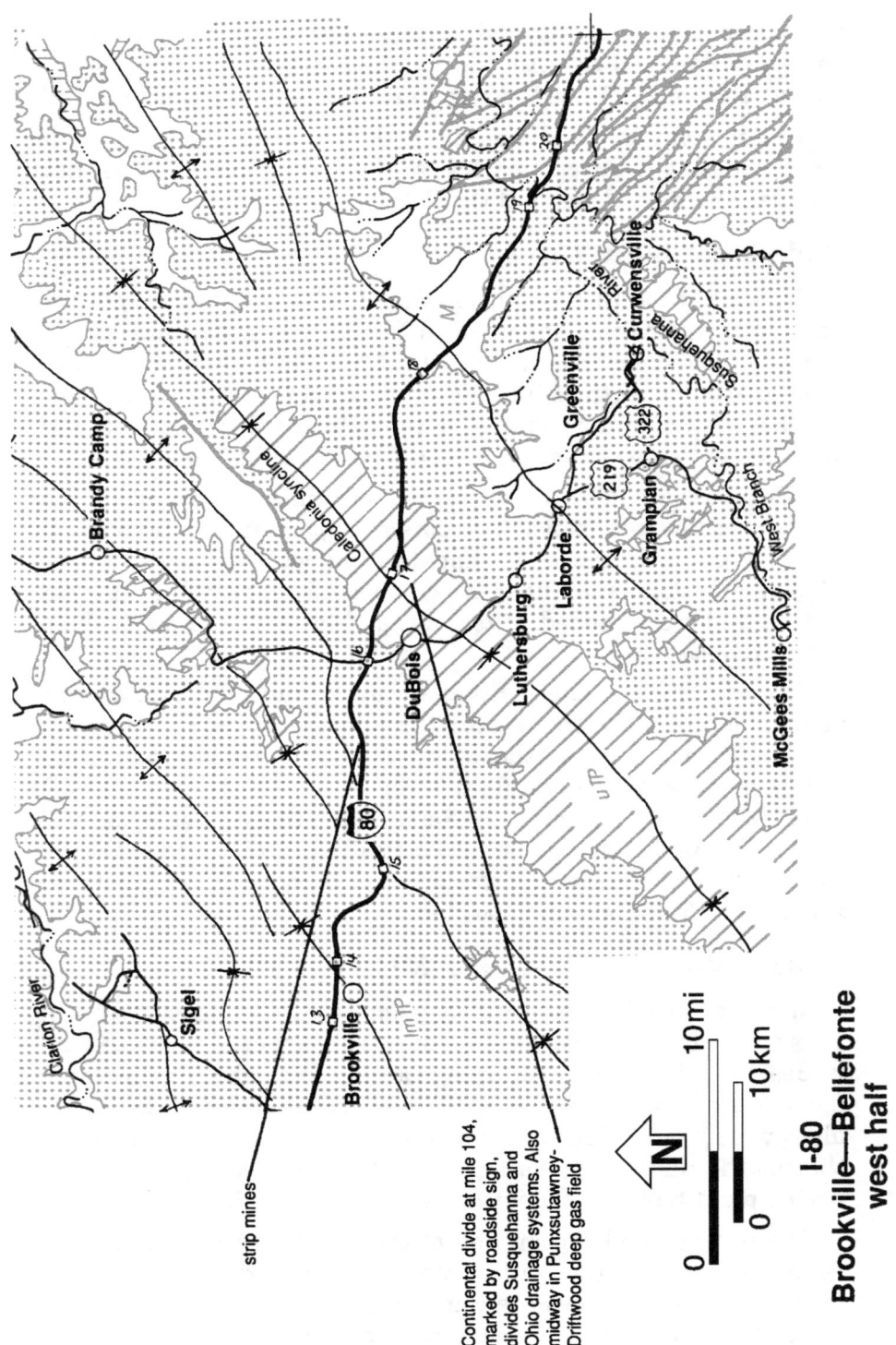

I-80
Brookville—Bellefonte
west half

I-80
Brookville—Bellefonte
82 mi./132 km.

This route crosses the eastern part of the Allegheny Plateau through roadcuts that expose many Pennsylvanian Pottsville, Allegheny, and Conemaugh beds, some with coal seams. This is part of Pennsylvania's bituminous coal region; numerous strip mines dot the landscape. Plateau-style, large-scale folding is more pronounced here than in the region west of Brookville. Where the highway cuts deeply into anticlines, it repeatedly passes down the stratigraphic section through older rocks and then back up again.

Between miles 78 and 81, you cross the Shawmut syncline, between miles 81 and 86, the Sabinsville anticline. You'll see Allegheny dark shales and light brownish siltstones and sandstones, but no coal. several exposures of coal-bearing Allegheny sequences, called cyclothems, appear between miles 87 and 97.

Cyclothems

Repeated coal-bearing sedimentary sequences, or cyclothems, are cyclic alternations of marine and non-marine sediments formed in response to wide, gradual, back and forth shifts of the shoreline of a shallow inland sea. The coals rest on non-marine sandy shales and clay deposited by streams. The coal starts as deposits of plant matter, peat, that accumulate in coastal swamps and then buried under marine muds and sand as the sea transgresses. The vegetable debris is slowly converted to brown lignite, then coal. Meanwhile, clay muds become mudstone or shale; calcareous muds become limestone; and sand becomes sandstone.

In outcrop, coal cyclothems are generally thin bedded, and each bed has its distinctive color and resistance to erosion. Coal is almost invariably black and very easily eroded. Shale is typically gray or black, thinly laminated and also easily eroded, tending to break down to a rubble of small chips. The sandier shales are paler, less thinly bedded, and stand up better to erosion. The limestones generally are pale gray rocks in thicker beds, ledge formers that may show pockets formed by solution.

Pennsylvanian formations contain as many as thirty individual coal seams, representing as many different marine transgression-regression cycles.

• You see several strip mines between miles 92 and 97. Watch for the excellent outcrop of Conemaugh beds at mile 97, just east of DuBois exit 16, with thick layers of tan sandstone overlying coal and shale. Exit 17 to Pennfield is on the axis of the Caledonia syncline, one of the major folds of this region. It has been traced more than 200 miles in a northeast-southwest direction. You see Conemaugh beds in the core of the fold and the underlying Allegheny beds east and west of the core.

The column on the left shows two typical coal cyclothems, while the graph on the right shows how the component sedimentary rocks are interpreted in terms of environment of deposition. —Adapted from Stanley, S.M., 1986, *Earth and Life Through Time*, W.H. Freeman and Co.

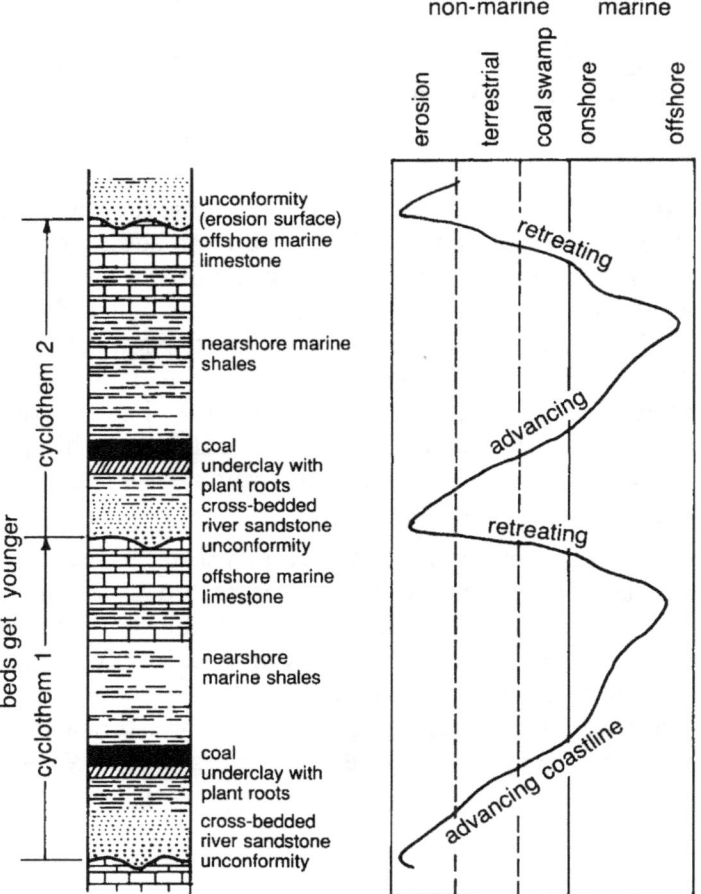

A sign at mile 104 marks the continental divide. Streams on the western side empty to the Allegheny River, then through the Ohio and Mississippi rivers to the Gulf of Mexico. Those on the eastern side flow to the Susquehanna River and then to the Atlantic Ocean by way of Chesapeake Bay. The eastern drainage network has long meandering segments that drain northeast along synclinal axes and then turn southeast to cross anticlines through water gaps. Most tributaries flow northwest or southeast off the flanking anticlinal ridges. These, and the strong northwesterly alignment of meanders in the trunk streams, especially in the west branch Susquehanna, reflect a dominant northwest-trending fracture pattern in the bedrock. By contrast, streams of the western drainage system form a dendritic pattern, like the branches of a tree, reflecting little or no apparent control by bedrock structures. On a regional scale, the stream patterns reveal much about the structure of the underlying bedrock.

At mile 104, you are about in the middle of the Punxsutawney-Driftwood Gas Field where deep wells produce from lower Devonian sandstone. Many depleted gas fields are now being used as natural reservoirs to store gas piped in from western states. The gas is pumped underground in the summer, and out in the winter.

You see Pottsville beds between miles 109 and 111. Exit 18 to Otocsin is almost on the axis of the Chestnut Ridge anticline that continues from northeast to southwest across the state, a distance of well over 300 miles. You cross the axis of the adjacent Clearfield syncline about nine miles farther east, near mile 120. That spacing gives some idea of the scale of plateau folding.

The tan, massively bedded sandstone on the highway divider at mile 111 is lower Mississippian Burgoon formation. In most places, you can recognize this unit by its striking, large cross-beds. It is an important formation in geologic mapping throughout the Allegheny Plateau, conspicuous on aerial photographs because of its light color and resistance to erosion. Its base is sharply defined against the greenish gray, finer-grained, slabby sandstones, siltstones, and shales of the Rockwell formation, which is transitional to the underlying upper Devonian Catskill formation and contains the boundary between Devonian and Mississippian rocks. The upper contact of the Burgoon formation with the reddish siltsones and sandstones of the upper Mississippian Mauch Chunk formation, is also sharply defined. The Burgoon and Rockwell foramtions are equivalent to the upper and lower parts, respectively, of the Pocono formation in northeastern Pennsylvania.

Ledges of sandstone project from soft Allegheny shales at mile 120 on I-80.

The Rockwell formation is at least partly a marine deposit laid down late in the career of the Catskill Delta, when seas alternately flooded and receded from this region. By contrast, the overlying Burgoon sandstone is a braided river deposit laid down on a great coastal sandplain. It probably formed at the very end of Catskill Delta sedimentation just before the earliest pulses of Alleghenian uplift. If true, Africa and North America were soon to begin closing as the original sediments were laid down.

Just east of the Clearfield Exit 19, you cross the west branch of the Susquehanna River on a high bridge. The river has cut deeply into the plateau, exposing the Burgoon sandstone. It flows northeast here, parallel to the plateau folds and the Allegheny Front. the greatly elongated meander loops follow a set of northwest trending fractures.

In the ten miles between the Woodland Exit 20 and the Phillipsburg Exit 21, you cross the prominent Laurel Hill anticline. Exit 21 lies near the axis of the Houtzdale-Snow Shoe syncline, the easternmost major plateau fold. The upturned eastern limb of this fold is the backbone of the Allegheny Front. You see several exposures of Pennsylvanian Pottsville and Allegheny beds, some with thin coal seams. Strip mines in the Allegheny formation appear at miles 124, 125, and 132.

Between Phillipsburg and the Snow Shoe Exit 22, 14 miles, you follow the crest of the northeast-trending Allegheny Front that rises east of the road. This section mostly lies on Mauch Chunk and

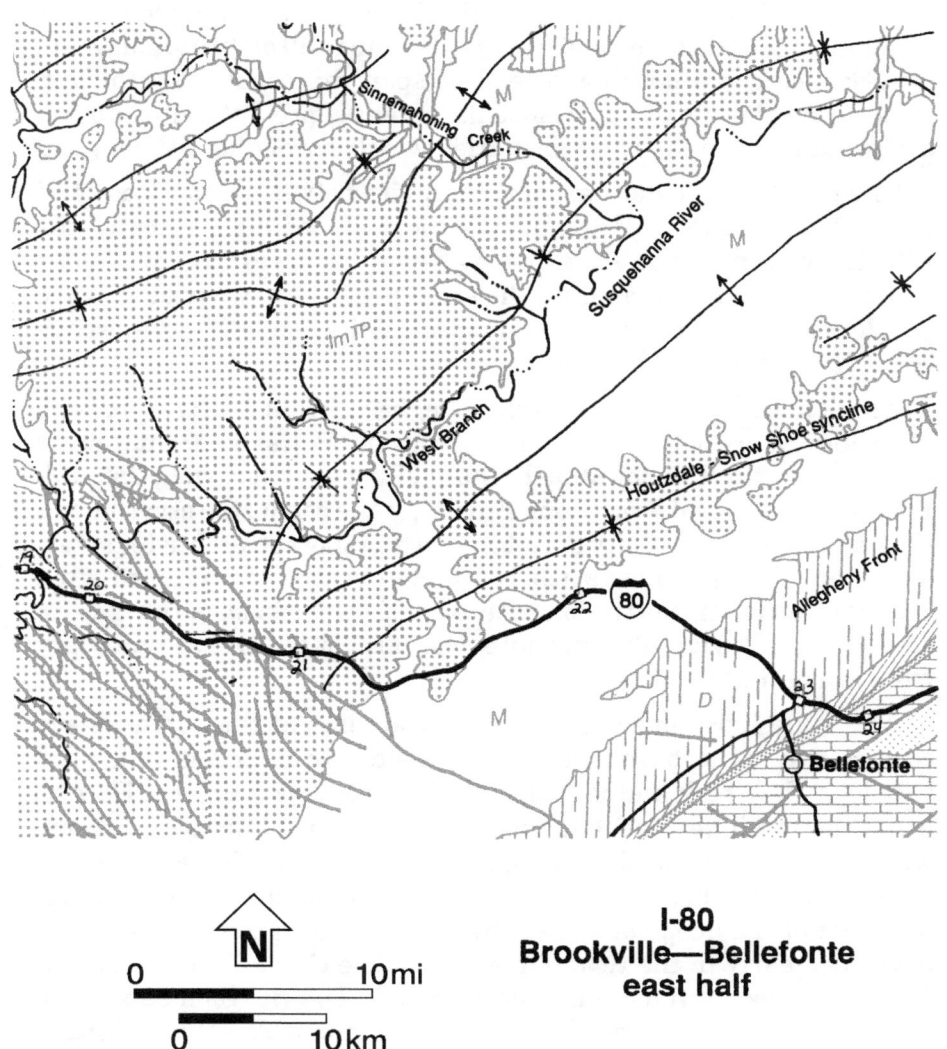

I-80
Brookville—Bellefonte
east half

Burgoon formations. Contact with the overlying Pottsville formation is never far away; it weaves back and forth in an intricate, stream-etched pattern. Snow Shoe is on the contact, three miles west of the scarp crest. The contact between the Pottsville and Mauch Chunk formations is a major unconformity where up to 30 or 40 million years of rock record are missing. The time span represented by the gap varies from place to place because the extent of pre-Pottsville erosion of the Mauch Chunk formation varied widely throughout the region.

The most spectacular part of the trip lies between Snow Shoe and the exit east of it at Milesburg — the 900-foot high scarp of the Allegheny Front. If you are traveling from west to east, you will enjoy a panorama that encompasses the scarp profile and wave after wave of ridges in front of it that distinguish the Valley and Ridge landscape from that of the Allegheny Plateau.

Think of the traverse across the Allegheny Front as a trip through 55 million years of geologic history, from the Pottsville formation at the top down through the Mauch Chunk, Burgoon, Rockwell, Catskill, Brallier, and Harrell formations. Finally, at the base of the escarpment, the middle Devonian Tully limestone. In this short stretch of highway, about ten miles, you cross a nearly complete sedimentary record of the collision between North America and Europe, bridge the gap to collision with Africa, and encounter the earliest positive sedimentary evidence of that collision.

The Tully limestone appears along the ramp at the northeastern corner of the Milesburg exit. This persistent thin formation is used widely in New York and Pennsylvania for geologic mapping. Here, it is not really a limestone, but a gray, fossiliferous, limey shale. Along with the gray, brown, olive Mahantango and black Marcellus shales below it and the black Harrell shale and gray Brallier siltstones above, it is part of the deep-water sequence deposited in the foredeep that developed early in the Acadian orogeny. The Catskill and Rockwell formations record the main Acadian uplift. The Burgoon sandstone was laid down in a tectonically quiet environment. The Mauch Chunk formation may record the onset of Alleghenian uplift; the Pottsville formation certainly does.

The first even-topped ridge east of the Allegheny Front is double-crested Bald Eagle Mountain with one ridge held up by lower Silurian Tuscarora quartzite, the other by upper Ordovician Bald Eagle conglomerate. The broad expanse of the Nittany Valley lies east of the ridge. Several small Tuscarora talus slopes are visible on the west side near the ridge top. I-80 crosses through the ridge between Exits 23 and 24 along Curtin Gap. In the gap, you see an excellent outcrop of

Vertical Tuscarora quartzite beds with shaley partings in Curtin Gap near mile 159 of I-80.

Tuscarora quartzite in vertical beds with shaley partings. The route through the gap traverses another 75 million years or so of geologic time, across the Silurian and Ordovician rock sections. The large quarry near Exit 24 is in middle Ordovician Bellefonte dolostone.

High-Calcium Limestone at Bellefonte

The Bellefonte area of Nittany Valley contains some of the purest high-calcium limestone deposits in Pennsylvania, mined in quarries and underground. Part of the upper Ordovician Benner formation, the beds crop out on both sides of the valley. They dip about 80 degrees to the northwest under Bald Eagle Mountain and 25 degrees southeast on the other side of the valley under Nittany Mountain, thus forming a large anticlinal arch. The limestone (calcium carbonate) is mostly converted to lime (calcium oxide) in kilns fired by pulverized bituminous coal with a limestone to coal ratio of about 3:1. Other products are hydrated lime used in glass-making, and crushed agricultural limestone.

Penns Cave

The Nittany Valley is floored by Ordovician limestones and dolostones, which dissolve in groundwater to form many caves and sinkholes. Penns Cave is a commercial attraction that differs from most others in the state in being partly flooded, so a trip through it includes a boat ride. The cave is east of Nittany Mountain on Penns Cave Road just off of Brush Valley Road, Pennsylvania 192, approximately 13 miles east-southeast of Bellefonte.

These thick Llewellyn beds at mile 181 of I-81 are sandstone, the black rock at the base is coal, and the thin-bedded layers above are shale and stiltstone. Another thin coal seam is near the top of the cut.

I-81
Wilkes-Barre—New York Border
67 mi./108 km.

Rock formations traversed by this section of I-81 are schematically illustrated in the Geologic Time Travel Map on page 197.

Between Wilkes-Barre Exit 45 and Exit 57E on the northern side of Scranton, you follow the floor of Lackawanna Valley. Exposed bedrock is mostly grayish conglomerates, sandstones, siltstones, darker gray shales, and some coal, all part of the Llewellyn formation, the principal coal-producer in eastern Pennsylvania. Many of the sandstones exhibit large cross-beds.

Over its 55-mile length, between Shickshinny on the south and Forest City on the north, the valley topography faithfully expresses the Lackawanna syncline. The valley sides are held up by the inward-dipping Pottsville sandstones and conglomerates, and the valley floor is underlain by the less resistant Llewellyn formation in the fold trough.

The Susquehanna River cuts through the Pottsville ridge and enters the valley from the northwest at Pittston, halfway between Wilkes-Barre and Scranton. It then turns southwest and follows the

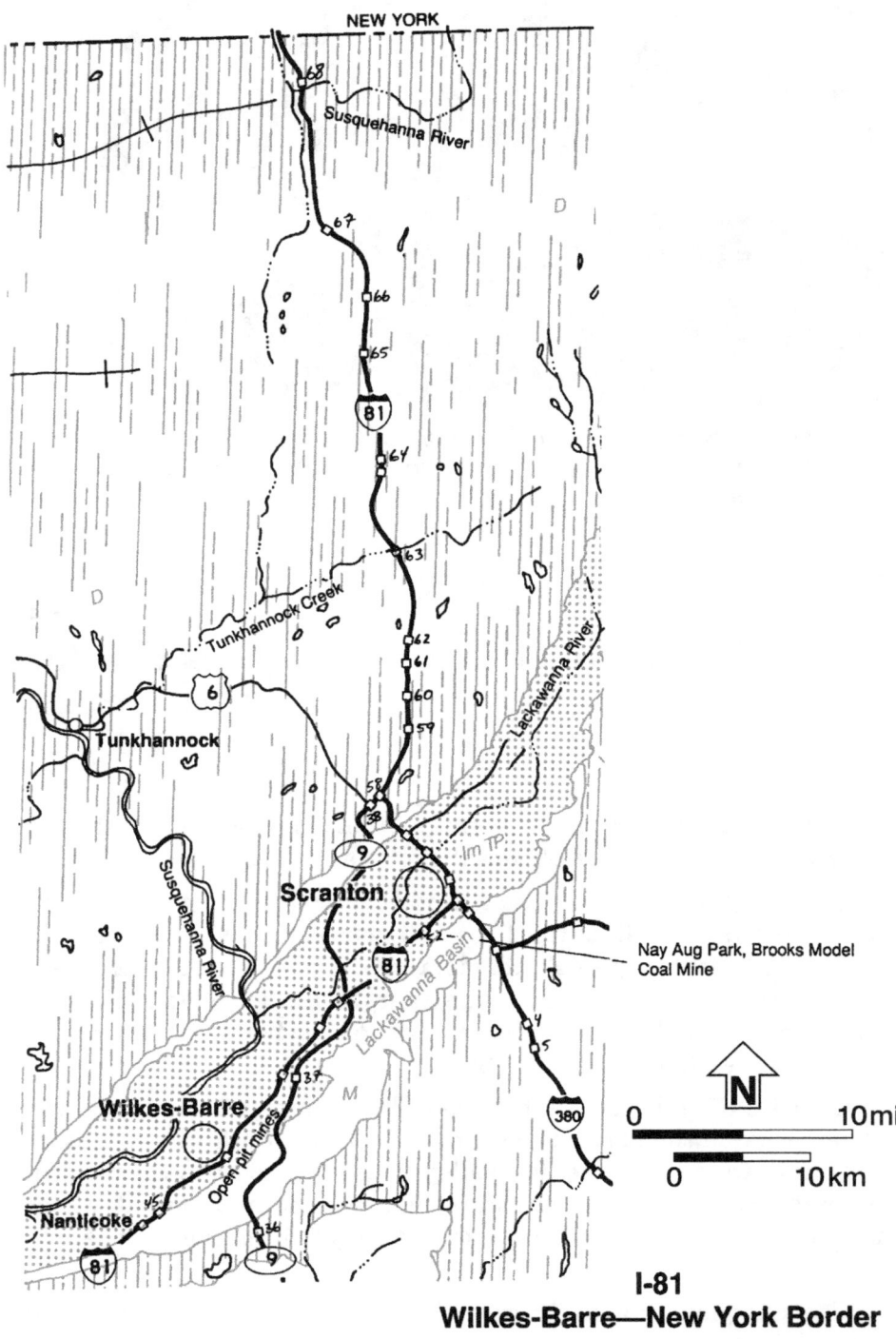

I-81
Wilkes-Barre—New York Border

valley to Nanticoke. There, it recrosses the Pottsville ridge and immediately turns southwest again to follow the soft Mauch Chunk rocks that separate the Pottsville sandstone from the Pocono formation of Shickshinny Mountain. At Shickshinny, the river turns south and passes through a gap into the next valley, carved from the core of the Berwick anticline. Finally it turns southwest again and follows the fold 40 miles to join the west branch of the river at Sunbury.

Thus, the river generally follows the trend of the bedrock structures except where it crosses hard ridges. Geologists now think this configuration, here and in other parts of Pennsylvania, results either from the effects of stream superimposition from an old erosion surface or from stream piracy.

The Pottsville and Llewellyn formations here correlate with Pennsylvanian soft coal-bearing formations of the Allegheny Plateau and were originally continuous with them. Folding during the Alleghenian orogeny subjected these rocks to intense deformation and slight metamorphism, converting the coals to anthracite. The associated folding pushed most of the rocks to high levels, where they were destroyed by erosion. The remaining coals are preserved only in the troughs of the synclines.

Old strip mines are all over the Lackawanna Valley; they even follow the Pottsville formation up the flanks of the ridges. Some of the old mines have been reclaimed, but most have not, so you will see many of them from the highway. Obviously, coal-mining played a major role in the settlement and development of the valley and the state.

Scranton Anthracite Museums

Three Scranton museums offer the interested visitor some of the anthracite geology and history of this region. The Scranton Anthracite Museum is in McDade Park near the southwestern corner of the city adjacent to Pennsylvania 9, the northeastern extension of the Pennsylvania Turnpike. The park is a rehabilitated strip mine site. Everhart Museum in Nay Aug Park near Exit 52 on the east side of the city has excellent coal displays and the Brooks Model Mine. Roaring Brook excavated an impressive small gorge through the park area, exposing gray to black shale and siltstones of the Llewellyn formation. The Scranton Iron Furnaces are on Cedar Avenue, between Lackawanna Avenue and Moosic Street, conveniently reached from Exits 52 or 53. The four restored stone smokestacks of blast furnaces were built in the 1840s and 1850s by George W. Scranton and associates as part of the Lackawanna Iron Company.

Upper Devonian Catskill beds at mile 222 of I-81, with strong vertical jointing oblique to the highway.

Of course, anthracite fired the furnaces, which catalyzed development of the city. In the 1860s, they were the second largest iron producer in the nation. Iron rails for the Erie Railroad were made here, and steelmaking began in 1875. The forges closed in 1902.

• You cross the northwestern limb of the Lackawanna syncline between exits 57W and 58, passing through narrow bands of Mauch Chunk, Pocono, and Spechty Kopf formations into upper Devonian Catskill formation. The Catskill formation prevails almost to the New York border. Roadcuts expose reddish to greenish gray sandstones, siltstones, and shales, mostly in flat- or nearly flat-lying beds, as you would expect in the generally mild folding of the Allegheny Plateau.

At mile 231, just a mile south of the New York border, you see flat-lying, olive-grey shales and siltstones of the Trimmers Rock formation exposed where the Susquehanna River eroded through the overlying Catskill beds. Their dark color and fine-grain size signifies a deep-water environment of deposition. They are, in fact, flysch deposits of the basin that formed as the continental crust warped down during the initial stages of the Acadian orogeny. The lighter reddish to greenish-gray and coarser-grained Catskill rocks exposed between here and Scranton are largely river and shallow marine deposits of the great Catskill Delta that issued from the Acadian Mountains during and after the main uplift.

This is the Great Bend of the Susquehanna River. The headwaters region of the river is in New York near the Onondaga-Helderberg scarp at the northern rim of the Allegheny Plateau. The main branch drains from Otsego Lake and flows generally south to the Pennsylvania border eight miles east of I-81. It penetrates four miles into the state, then bends sharply west, to finally flow back into New York alongside I-81. From there, it traces a course west past Binghamton to Waverly, New York, about 35 miles from I-81, where it re-enters Pennsylvania and joins the Chemung River. Then it winds its way southeast across the plateau in deeply entrenched meanders to Scranton.

Great floods of water went through the Susquehanna Valley in the waning stages of the last ice age. As the glacier margin receded northward, meltwater filled valleys to the south of it, including the Susquehanna Valley, which became the major outlet for the forming Finger Lakes. At a later stage, when the Mohawk Valley was uncovered, the forming Great Lakes and the Finger Lakes drained through it, leaving the Susquehanna River much diminished. At an even later stage, the St. Lawrence River became the main drainage route.

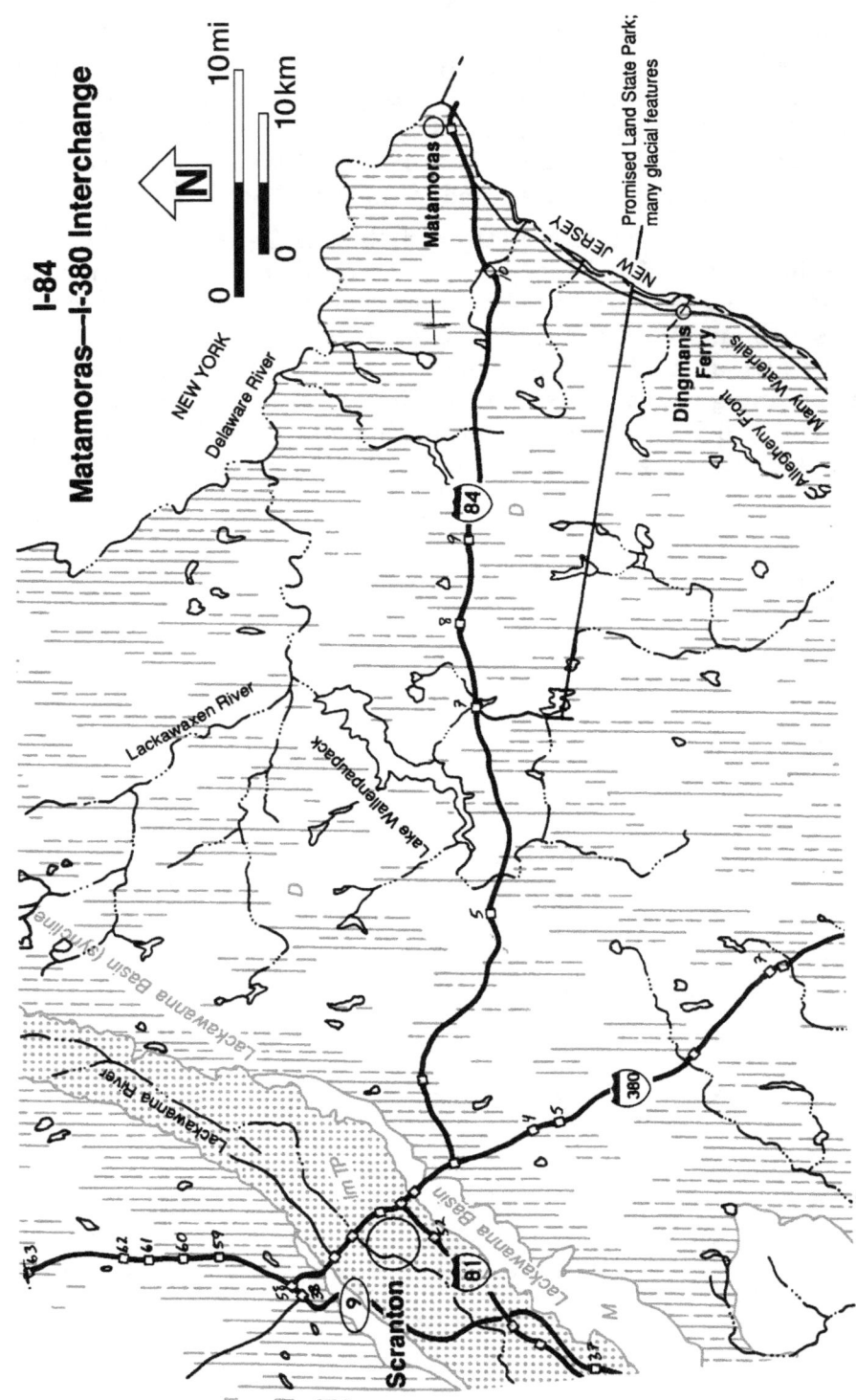

I-84
Matamoras—I-380 Interchange
48 mi./77 km.

Between Exit 11 at Matamoras and Milford Exit 10, seven miles, I-84 follows the Delaware River along the Port Jervis trough. A pronounced high scarp on its northwest side, part of the Allegheny Front, is carved from the middle Devonian Mahantango group, exposed in many places along the road at the base of the scarp. The rock is mostly gray, brown and olive shale, siltstone and sandstone. The scarp is the site of many marvelous waterfalls where streams cross resistant sandstone beds. On the other side of the trough, just over the state line in New Jersey, is Kittatinny Mountain, a ridge of sandstone and conglomerate of the lower Silurian Shawangunk formation. Farther north in New York, the same ridge is the backbone of the Shawangunk Mountains. Throughout most of Pennsylvania, the terms Tuscarora and Clinton are used for lower and upper parts of the Shawangunk formation. Tuscarora quartzite is an extremely resistant rock that forms high ridges, which continue for hundreds of miles.

The Delaware River descends the Allegheny Front from the northwest and enters the trough at Port Jervis, New York, then follows it southwestward to Stroudsburg. At Stroudsburg, the river bends sharply southeast and cuts right through Kittatinny Mountain in the spectacular Delaware Water Gap.

Between Milford Exit 10 and Lords Valley Exit 9, 12 miles, you cross the Allegheny Front. At the base of the hill you see olive gray siltstone and shale of the Trimmers Rock formation that overlies the Mahantango formation. The dark color, fine texture and marine fossils of the Mahantango and Trimmers Rock formations indicate that they formed in deep, poorly oxygenated waters. They are flysch sediments deposited in a basin that developed during the initial phase of the Acadian orogeny.

Above the Trimmers Rock outcrops, you see Catskill formation, a succession of sandstones, siltstones, and some shales of generally coarser texture and lighter color than those in the Trimmers Rock formation. Reddish and greenish gray colors are common and in places interlayered to produce an attractive banding. These are river and shallow marine deposits of the Catskill Delta that signal the collision of North America and Europe and the main phase of the Acadian orogeny. The Catskill formation underlies the rest of the route to the I-380 interchange.

The flat plain near Exit 9 is typical of much of the glaciated Allegheny Plateau in Pennsylvania and New York. Some parts are lake plains that were flooded by meltwater during the last deglaciation. More lakes exist in the Poconos than in any other part of Pennsylvania. Even more existed just after the last ice age, but many have since filled with sediment. Many of the lakes formed as kettles where blocks of ice incorporated in the glacial deposits finally melted.

Several of the Catskill red sandstone beds between Exits 9 and 7 are cross-bedded, and some have thin greenish interbeds. Many more such exposures appear between Exits 7 and 5, some with more gray, rather than red, sandstone colors.

Colors of the Catskill Formation

The reddish and greenish colors of Catskill beds both come from iron compounds incorporated in the sediments. Each suggests an environment of deposition. Color, therefore, is a key to geologic history, an important element of the geologist's window on the past. Reddish colors come from the iron oxide, hematite, formed in highly oxidizing environments as, for example, where sediments accumulate on land. Greenish colors come from iron compounds formed in oxygen-poor settings, and may indicate marine deposition in more stagnant water. Alternating reddish and greenish beds may thus suggest fluctuating conditions of submergence and emergence. This corresponds well with what we know about the late Devonian period in this part of the world — the Acadian Mountains were rising to the east and were shedding sediments westward, gradually displacing a vast shallow inland sea and forming the Catskill Delta. A gentle sediment plain leading down to the shore would, under these conditions, be sensitive to even slight changes in sea level and would be submerged when the level rose.

Promised Land State Park

This park, south of Exit 7, is an excellent place to see some of the features typical of the glaciated Pocono Plateau. Catskill formation outcrops are generally blocky, creased by parallel sets of open vertical fractures or joints, that cross-cut the bedding. The soil nearly everywhere is littered with blocks plucked from these outcrops by glacial ice.

Blocks separated along joint, bedding and cross-bedding surfaces were carried away from the outcrops in the slow-moving glacier, then dropped in jumbled disarray as glacial erratics when the ice melted. Many of the blocks were worn down as they scraped against each other and the bedrock during transport, producing polished and striated surfaces, which are poorly preserved except where they have been protected from weathering under a cover of soil. Surfaces exposed to air, water, and temperature extremes quickly roughen as the rock disintegrates, grain by grain. Mounded chaotic deposits of glacial till, a mixture of boulders, gravel, sand, silt, and clay, are a common sight in the park.

The many potholes along the east branch of Wallenpaupack Creek were carved during flooding that accompanied glacial recession. The four lakes of the park: Promised Land, Lower, Bruce, and Egypt Meadow are all probable kettle lakes that formed where huge blocks of ice calved from the receding ice front and were partly buried in glacial outwash sediments before they melted. Bruce Lake is a good example of a peat bog, where vegetation has slowly filled in the depression during the past 18,000 or so years.

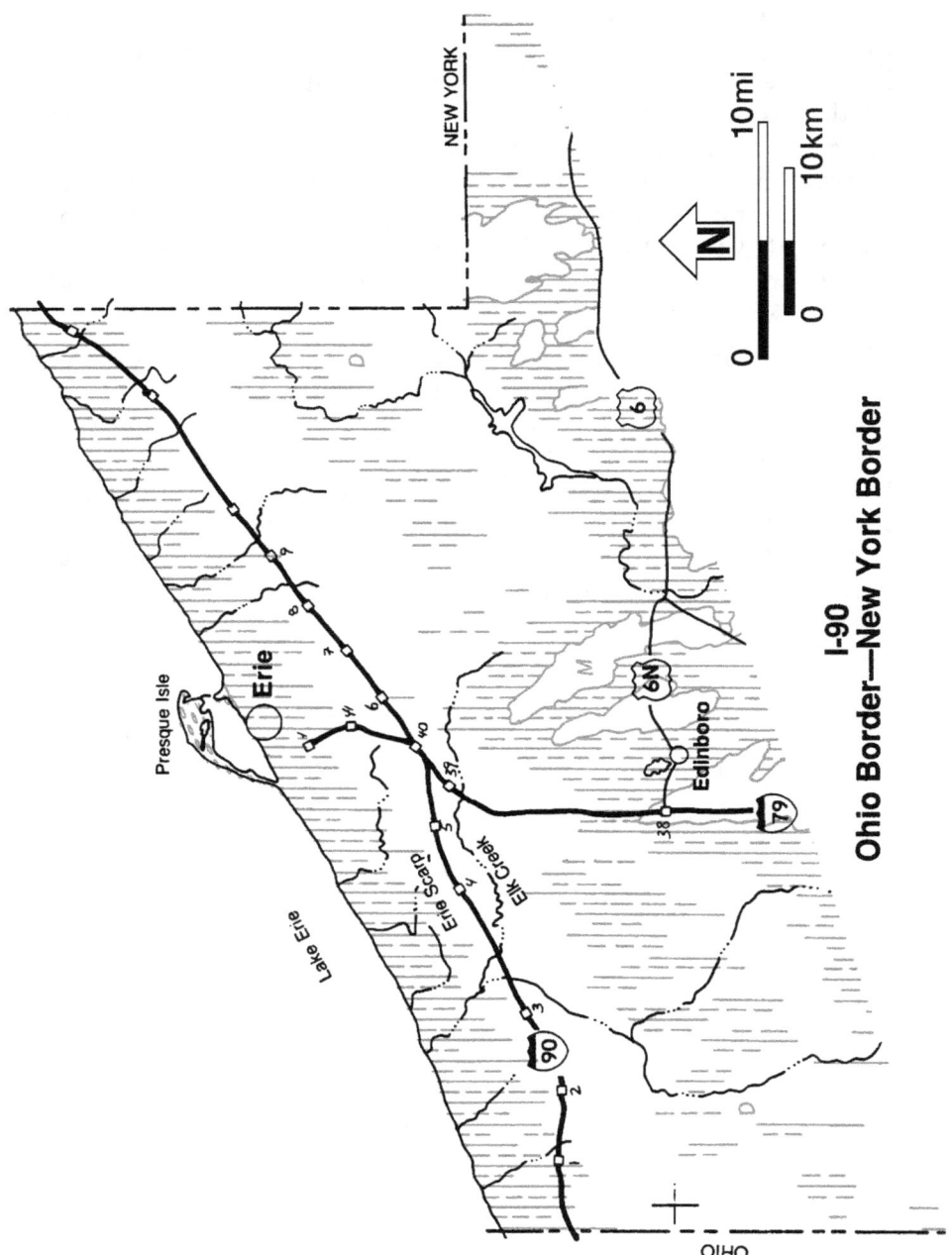

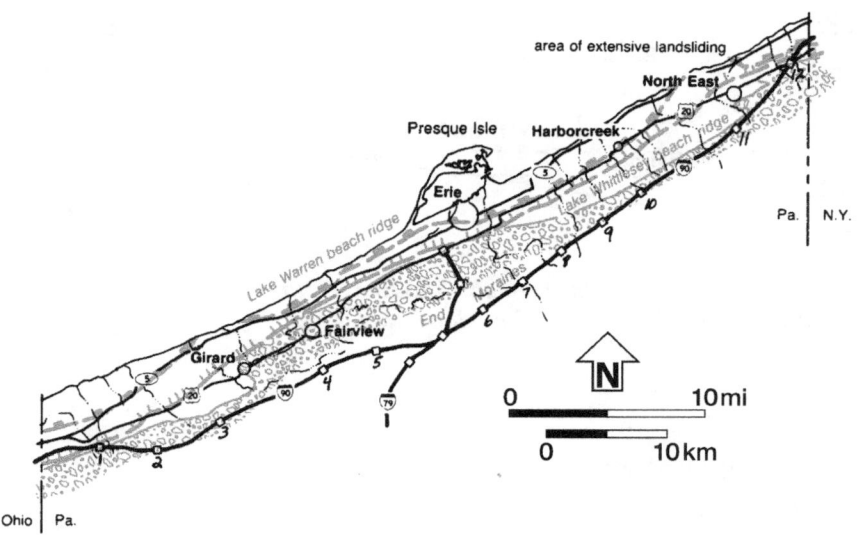

Beach ridges of glacial lakes Whittlesey (earlier) and Warren (later) near I-90. The poorly preserved pre-Whittlesey beach of Lake Maumee is not shown. —Adapted from Schooler, E.E., 1974, Pennsylvania Geological Survey General Geology Report 64.

I-90
Ohio Border—New York Border
46 mi./74 km.

This short route crosses Erie County from southwest to northeast close to the base of the Erie scarp, or rises onto the scarp, with the gently sloping Erie Plain stretching away to the northwest. The beach ridges of Lakes Whittlesey and Warren, Erie's larger late-glacial predecessors, lie northwest of the highway all the way except near the New York border. They are visible southeast of the road. All the small streams that cross the highway on their way to the lake are entrenched, especially near the lake. Their cutting is a consequence of the subsiding lake level, from that of Lake Whittlesey to the present level of Lake Erie. In places, the streams carved small bedrock gorges that expose flat-lying beds, mostly of shale belonging to the late Devonian Canadaway and Conneaut groups.

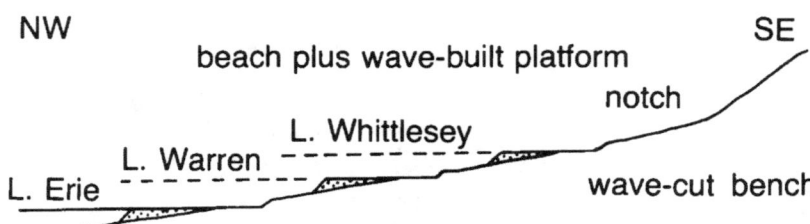

Schematic cross-section through shoreline features of glacial lakes Whittlesey and Warren, and modern Lake Erie.

Waves, Longshore Drift, Jetties, and Groins

Lake Erie beaches, like most beaches, are not static; they are rivers of sand flowing along the shore. Along the Pennsylvania shore, the beach flows ever northeastward. At any one point, the sand may stay a while, but it will move on. Sand comes from western beaches and goes to eastern beaches. It originates from real rivers that bring sediment to the lake, and from wave erosion of bluffs or seacliffs. Wave action moves sand northeast, parallel to the shore. The waves, generated by the prevailing winds, move onto the shore obliquely eastward, carrying sand up the beach. The backwash carries sand straight down the slope of the beach. The net result is longshore travel as each new wave moves the sand a little farther.

Waves that approach the south Erie beaches from the northwest move sand to the east in jumps shown by the barbed arrows. Groins and jetties act as dams to this drifting sand, which thus piles up against them.

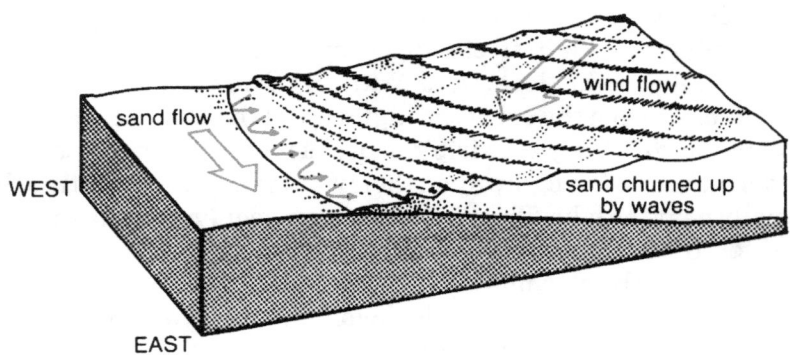

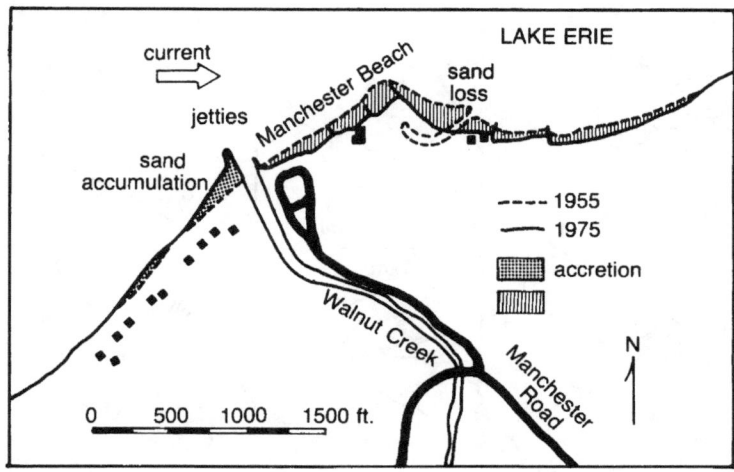

Shoreline changes near the mouth of Walnut Creek since construction of jetties. —Adapted from Delano (1987, p. 67)

Presque Isle peninsula is eloquent testimony to long-term northeastward drift along the Pennsylvania shore. Historically, the whole six-mile long peninsula has migrated northeastward as waves erode its western side and deposit material on its eastern end. Its shape clearly indicates northeastward longshore currents.

At Manchester Beach, just west of the city of Erie at the mouth of Walnut Creek, the Pennsylvania Fish Commission has a boat-launching site that interrupts the natural flow of sand. In 1972, the commission built a long jetty on the western side of the creek mouth to prevent the channel from clogging with sand, and also dredged the channel to facilitate boat launching. They built an additional short jetty on the eastern side of the creek. The jetties interrupted eastward longshore drift of sand, and that almost immediately increased the erosion of beaches to the east. Groins were built there in 1973 and 1974 in a futile attempt to stem the erosion. The modifications drastically altered the shoreline in an instant of geologic time, as shown on the before and after map, derived from the 1957 and 1975 versions of the Swanville, Pa. 7.5-minute topographic quadrangle map. Jetties and groins are very real dams to the river of sand. Here, the long jetty did the most damage, causing sand to pile up on its windward side into an exceptionally wide beach. It virtually cut off the supply to the eastern beaches. That caused the shoreline to recede east of the jetty despite the groins. In fact, only the two easternmost groins appear to have captured any sand at all. In this case, the lion's share of sand must come from longshore drift. If Walnut Creek were bringing much sand to the shore, the eastern beaches would continue to receive a goodly supply despite the jetty.

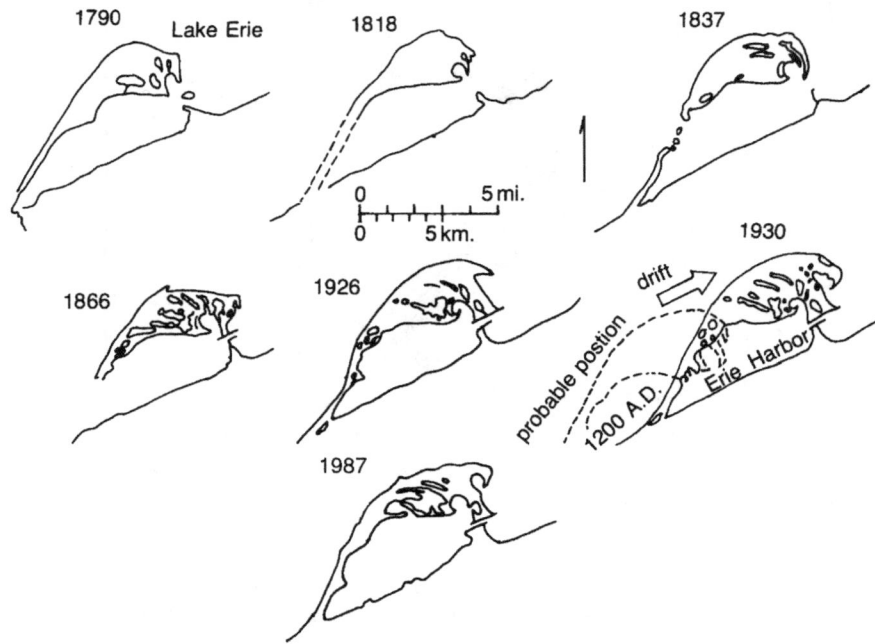

Growth and migration history of Presque Isle. —Adapted from Jennings, O.E., 1930, Perigrinating Presque Isle, Carnegie Magazine, v. 4, p. 171-5.

Presque Isle

Presque Isle is a recurved sand spit, the only major depositional feature that projects from the otherwise monotonously smooth south shore of Lake Erie. Its existence has been fortunate for Pennsylvania because it encloses a harbor called Presque Isle Bay that allowed Erie to develop into a port and manufacturing center. During the War of 1812, the harbor provided shelter for Commodore Oliver Hazard Perry. It was here that he built the six ships with which he defeated the British fleet in the fall of 1813, and where he returned with his wounded. In 1815, two captured British ships, plus two of Perry's original ones, were deliberately sunk in Misery Bay, near the tip of the spit. The ship Niagara was raised in 1913 and reconstructed; it now stands as an historical monument by State Street in Erie.

The peninsula that offered Perry protection to build his ships also presented a final obstacle to launching them. A sand bar at the eastern entrance to the bay prevented two relatively deep-drought brigs from moving through. Perry solved the problem by floating the ships higher with the help of empty tanks, called camels, attached to their sides.

Today, constant dredging maintains a deep channel at the bay entrance for ships to pass to and from the city docks.

The tendency of drifting sand to close this opening is natural and probably cannot be totally eliminated. As sand moves into the entrance, it gradually and inexorably changes the shape of Presque Isle. Its shape shows that the prevailing currents move from southwest to northeast, carrying the sand with them. This sand transport has, over the years, increased the size of the bulging eastern end of the spit, while shifting the whole peninsula eastward. The peninsula apparently migrates by repeated development and accretion of beach ridges at the eastern end, coupled with erosional removal of earlier deposits along the western edge.

The existence of Presque Isle is probably related to a 12-mile-long, shallow, sand-covered morainal ridge that crosses the lake to the Ontario side and divides the central and eastern lake basins. The ridge connects with Long Point on the opposite shore, a spit similar to, but larger than, Presque Isle. Whatever its beginnings, the peninsula has perpetuated itself by providing a catchment for the northeast-drifting sands.

Erosion and Landsliding on the Erie Shore

The southern shore of Lake Erie is a region of sudden, dramatic, and costly geologic change. Much of the Pennsylvania shore has high bluffs composed of unconsolidated, or weakly consolidated glacial gravels, sands, silts, and clays. Wave erosion at their bases maintains the steepness of the bluffs, and makes them very unstable and subject to landsliding. This problem is particularly nettlesome along a section of shore two and one half miles west of the village of North East, eight miles northeast of Erie, where the bluff is 160 to 170 feet high.

As waves undercut the bluff, the upper part caves in. Even though the material is unconsolidated, the initial breakdown often drops large blocks that break off along joints like those that occur in solid rock. The blocks eventually break down to form a pile at the base of the bluff. Along with other landslide and slope wash debris, this prevents further wave erosion there — at least for a while. These erosional processes are most rapid when the lake level is high during storms, and during the record high-water years of 1976 and 1987.

The bluff is higher in this section of shore than to the west because one of the Lake Warren beach ridges intersects the shore here. Farther west the ridge traverses inland where it is marked by a 50-foot scarp.

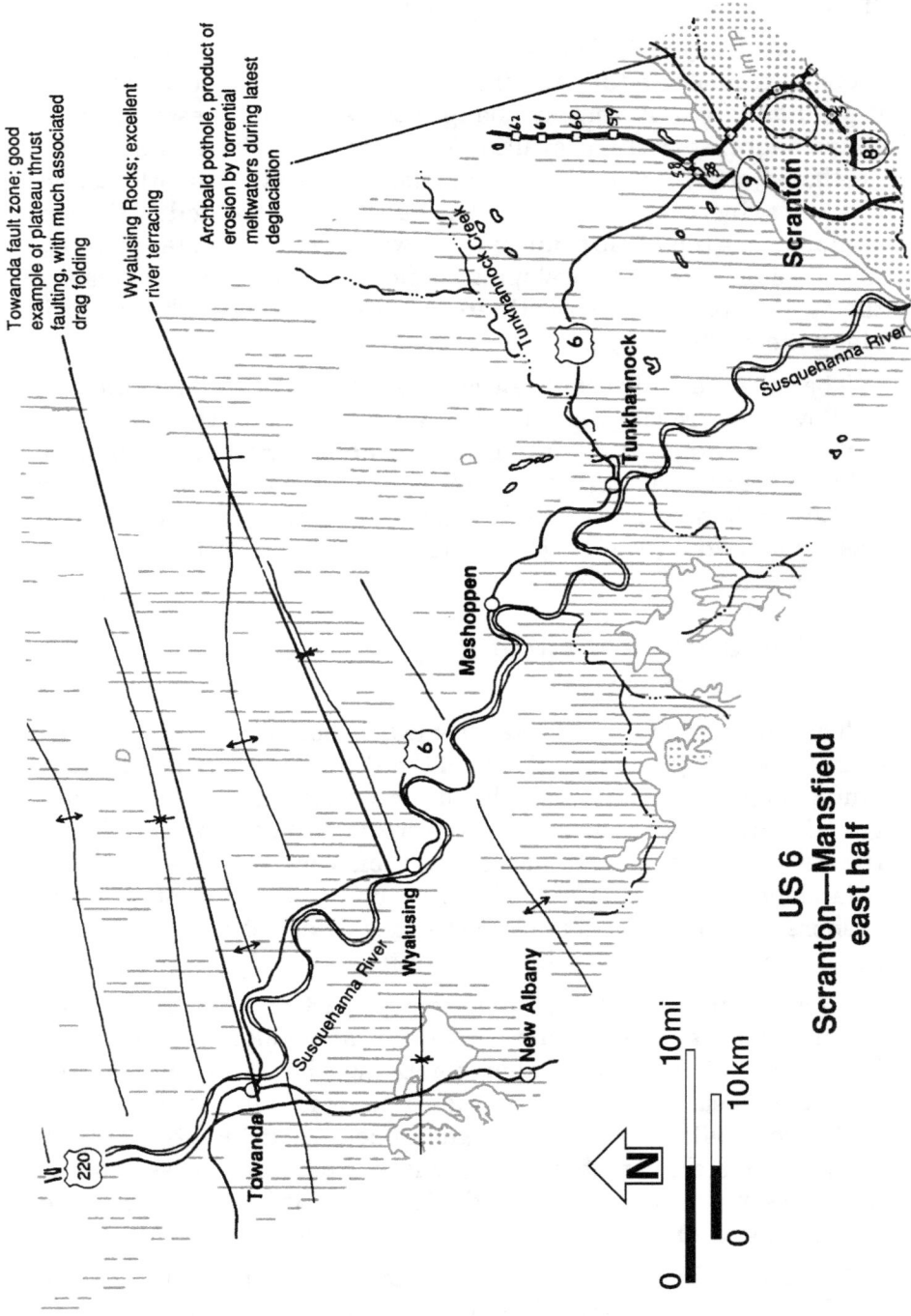

US 6
Scranton—Mansfield
93 mi./150 km.

This route crosses the northern part of the Allegheny Plateau in which the most important bedrock ranges in age from late Devonian to Pennsylvanian. The plateau beds are mostly flat-lying or gently folded into broad, open anticlines and synclines.

Scranton is not on the Allegheny Plateau; it is just east of the Allegheny Front, nestled in the Lackawanna Valley, in the deep Lackawanna syncline that houses the Northern Anthracite Field. The strip mines and waste rock piles that scar the region are reminders of a once-thriving hard-coal industry.

The road to Archbald Pothole State Park crosses the northwestern limb of the Lackawanna syncline over Pennsylvanian Llewellyn beds, the principal coal-producer of the region. The underlying, more resistant Pottsville beds crop out upslope, forming a modest ridge called Bell Mountain at the plateau margin. A similar ridge, Moosic Mountain, marks the opposite limb of the syncline, clearly visible across the valley. Topography here, unlike that in many other parts of the Valley and Ridge province, faithfully expresses the underlying bedrock structure — low elevations over downfolded beds, high elevations over upended, resistant beds.

Archbald Pothole

Potholes in general are carved where sand and gravel grind against bedrock in swirling eddies. Archbald Pothole apparently formed during glacial retreat about 15,000 years ago at a place where torrents of meltwater broke through a crevasse and fell with great force, probably many tens or hundreds of feet, to the rock surface below, creating a powerful whirlpool. Such streams normally contain much sediment released from the ice as it melts, and thus may provide a copious supply of grinding material to excavate the bedrock. Archbald Pothole is 42 feet wide at the brim, and tapers to a depth of 38 feet, making it one of the largest in the world. Its large size indicates that the conditions for its formation persisted for a long time, perhaps hundreds of years.

• Along the 18 miles between Clarks Summit and Tunkhannock, you are in plateau country some 700 feet higher than Scranton, but with markedly lower relief that reflects the gentle bedrock folding. No

Swirling, sediment-charged meltwater from the wasting latest ice age glacier carved the enormous Archbald pothole, which measures 42 feet wide at the top by 38 feet deep, into Pennsylvanian Llewellyn sandstones.

well-defined scarp marks the Valley and Ridge to Allegheny Plateau transition here, as in much of the state; but the change from rugged to gentle landscape west of Scranton is remarkable. Most of the plateau relief results from geologically recent stream dissection. Over the western half of the way, the route follows stream valleys cut 300-400 feet into the plateau.

Between Factoryville and Bardwell, four miles, you follow the southern branch of Tunkhannock Creek and in the six miles between there and Tunkhannock, you follow the main branch of Tunkhannock Creek. Tunkhannock is an Indian name meaning "little creek;" the village of the same name is where the main branch joins the Susquehanna River. In this stretch, upper Devonian Catskill formation, consisting of nearly flat beds of sandstone, siltstone, and shale appears in numerous cuts.

The Susquehanna River snakes across this section of plateau in curving meanders up to three miles wide that show no sign of bedrock influence on their form, which suggests flat or nearly flat strata. This is a trademark of much, but not all, plateau stream dissection. It contrasts strikingly with Valley and Ridge stream patterns, where, the intense folding offers alternating hard and soft rocks, and the stream drainage prefers the soft. Stream patterns on folded rocks tend to trace the underlying bedrock structures. The Tunkhannock-Scranton and Scranton-Wilkes-Barre segments of the Susquehanna River are a fine example. At Scranton, or more accurately, Pittston, seven miles southwest of Scranton, the meandering abruptly ends; the river continues almost ramrod-straight to the southwest for 23 miles along the trough of the Lackawanna syncline, before turning south to breach the hard ridge of Penobscot Mountain that defines the southeastern limb of the fold.

The route follows the Susquehanna River for 58 miles northwest of Tunkhannock, although with a much straighter course that affords intermittent, but inspiring, river views. This historic route illustrates the strong influence bedrock exerted on human activity in Pennsylvania during the 18th century and earlier, when there were few roads and no railroads. Transportation was by foot, horse, wagon or boat, and all by the path of least resistance. That meant, in most cases, rivers or river valleys. The Indians followed these natural paths and settled along them long before white men came. During the American

Gentle terracing on the inside of a large Susquehanna River meander at Azilum results from downcutting by the river as the meander shifted sideways.

Revolution, the famous Sullivan's March along the old Susquehanna path followed their example. Several historic markers by the highway commemorate this event.

Sullivan's March was a major component of the Sullivan-Clinton Campaign, one of the most ambitious American campaigns of the Revolutionary War. Its purpose was to crush the Indian-Tory frontier menace in New York by destroying Indian villages and crops upon which the British depended. George Washington placed General John Sullivan in command. At Fort Wyoming, Pennsylvania, near present-day Wilkes-Barre, Sullivan organized the main force of nearly 4000 troops, several cannon, 1200 pack horses, 800 beef cattle, and a flotilla of boats. Between July 31 and August 11, 1779, he marched this army up the Susquehanna River to Tioga Point at the junction of the Chemung River, where he built Fort Sullivan. Meanwhile, General James Clinton, second in command and brother of Governor George Clinton of New York, organized another force at Canajoharie, New York, and proceeded up and over the Onondaga-Helderberg escarpment to Otsego Lake at the headwaters of the Susquehanna River. From there, he came down the valley to join Sullivan at Tioga Point. Fort Sullivan was the base for all later incursions into the Indian territory of western New York. Sullivan terminated the campaign at the end of September, having burned some 41 Indian villages and destroyed all of their crops.

Between Tunkhannock and Towanda, 54 miles, the highway offers several grandstand views of the winding Susquehanna Valley. If you

Flat-lying upper Devonian Catskill beds of sandstone and shale on U.S. 6 near Wyalusing.

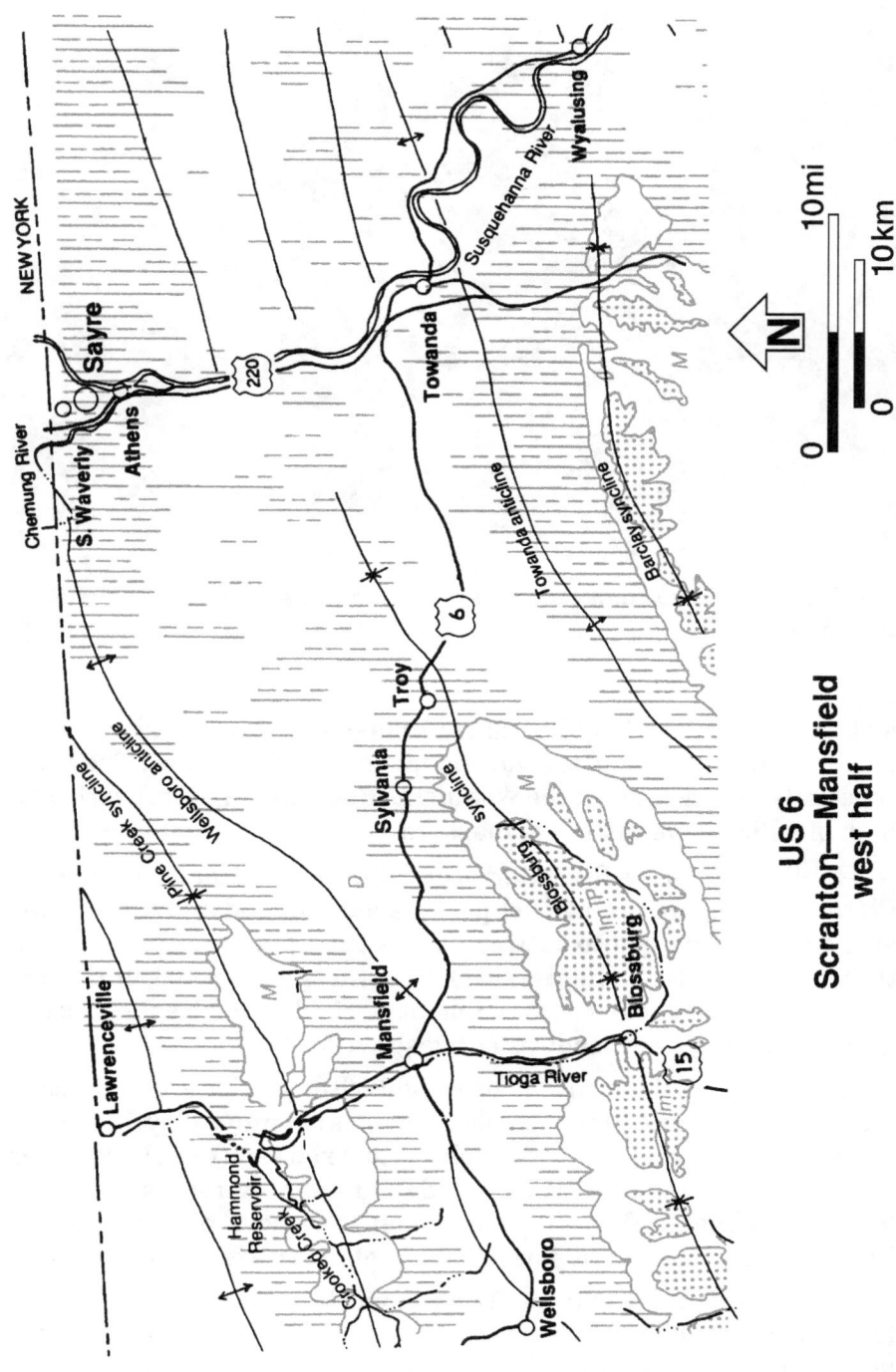

US 6
Scranton—Mansfield
west half

Flat-lying upper Devonian Catskill sandstones and shales near Mansfield on U.S. 6.

look carefully, you should be able to make out terraces above river level on the insides of the meanders in several places. They are especially well-defined at the Wyalusing Rocks and Azilum overlooks, both of which have historical markers.

The Susquehanna River terraces, like those of many Allegheny Plateau rivers, are products of post-glacial rebound and consequent erosion of the rivers into outwash deposits that had accumulated as the ice sheets melted. The margin of the melting ice receded northward as swollen rivers choked with sediment washed through valleys to the south, dropping their loads as outwash on the way.

After removal of the enormous weight of the glacier, the depressed land rebounded, and the diminished streams cut into valley fill as they meandered back and forth. Floodplains marked each new, lower river level. The terraces that you see today are simply remnants of the former floodplains that the river failed to remove by sideswiping. Thus, the terraces record part of the the glacial history of this region.

Most of the bedrock between Tunkhannock and Towanda is Catskill formation, and it is well-exposed. One particularly large exposure near the Wyalusing overlook consists of reddish to greenish-gray sandstone, siltstone, and shale. The distinctively olive gray shale, siltstone, sandstone, and conglomerate of the underlying upper Devonian Lock Haven formation crop out at lower elevation near the river. The road traverses this latter unit exclusively in the nine miles between Rummerfeld and Towanda. Its exposure here is due to erosion of the northeast-trending Towanda anticline — one of the large and gently-folded variety that wrinkle the plateau. The traverse from southeast to northwest takes you across the southeast limb of the fold, passing from younger to older rocks — from the Catskill to the Lock Haven formations — the normal outcrop pattern for eroded anticlines. Towanda lies about on the axis of the fold. The flanking Barclay syncline to the southeast and Blossburg syncline to the northwest form hills capped by Mississippian and Pennsylvanian formations — younger rocks.

Between Towanda and Mansfield, 35 miles, you go east-west oblique to fold axes. At halfway, Troy is on the Blossburg synclinal axis. Pennsylvanian Pottsville and Allegheny beds crop out in the fold trough ten miles to the southwest. Mansfield, 17 miles west of Troy, is about on the axis of the Wellsboro anticline, back in the olive gray sandstones, siltsones, shale, and conglomerates of the upper Devonian Lock Haven formation.

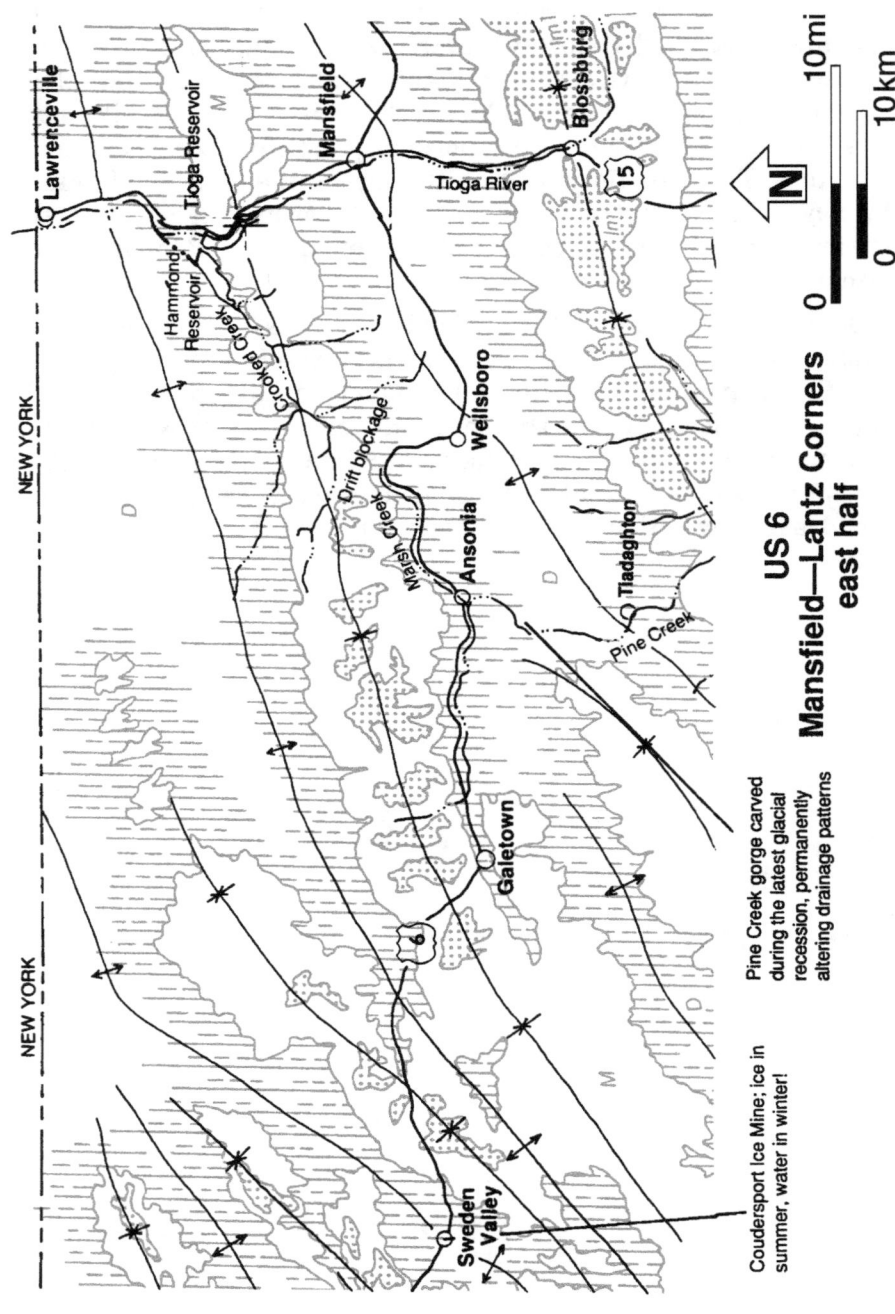

US 6
Mansfield—Lantz Corners
104 mi./168 km.

The 13-mile drive between Mansfield and Wellsboro reveals several lovely views of the valley country but few outcrops. Between Wellsboro and Ansonia, eight miles, the route follows the sinuous course of Marsh Creek along a heavily wooded valley. Ansonia lies at the northern end of Pine Creek Gorge, one of Pennsylvania's most beautiful natural features.

Pine Creek Gorge - The "Grand Canyon of Pennsylvania"

Pine Creek Gorge extends south nearly 50 miles downstream to Waterville, at the Little Pine Creek juncture. Best and most accessible views of the gorge are a few miles south of Ansonia, at Colton Point State Park on the west rim and Leonard Harrison State Park directly opposite on the east rim. The depth of the canyon varies from 600 to nearly 800 feet; the rim to rim width of the first eight miles south of Ansonia is remarkably uniform, averaging about 4000 feet. In aerial view, it looks as if it were gouged into the plateau by a giant router like those used to carve letters in wooden signs. South of Tiadaghton, the canyon widens considerably and becomes more irregular, with gentler slopes and better-developed tributaries. Depth increases southward to a maximum of 1450 feet at Waterville.

The present configuration of Pine Creek Gorge, and the course of the creek result from drainage changes at the end of the last ice age.

West of Ansonia, the pre-glacial river was probably similar to the modern one, which flows east. However, the pre-glacial river continued northeast from Ansonia along the large valley now occupied by tiny Marsh and Crooked creeks, to Tioga, then went north along the Tioga Valley to the Chemung River. A small tributary descended northward from a pass a few miles south of Ansonia, and joined the main stream there. The pass separated the drainage from another south-flowing system.

Pine Creek Gorge, Pennsylvania's "Grand Canyon;" looking north from Colton Point State Park on the west rim.

Enter glaciation. For a time the Marsh-Crooked Creek valley must have been blocked by ice, and meltwater ponded there until it overtopped the pass and poured through to join the southern stream system. Long-continued flow through the breach eventually destroyed the pass as it carved out Pine Creek Gorge. When the ice finally melted out of the Marsh-Crooked Creek valley, the new southern route through the gorge was well-established.

Of course, another much odder geologic story is written in the rocks of the canyon walls. All of the rocks of the northern part of the canyon belong to the upper Devonian Catskill formation, which crops out in the lower canyon all the way to its mouth near Jersey Shore, at the Susquehanna River. South of Tiadaghton, the creek alternately cuts across broad open synclines and anticlines that form ridges and valleys, respectively. Pine Creek Gorge is deepest across the synclines, where the upper Devonian to lower Mississippian Huntley Mountain formation is the rim rock, and the lower Mississippian Burgoon sandstone and Pennsylvanian Pottsville beds are not far back from the rim.

The Huntley Mountain formation is transitional between the Catskill and Burgoon formations, and contains the boundary between

Devonian and Mississippian rocks. It is dominantly composed of flaggy, greenish-gray sandstones and minor red shales. The lower sandstones are similar to the underlying red Catskill sandstones; the upper sandstones resemble those of the overlying Burgoon formation.

Nearly all of the exposed rocks are terrestrial or shallow marine. The Catskill formation consists of sediments carried by streams from the Acadian Mountains, that once stood where you now see the Atlantic coastal plain. The Huntley Mountain formation represents a similar environment, except that the streams probably carried heavier loads of sediments. It may mark the first pulse of Alleghenian mountain building, which followed closely on the heels of the Acadian mountain building.

Pine Creek is part of the Seneca Trail long used as a trade and travel route by the Indians between the Genesee and Susquehanna rivers. South of Ansonia, the trail branched, with one branch following the creek, and the other going overland to rejoin the creek at Blackwell.

• An 18-mile scenic drive south of the highway on Pennsylvania 44 between Carter Camp and Sweden Valley takes you high up on the plateau through Susquehannock State Forest. It begins from Pennsylvania 144, ten miles south of Galeton. Cherry Spring tower and an overlook offer panoramic views of the part of the plateau dissected by Sinnemahoning Creek, a tributary to the west branch of the Susquehanna River. The plateau is deeply creased by the streams, and heavily wooded. Between Sweden Valley and Galeton, 19 miles, you pass over Denton Hill, 2424 feet, one of the major stream divides.

Coudersport Ice Mine

This oddity is in the ridge directly southwest of Sweden Valley, where ice appears in spring each year and continues through the hot summer, but disappears in winter! The ice forms in a vertical shaft eight to ten feet wide and 40 feet deep. It takes various forms, often appearing as huge, crystal-clear icicles, up to three feet thick and 25 feet long. One theory for this puzzling and seemingly backwards phenomenon has cold winter air descending into openings in the Lock Haven formation, and expelling into the shaft the warm air that accumulated in them during the summer, which melts the ice. The cold air remains stored in the rocks during the summer, freezing groundwater that seeps into the shaft.

• The junction between US 6 and Pennsylvania 872, two miles east of Coudersport, was originally called Lymansville. For many years it was the northern terminus of the Jersey Shore Pike, a major trade

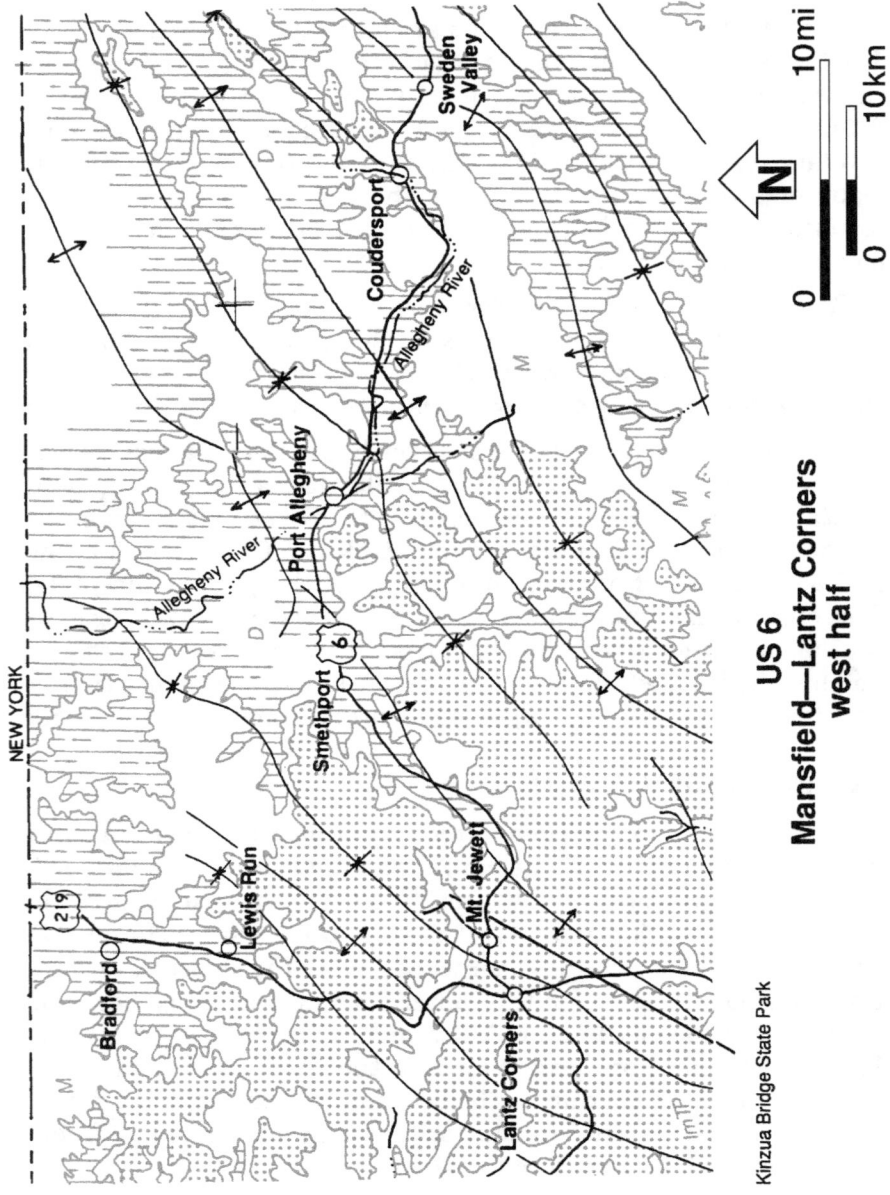

US 6
Mansfield—Lantz Corners
west half

route for early settlers in the nineteenth century. This packtrain and wagon route followed Sinnemahoning Creek to its junction with the west branch of the Susquehanna River at Jersey Shore west of Williamsport.

From Coudersport to Port Allegany, 16 miles, US 6 follows the Allegheny River which then flows northward to Olean, New York, goes around Allegany State Park, and then comes back into Pennsylvania. This looping bend, the Salamanca Re-entrant, roughly parallels the southern limit of ice-age glaciation.

Unglaciated Landscapes

The unglaciated landscape of the Salamanca Re-entrant between Port Allegany and Warren differs subtly from that of the glaciated regions. It lacks, for example, glacial till and erratics, but does contain outwash sediments derived by stream reworking of drift of all kinds in the bordering glaciated regions. Natural lakes are almost non-existent, while they are numerous in the northeastern Pennsylvania glaciated region and moderately so in the northwestern region. Other differences are more subtle: narrow, instead of broad valleys, and ridges less rounded because they have not been reshaped by overriding ice.

Allegheny Portage

Port Allegany is at the northern end of a 23-mile portage along the Sinnemahoning path through the Allegheny Mountains; which the Indians probably established before the 17th century. The southern end of the portage was at the present site of Emporium Junction, on Sinnemahoning Creek. Pennsylvania 155 now follows the original route between the two points the Indians called "canoe places," following Portage Creek and the Driftwood branch of the Sinnemahoning. From Emporium Junction, the old trail continued southeast along the present route of Pennsylvania 120 to Keating, on the Susquehanna River. The portage thus connected the Allegheny and Susquehanna rivers, which were major travel corridors, first for the Indians and then for white settlers.

• Several good views over the plateau country appear in the ten miles between Port Allegany and Smethport. Red and green sandstone and shale of the Catskill formation are exposed in a few large cuts along the way. Smethport is on Potato Creek, a north-flowing tributary to the Allegheny; the heavily wooded segment of highway between Smethport and Lantz Corners follows its tributary, Marvin Creek.

Kinzua Viaduct.

Kinzua Viaduct

A few miles northeast of the village of Mt. Jewett is Kinzua Bridge State Park and the Kinzua Viaduct over Kinzua Creek, a reminder of the historical importance of coal in this region. The bridge was built in 1882 for a branch of the Erie Railroad to ship coal north. It was then the world's highest rail viaduct at 301 feet and longest, at 2159 feet Rebuilt of steel in 1900 to carry heavier loads, it was in service until 1959.

US 6
Lantz Corners—Meadville
112 mi./181 km.

Between Lantz Corners and Warren, 36 miles, the topography is neither rounded nor deeply dissected. You see Devonian, Mississippian, and Pennsylvanian rocks including several conglomerate beds. This section of Pennsylvania is oil and coal country.

Kinzua Dam

Kinzua flood control dam, six miles east of Warren via Pennsylvania 59, spans an exceedingly narrow segment of the Allegheny Valley just west of a sharp turn called Big Bend. This is an ideal dam site, with a narrow bedrock valley that widens immediately upstream to provide a reservoir of much greater width. The dam is like the proverbial "cork in the bottle." The dam is a combination earth and concrete structure anchored to bedrock, which is part of the upper Devonian Catskill Delta. Watch for the horizontal layers of sandstone, siltstone, shale, and some conglomerate of the Chadakoin (lower) and Venango (upper) formations. The huge cut near the south end of the dam is in the Venango formation.

This area offers a rare and dramatic view of geologic history, best visualized from overlooks atop Coal Knob ridge immediately east of the dam. Continue on Pennsylvania 59 for three miles from the dam, and turn right just before the Cornplanter Bridge onto Longhouse

Upper Devonian Venango shales by Kinzua Dam.

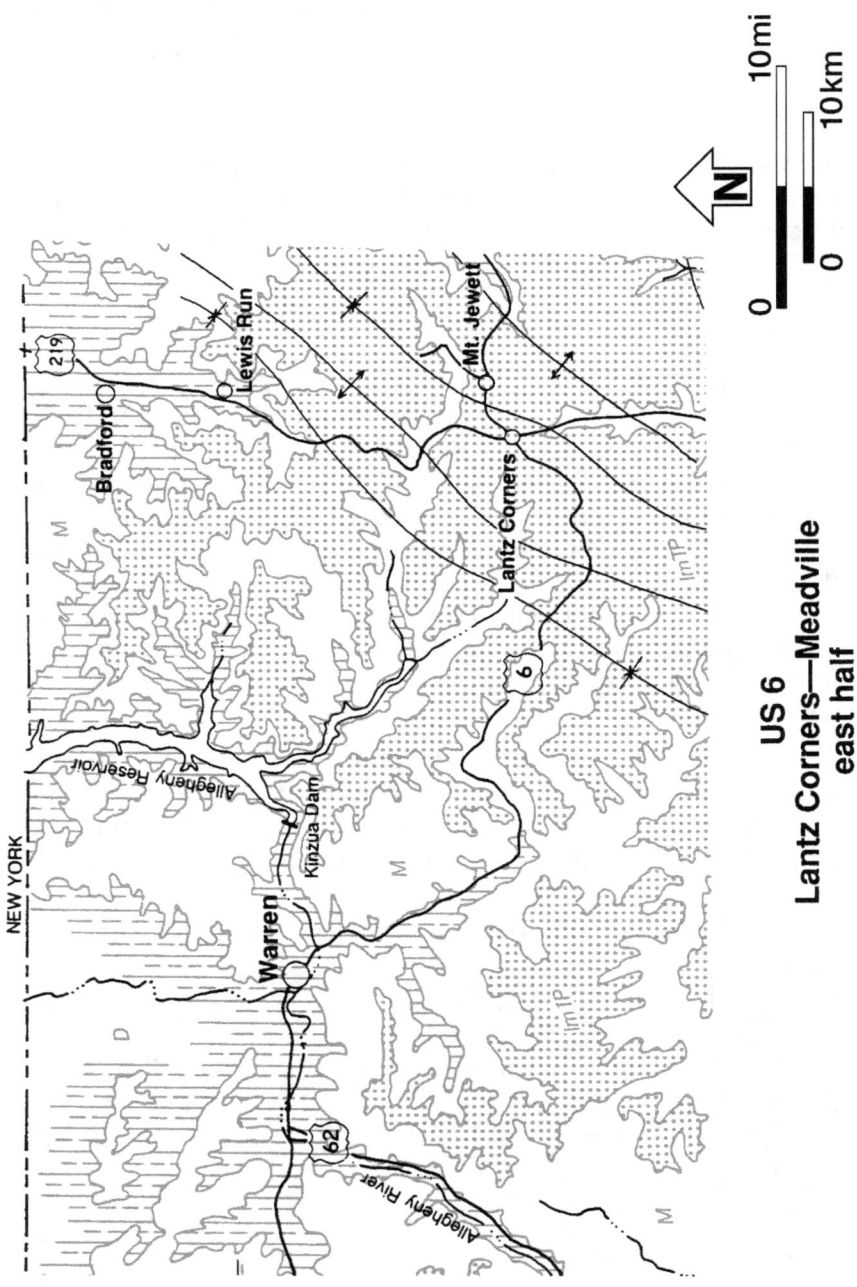

US 6
Lantz Corners—Meadville
east half

Scenic Drive, then right again after 1.5 miles, where the road forks. Continue to the top of the ridge to a picnic area and paved trail to Jake's Rocks, and a loop drive with two overlooks. The southernmost overlook affords the best view, a panorama encompassing the southern end of the reservoir, the dam, and the narrow valley downstream from it.

Imagine the scene of about 20,000 years ago. Glacial ice is just visible beyond the hills to the west. You can't see the ice to the north because it is too far away, near Allegany Park in New York. The lake is banked up against the ice on the north and high ground elsewhere. In place of the dam is Kinzua Pass, a saddle, through the ridge leading to the headwaters of a river that runs down to the west on the other side. Waves lap against the shore close to the pass driven by the cold winds descending from the ice sheet. Behind you, on the other side of your ridge, a long arm of the lake reaches southward up tributary Kinzua Creek valley. It is summer, yet icebergs drift on the lake; snow dusts the barren ridges; you shiver in the arctic scene.

Now take a few millenial leaps forward, one at a time. After the first thousand years, the scene hasn't changed much, except that the ice front has melted back enough that it is no longer visible in the west.

The Allegheny Reservoir now occupies approximately the same position as glacial Lake Carll, but it's probably not as deep.

Kinzua Dam spans a bottleneck in Allegheny Valley, where Kinzua Pass served as a drainage divide in pre-glacial time. Glacial Lake Carll overtopped the pass and wore it down, and thereby drastically changed the Allegheny River system.

Most significantly, the lake has risen with the added influx of meltwater; great torrents of water pour through a small canyon cut into Kinzua Pass.

Another thousand years. The lake has dropped because the slash through the pass is deeper. The same is true for the next few millenia — the canyon gets deeper and deeper, and the lake smaller and smaller. After several thousand years, the lake is gone and the modern Allegheny River meanders south over a broad plain of lake sediments, then turns abruptly west at Big Bend to flow through the narrow canyon where the pass was.

Long before the ice age lake filled the valley where you now see the Allegheny Reservoir, there existed a pre-glacial Allegheny River system very different from the present. The middle branch descended from this side of Kinzua Pass and flowed north to the basin that was later to become Lake Erie. The west branch went down the other side of the pass to the present site of Warren, then turned north to join the ancestral Allegheny River near Randolph, New York. The main, or east branch descended from Pennsylvania into Olean, New York, then looped around Allegany State Park to join the middle branch near Steamburg, New York.

The modern river system empties to the Gulf of Mexico via the Ohio and Mississippi rivers, instead of to the Atlantic Ocean through the St. Lawrence River. Flow in the west branch, now Conewango Creek, was reversed, as was that of the middle branch of the old river, now the main branch of the new river. In New York, the lower reach of the pre-glacial river, now Cattaraugus Creek, is separated from the Conewango headwaters by the Gowanda moraine; another moraine blocks the old junction at Steamburg.

You can't, of course, see all of this from the overlook. But with the reservoir in place of the old lake, a little imagination and knowledge of geological processes will give you a unique "window on the past."

Jake's Rocks

The caprock of Coal Knob ridge is massive conglomerate of the Pennsylvanian Pottsville group. Jake's Rocks, at the north end of the ridge, are a rock city typical of this and similar formations of Pennsylvanian and Mississippian ages in western Pennsylvania and southwestern New York. The buildings of these fascinating cities are

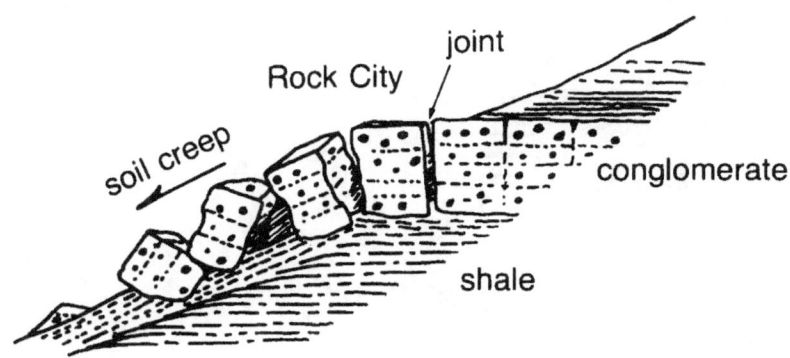

Schematic cross-section of a thick bed of conglomerate over shale illustrating ideal conditions for formation of rock cities.

giant, generally rectangular, blocks leaning this way and that in disarray; and the streets are open passageways, crawl spaces, and tunnels along the fractures that separate them. Rock cities develop from flat-lying, massive beds of resistant conglomerate or sandstone exposed on hillslopes above much weaker beds of shale or siltstone. Widely spaced vertical fractures called joints outline the blocks.

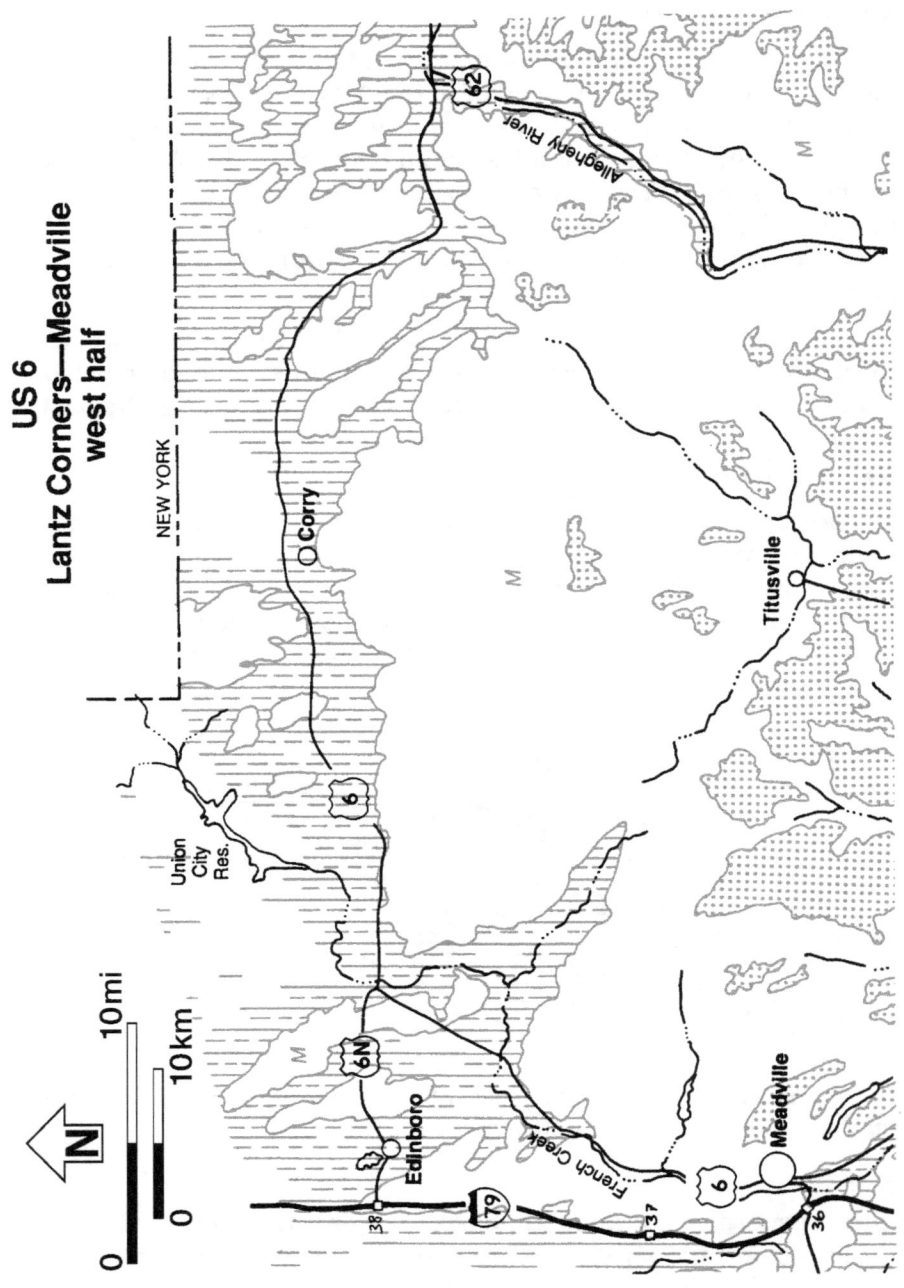

Jake's Rocks, a rock city in Pennsylvanian Pottsville sandstone and conglomerate on Coal Knob ridge near Kinzua Dam.

Erosion of the hillslope proceeds more rapidly in the weak beds than in the conglomerate, leaving it overhanging so the blocks eventually tumble. Frost heave and root-pry in the joints aid the process, particularly in the early stages of separation. In time, some buildings may creep far from their source. The rock cities are fun to explore and climb around, natural playgrounds.

• Warren is at the junction of the Conewango and Allegheny rivers, also at the margin of the western glaciated region of Pennsylvania. Therefore, the topography from here west is more rounded, reflecting glacial scour; glacial till is visible in some road banks. Between Warren and Irvine, five miles, the route follows the Allegheny River downstream. From Irvine the river continues south, eventually to join the Monangahela River at Pittsburgh to form the Ohio River.

The countryside in the long 76 miles between Warren and Meadville, is gently rolling farmland. Bedrock is principally Pennsylvanian and Mississippian formations, but it is poorly exposed. Oil and gas wells dot the landscape. The lesser relief is in part due to glacial erosion and deposition and in part to the great distance from the Allegheny Front. Stream dissection is most active in the face of the scarp and lessens with increasing distance from it. Drumlins, eskers, and kames formed by glacial deposition add detail to the landscape. Kettles and kettle lakes exist, but not as abundantly as in the Pocono Mountains.

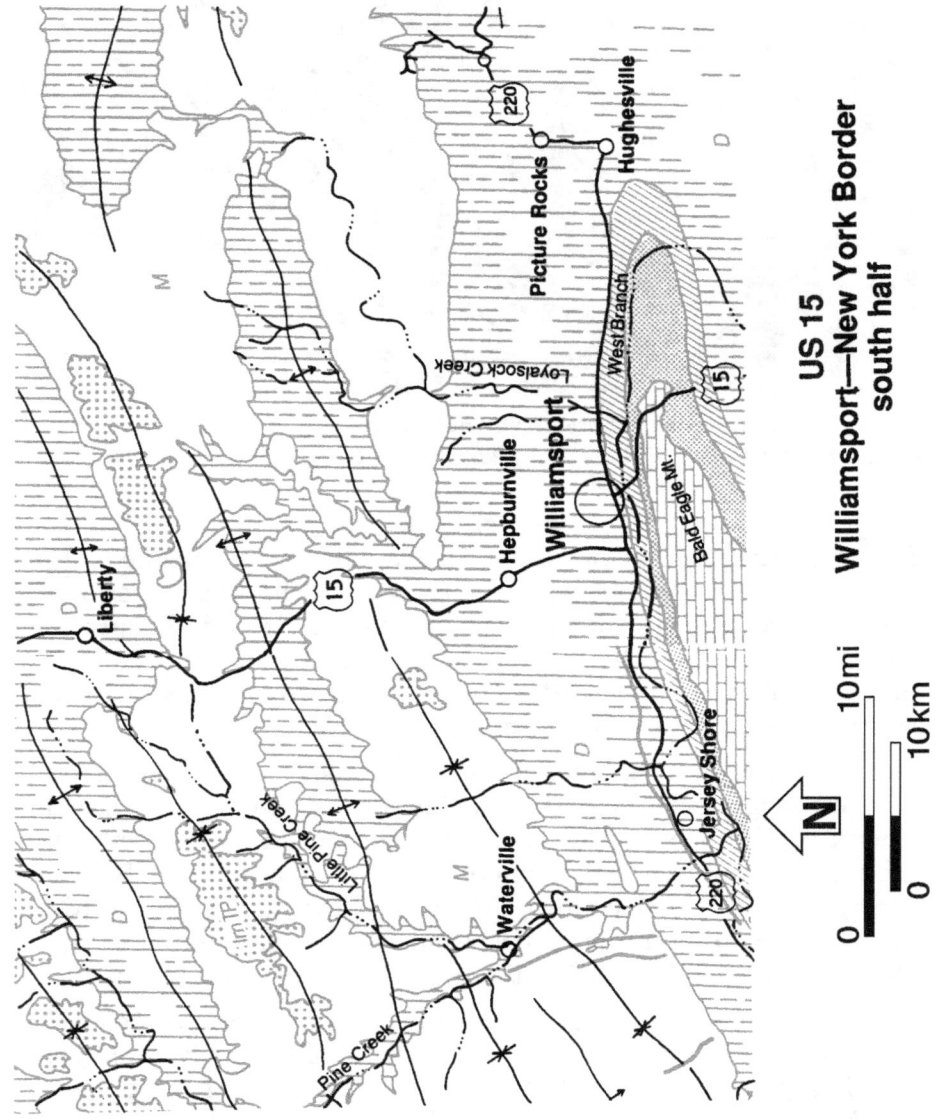

US 15
Williamsport—New York Border
60 mi./97 km.

This segment of US 15 crosses the Allegheny Plateau. The Allegheny Front rises 1300 feet in a gentle slope just north of Williamsport. As you travel north from Williamsport, you climb the section of strata that make up the scarp, most of which are poorly exposed. Near the river, you cross dark shales of the middle Devonian Mahantango and upper Devonian Harrell and Brallier formations. Above those are the sandier, grayish rocks of the upper Devonian Lock Haven formation. Contact with the overlying Catskill formation is at Hepburnville, five miles north of the river. The hamlet of Powys, at ten miles, is on the east-west line of the rim rock of lower Mississippian Burgoon sandstone.

Rocks of the Allegheny Front encompass a nearly complete stratigraphic record of Acadian mountain building. The dark shales at river level are deep-water sediments deposited in the Acadian foredeep basin, which formed in the earliest stage of the collision between North America and Europe. The Lock Haven formation records the gradational change to the dominantly terrestral conditions under which the Catskill sediments were laid down during and after the main Acadian uplift. The mountains were then rising about where the coastal plain now lies. The Huntley Mountain formation contains the latest Acadian sediments, and the Burgoon sandstone represents seashore and coastal plain sedimentation.

All of the beds dip into the Allegheny Front. At the base, the dip is steep, like that of the Tuscarora quartzite on the north flank of Bald Eagle Mountain across the river. The angle gradually lessens upslope to a gentle dip at the crest. The beds are on the north limb of a gigantic, eroded anticline, the Nittany Arch, of which Bald Eagle Mountain is the core. The fold formed during Alleghenian mountain building above a thrust fault ramp where the flat-lying sole thrust, which cuts Cambrian strata about 30,000 feet under the Valley and Ridge province, ascends at a steep angle and then levels off in weak Silurian shales about 10,000 feet beneath the plateau. The beds carried on the sole thrust arch over the top of the ramp.

Originally, the Burgoon sandstone of the scarp crest arched thousands of feet over Bald Eagle Mountain, but erosion destroyed the whole top of the fold, including the entire section between the Tuscarora and the Burgoon formations. Furthermore, east and west of US 15 along the Allegheny crest are erosional remnants of upper

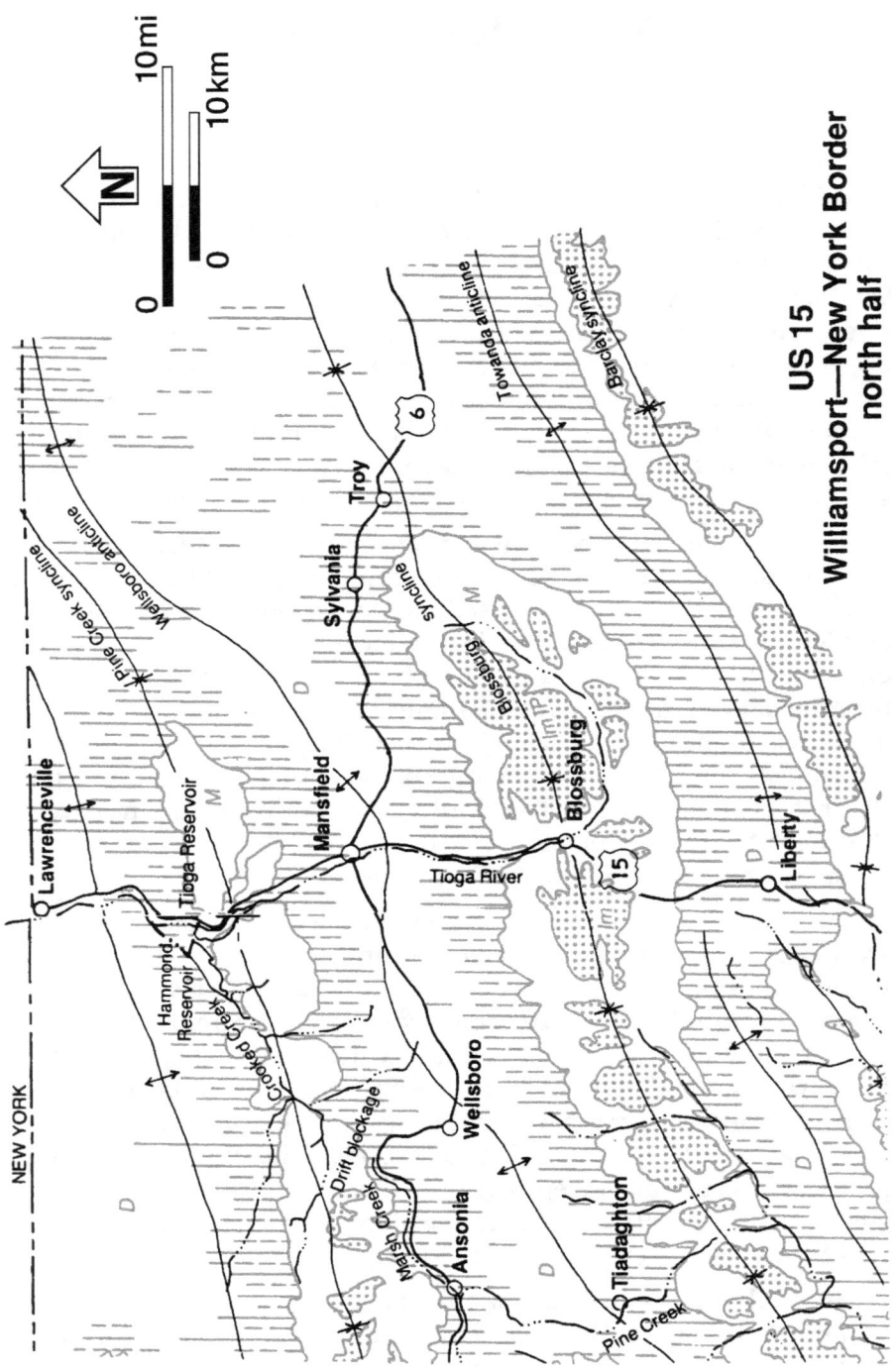

Mississippian Mauch Chunk and Pennsylvanian Pottsville formations that lie on the Burgoon sandstone. They also were part of the arch.

The west branch of the Susquehanna River now follows the belt of weak Silurian and Devonian shales that lie stratigraphically above the Tuscarora quartzite. Imagine an earlier time when the river, at a higher elevation, banked against the inclined Tuscarora beds on the northern flank of Bald Eagle Mountain. Unable to cut into the hard quartzite, it gradually slid down the dip as it stripped the soft overlying beds, eventually bringing it to its present niche.

US 15 passes through a wind gap near the eastern end of Bald Eagle Mountain. This was a spillover channel for glacial meltwater. Old moraine deposits high on the sides and end of the ridge indicate the position of the ice front during at least two glaciations. On both occasions, the ice blocked the Susquehanna drainage around the end of the ridge, so meltwater ponded to the west between ridge and ice. When it got deep enough, the water spilled out through a shallow pass on the ridge — at the site of the wind gap — and, in time, carved it deeper. When the ice melted away from the ridge end, the lake drained around it and abandoned the gap, establishing the present river course.

In the 50 miles between Powys and the New York border at Lawrenceville, you cross several alternating, northeast-trending anticlines and synclines of the very large, open variety that gently corrugate the plateau. This is a region of pronounced topographic inversion, where structural highs or anticlines, form valleys; and structural lows or synclines, form ridges. This relationship characterizes nearly all of Pennsylvania's plateau country.

You see mostly Catskill formation, dominantly reddish, sometimes brick red or maroon, to greenish-gray, cross-bedded sandstones and crumbly shales. Where the road crosses synclines, as at Blossburg,

Red to gray, cross-bedded sandstones and dark shales of the upper Devonian Catskill formation on U.S. 15 near Liberty.

you see massive Huntley Mountain sandstones, distinguished from those of the underlying Catskill formation by their darker greenish gray to olive gray color. Loose blocks of buff sandstone are from the Burgoon formation that crops out farther upslope.

East and west of US 15 at Blossburg, the even younger Pennsylvanian Pottsville beds are preserved in the core of the Blossburg syncline, but you don't see them by the highway.

Rocks of the Lock Haven formation appear where the road crosses anticlinal axes, as at Mansfield, on the axis of the Wellsboro anticline, and at Tioga Junction, on the axis of the Sabinsville anticline.Lock Haven beds are beautifully exposed in a huge cut six miles south of Lawrenceville, on the side of a hill in view of Tioga Dam. Here they consist of thick-bedded, light gray, fine-grained sandstone that contains tiny flecks of sparkly mica, and interbedded dark gray shale that breaks to a chip rubble.

The Tioga Dam is one of two at this site; the other is Hammond Dam. An extremely narrow divide near the dams separates the reservoirs.

At Tioga Junction, three miles south of Lawrenceville, you cross deep gas fields that produced from the lower Devonian Oriskany sandstone on the axis of the Sabinsville anticline. They no longer produce, but are used to store gas imported from other states. The gas is pumped into the natural sandstone reservoir deep undergound in summer when demand is low and pumped out again in winter.

Tioga Reservoir dam from U.S. 15. Twin Hammond Reservoir dam is just over the low ridge with the tower on it.

US 62
Franklin—Junction US 6
53 mi./85 km.

This route is neither heavily traveled nor blessed with many great views or outcrops, but it is geologically important. It follows the Allegheny River approximately parallel to the glacial boundaries, which lie several miles to the northwest along much of the way.

The direction of river flow is opposite the pre-glacial direction, which was north into the Lake Erie basin. Now the river flows south to join the Monongahela at Pittsburgh, forming the Ohio River which empties to the Mississippi and the Gulf of Mexico. This momentous alteration of the drainage network happened during the last ice age, when ice blocked the northern escape routes and meltwater had nowhere to go but south. It spilled over passes, destroying them in the process, and followed existing valleys. Certain visible features of the modern Allegheny Valley record this history: deep incision into the plateau bedrock, deeply notched tributaries, and stretches of valley outwash deposits that are now terraced from post-glacial erosion.

US 62 cuts through the region of shallow oil fields where the oil industry began with completion of the first commercially successful well at Titusville in 1859. The region of shallow gas fields begins southeast of the route. Shallow oil and gas come principally from upper Devonian sandstones. They probably originated at greater depth in dark middle Devonian shales deposited in the Acadian foredeep basin. Most dark marine muds contain much organic debris that may convert to hydrocarbons as deep burial compresses and heats them. Burial also squeezes oil, gas, and water out of the shale; they migrate upward through the pore spaces of the overlying rocks, stopping when they reach another tightly compacted, impermeable shale. You will not see many wells right by the highway, but thousands are nearby, most of them abandoned.

The city of Franklin is where French Creek joins the Allegheny River. Its strategic position made this the site of several forts during the 18th century, including French Fort Machault, built in 1753, British Fort Venango, built in 1760 and destroyed in the Pontiac War of 1763, and American Fort Franklin, built in 1787. A blacksmith drilling for water in 1860 struck oil, and the city's economy has been largely oil-based since.

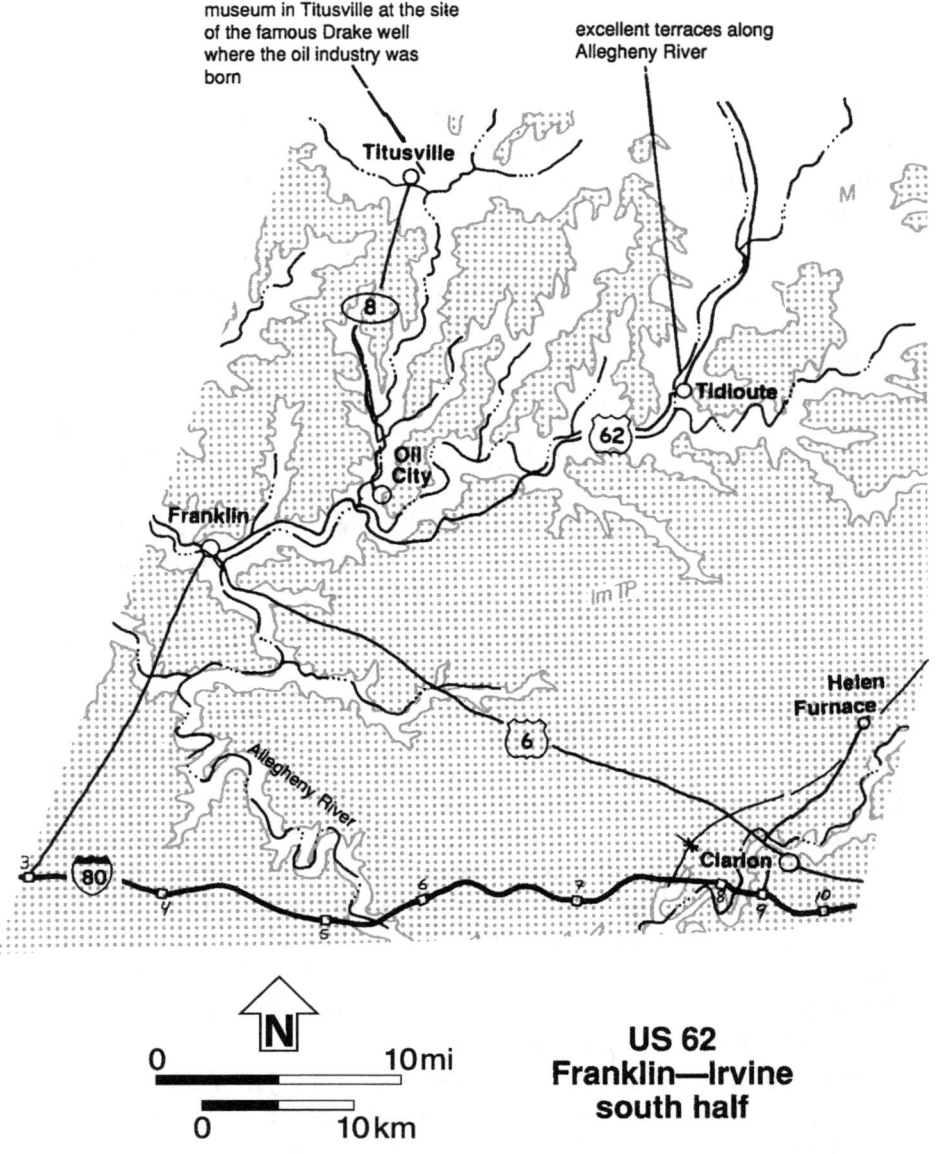

Pump-jack on Pennsylvania 36 near Titusville

The route crosses the Allegheny River in Franklin, beside a high scarp where you see yellowish gray siltsones, dark gray shales, and some thin sandstones of the lower Mississippian Cuyahoga formation. The same unit appears in river cutbanks for many miles upstream. The regional caprock on the plateau is lower Pennsylvanian Pottsville conglomerate and sandstone. The river and its short, deep tributaries have cut through it and the underlying lower Mississippian Shenango formation into the Cuyahoga beds. The boundary between the Shenango and Pottsville formations is a buried erosion surface that separates parallel beds about 30 million years apart in age. The layers all dip very gently southward, but more steeply than the river gradient; so the strata exposed by the river get older upstream. This is readily apparent in the outcrop pattern shown on the map.

Typical gas well and storage tank of northwestern Pennsylvania

Reconstructed Drake well at Titusville.

The eight miles of picturesque road between Franklin and Oil City, follows close to the river. Oil City, at the junction with famous Oil Creek, became a boom town almost overnight after the completion of the Drake Well in 1859 prompted wildcat drilling at a furious rate. Oil Creek valley quickly became the busiest place on the continent, lined with hundreds of derricks; seventeen million barrels of oil were shipped from here to Pittsburgh between 1860 and 1870.

In the 15 miles between Oil City and Titusville, Pennsylvania 8 follows Oil Creek for about four miles. Most of the old wells are abandoned and no longer visible, but one just south of Rouseville, McClintock No. 1, produced continuously for more than 100 years.

The reconstructed derrick and engine house of the Drake Well are now attractions of a lovely memorial park at Titusville, along with many outdoor displays of early oilfield equipment, an indoor museum, library, and theater. The park makes the extra miles to Titusville worthwhile.

You cross to the south side of the Allegheny River at Oil City. Between Oil City and Tionesta, 18 miles, the road veers south of, then returns to the river, crosses to the north side at Hunter, then crosses back again at Tionesta. In the traverse away from the river, you pass over Pottsville conglomerate on top of the plateau, some 500 feet above

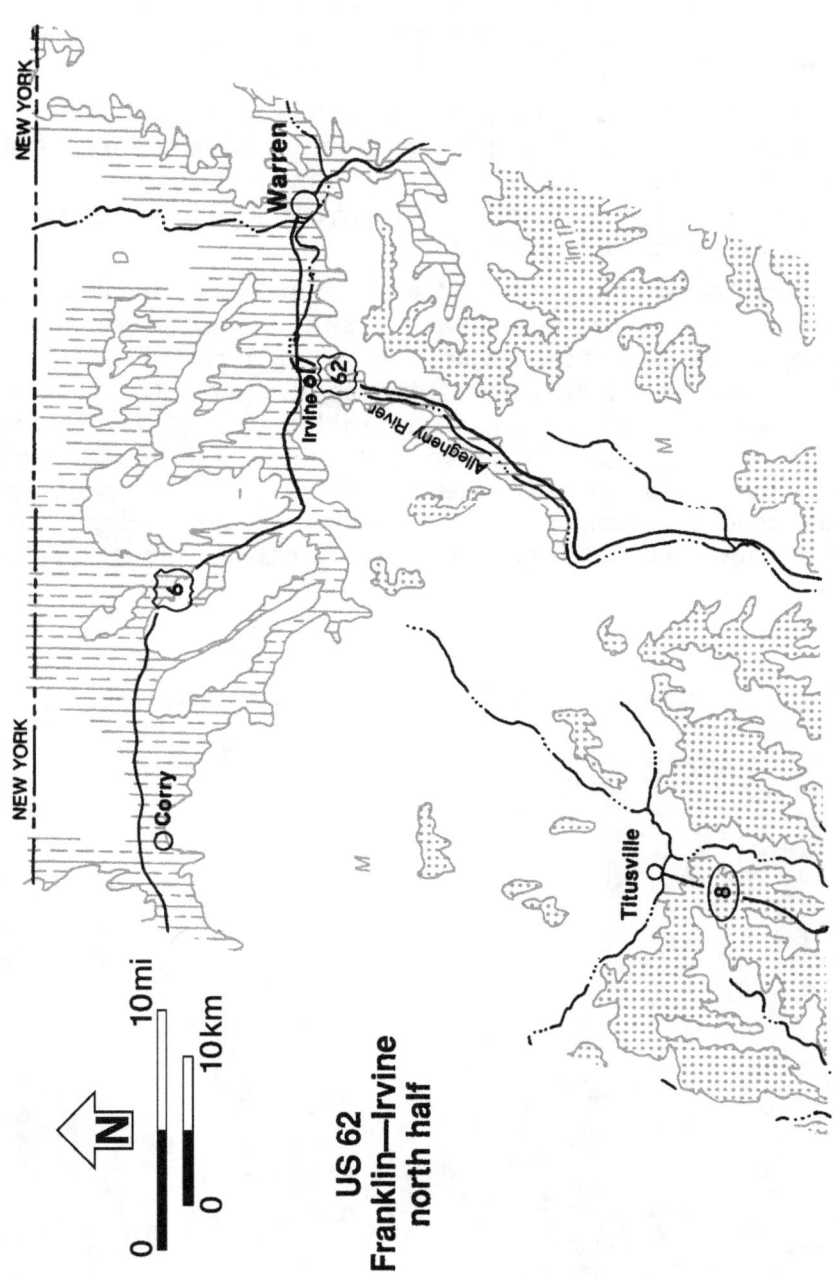

the river. Next to the river, you pass Mississippian strata. The beds now lie almost as flat as the original sediments were laid down in a shallow inland sea about 360 million years ago. They were only slightly folded as they rose to their present elevation.

River terraces are well preserved near President, Hunter, and Tionesta. The latter village name comes from Tionesta Creek, which joins the Allegheny here; it is an Iroquois Indian term that means "it penetrates the land." Tionesta Reservoir, over a mile upstream, takes advantage of this deep penetration.

Near Tionesta you see lower Mississippian redbeds that lie beneath the Cuyahoga formation exposed at Franklin. Light gray shales and flaggy sandstones of the Venango formation appear near Tidioute. This formation contains some thick conglomerates, loose blocks of which are locally mixed with soil and visible by the highway. This area was on the thin edge of the Catskill Delta, far from the rising Acadian Mountains, and normally received only very fine sediments. The sandstones and conglomerates probably originated during times of low sea level, when streams could carry the coarse sediments farther.

View of entrenched Allegheny River from Tidioute overlook, near Pennsylvania 62.

You can get a nice view of the river and the plateau from Tidioute overlook. Cross the bridge from Tidioute to the south side, turn west, and almost immediately after, turn sharply back to the left onto Pennsylvania 337. Follow this road for 1.2 miles to the overlook. Perhaps the most striking aspects of this scene are the evenness, or accordance, of the upland divides, and the depth to which the Allegheny River cut into the plateau. The even ridge crests throughout the Allegheny Plateau may be remnants of an old erosion surface.

Between Tidioute and US 6 junction, 16 miles, the route continues alongside the river over Venango beds. Contact with the overlying Mississippian strata is never far upslope on either side of the river, indicating the very gentle dip of the bedding. Near the US 6 junction you cross the basal contact of the Venango formation into the underlying upper Devonian Chadakoin formation, but there are no roadcuts.

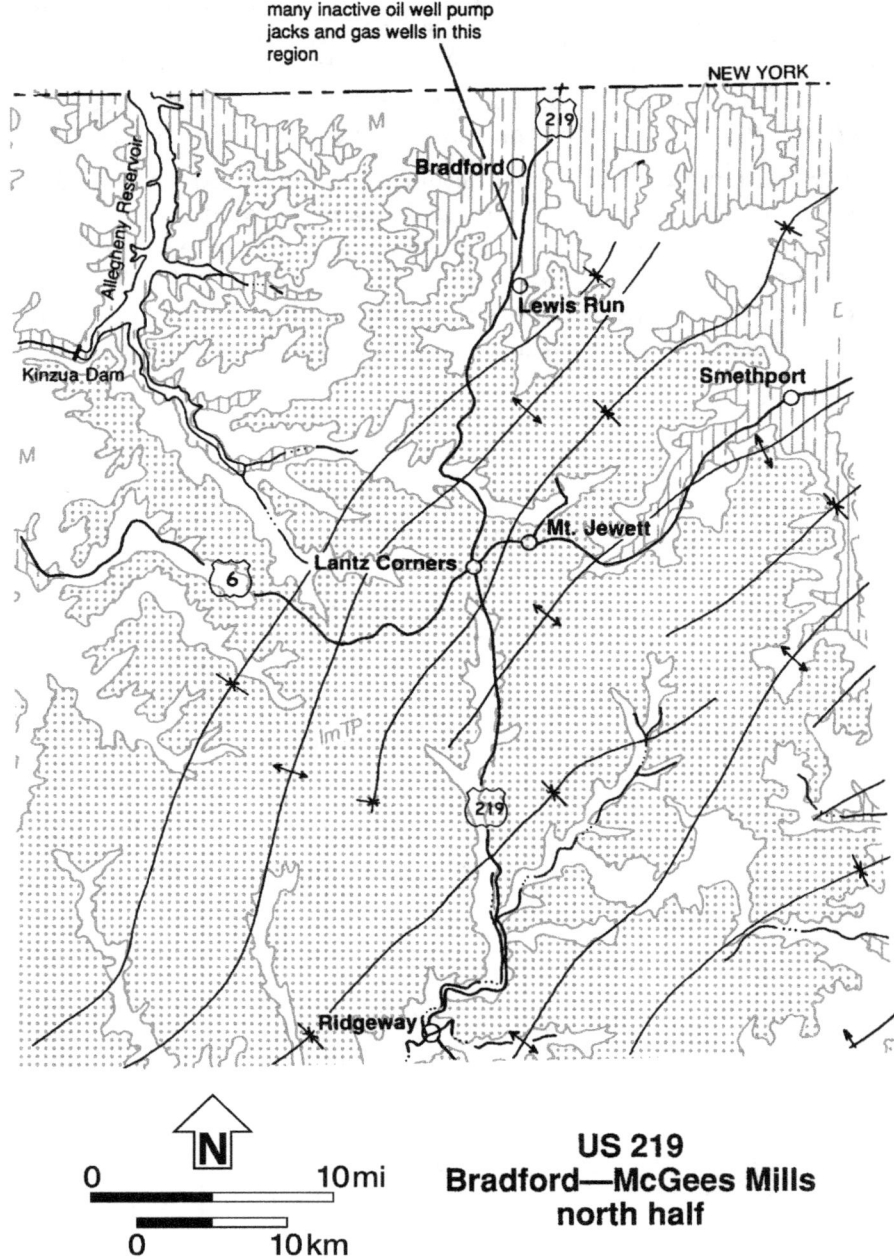

US 219
Bradford—McGees Mills
north half

US 219
Bradford—McGees Mills
104 mi./168 km.

South of Bradford for several miles, you follow the east branch of Tunungwant Creek, which cuts into the plateau strata exposing the upper Devonian Chadakoin formation. You see several road cuts in flat, grey-brown siltstone and some sandstone, interbedded with gray shale.

Several gas and oil wells are on the south side of Bradford. The old, rusting, and immobile pump jacks symbolize the dying oil industry in western Pennsylvania.

At Lewis Run, six miles south of Bradford, look for a low cut in reddish gray shale and siltstone of the upper Devonian Catskill formation that overlies the Chadakoin formation. The Catskill formation is relatively thin here because the original sediments were laid down near the edge of the Catskill Delta, far from their eastern source.

Catskill Delta

The Catskill Delta is really a complex of numerous deltas and fans built of sediments eroded from the Acadian Mountains which rose in late Devonian time as the northern Proto-Atlantic basin clamped shut, and Europe collided with North America. The axis of the range was probably somewhere along the line of the present coast, east of Pennsylvania, where the deeply eroded core rocks lie buried beneath coastal plain sediments. The sediments, carried west principally by streams, gradually displaced a vast inland sea, the Acadian foredeep basin, to form an enormous wedge-shaped deposit, thick in the east and thin in the west. Each new layer thickened the pile and extended the westward reach of the delta. In northwestern Pennsylvania, the Catskill Delta deposits, mainly represented by the Catskill formation, are thin because they are near its western margin.

• Between Lewis Run and Lantz Corners, 13 miles, the road crosses a segment of plateau and two valleys. Olive-gray shale and sandstone belonging to an interval from Catskill to lower Mississippian Shenango formation form the sides of each valley, but outcrops are rare. The overlying lower Pennsylvanian Pottsville sandstone, visible in a few exposures, caps the divides between. This outcrop pattern characterizes the Allegheny Plateau of northwestern Pennsylvania.

Outcrop Patterns of Northwestern Pennsylvania

For all practical purposes, the sedimentary strata of this section of plateau are like layers in a cake. Streams that carved the cake created a pattern of outcrops that follows the stream patterns. From the air, the drainage systems look like trees, with branches formed by tributaries, so we call the pattern dendritic. Dendritic drainage develops wherever the underlying material offers uniform resistance to erosion. As the streams erode their valleys they maintain the dendritic pattern. In the shallowest headwaters valleys only the uppermost layer crops out, while progressively lower layers appear at lower levels farther downstream. In its deepest part, the valley wall may display several flat-lying layers, one on top of the other. The layer contacts precisely follow the contours of the valley walls.

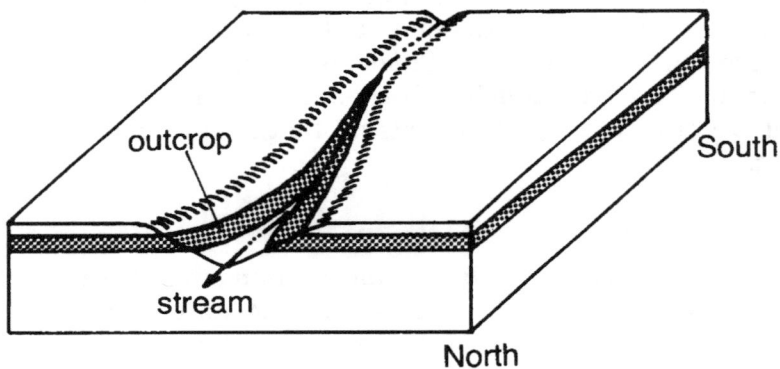

The combination of gentle southern dips and north-flowing streams in the Allegheny Plateau of northwestern Pennsylvania produces south-pointing finger-like outcrops of individual formations.

• The junction with US 6 at Lantz Corners is on the divide between Kinzua Creek on the north and Clarion River on the south. The underlying bedrock is Pottsville sandstone. Between Lantz Corners and Ridgeway, 24 miles, the road descends to and follows the Clarion Valley downstream. You see Oswayo and Shenango beds exposed in a narrow strip along the river beneath the plateau-capping Pottsville sandstone. This river does not penetrate lower strata as it progresses downstream because the layercake is slightly tilted — the beds dip gently south, and the river follows the dip.

In the nine miles between Ridgeway and Brandy Camp, you go over Boot Jack Summit on the axis of the Hebron anticline. This is the

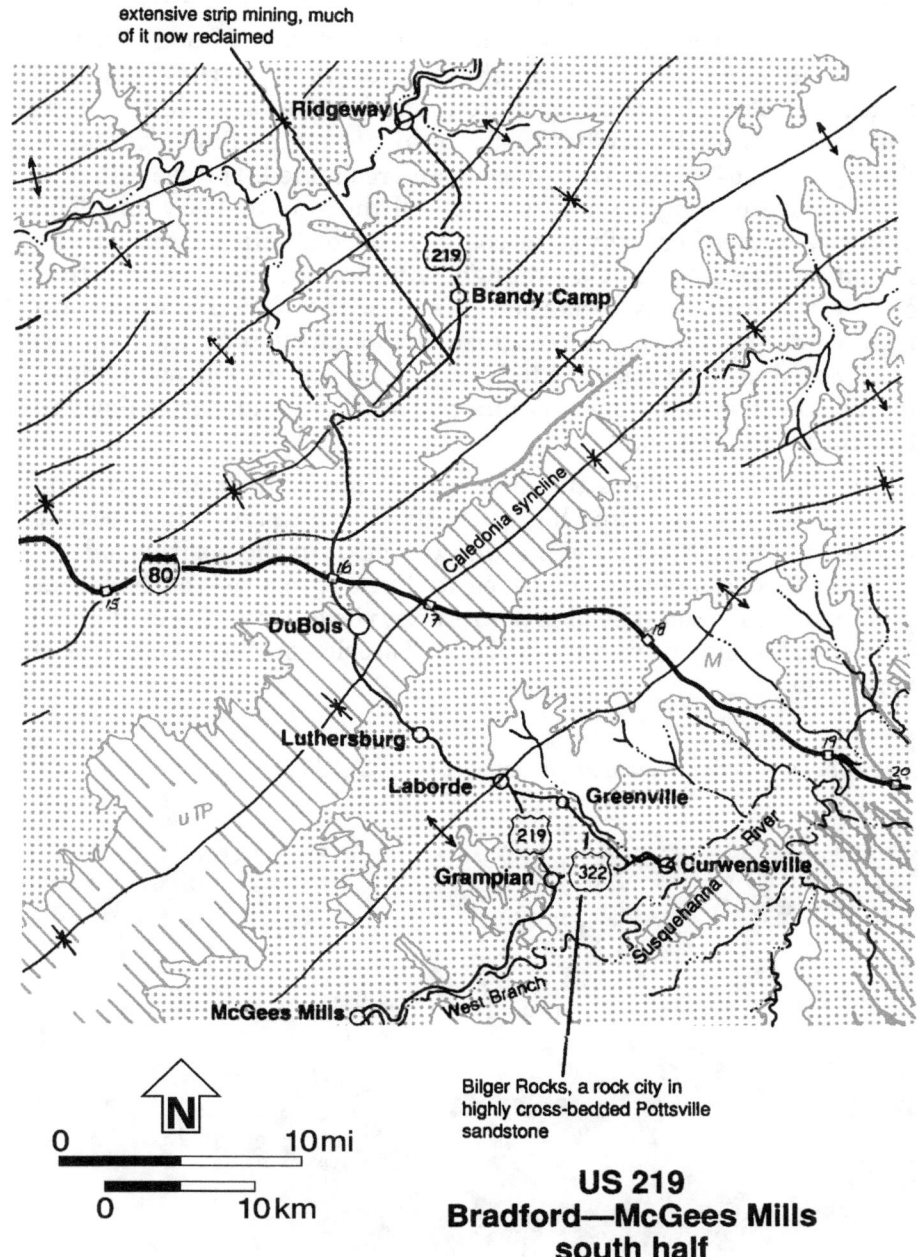

**US 219
Bradford—McGees Mills
south half**

continental divide that separates the Clarion River of the Ohio River system from Little Toby Creek of the Susquehanna system. At the summit are good views east and west of the plateau country.

Brandy Camp is on the axis of the Shawmut and St. Marys syncline, in middle Pennsylvanian Allegheny beds that overlie Pottsville sandstone. The Allegheny formation contains the valuable Freeport, Kittanning, and Brookville-Clarion coals; so strip mines abound in this region; most of them are inactive and reclaimed. You see evidence of extensive mining from here south all the way to Maryland. The highway crosses Pennsylvanian coal-bearing beds almost exclusively, including those of the Conemaugh and Monongahela groups that overlie the Allegheny formation. Many of the old strip mines are reclaimed, and difficult to recognize. Although current law requires that mined land be restored to its original form insofar as possible, many older mines in Pennsylvania still scar the land, particularly in the anthracite region.

This highway follows the slope of the land rather than cutting through the hills or riding high on valley fill. It's slow going because it goes up and down and winds and weaves. On the positive side, it affords the Roadside Geologist a feel for the landscape — for the extent of stream dissection. The high points offer views across the even ridge crests, which may be remnants of an old erosion surface of millions of years ago.

Small open pit coal mine near Brockport by U.S. 219.

You pass several strip mining operations between Brandy Camp and the I-80 junction. One open pit is near Brockport, five miles south of Brandy Camp, where reclamation follows stripping closely. Reclaimed mines can sometimes be recognized by smooth, regraded contours that lack outcrops and are covered with neat rows of newly-planted trees. In hilly country like that along US 219, erosion of mined-out lands may become a serious problem, and, of course, trees and grass help to prevent it.

DuBois, just south of I-80, is on the axis of the northeast-trending, Caledonia syncline, and the village rests on Conemaugh beds in its trough. DuBois is on the continental divide, with Susquehanna River drainage just north of it and Ohio River drainage south. You also cross the divide near Luthersburg on Coal Hill.

Between Luthersburg and McGees Mills, 22 miles, you cross the Chestnut Ridge anticline, where you see Allegheny and Pottsville beds. Rock cities are common in the Pottsville formation because it contains massive sandstones and conglomerates that tend to break into huge blocks.

Bilgers Rock

This rock city developed in highly cross-bedded sandstones of the Pottsville group. The sandstone is a single bed 20-25 ft. thick, broken along widely-separated joints. The outcrop is on the western bank of Bilger Run by Pike Township Route 203, four miles northwest of Curwensville, which is five miles east of Grampian on US 322. The accompanying map shows the location and route.

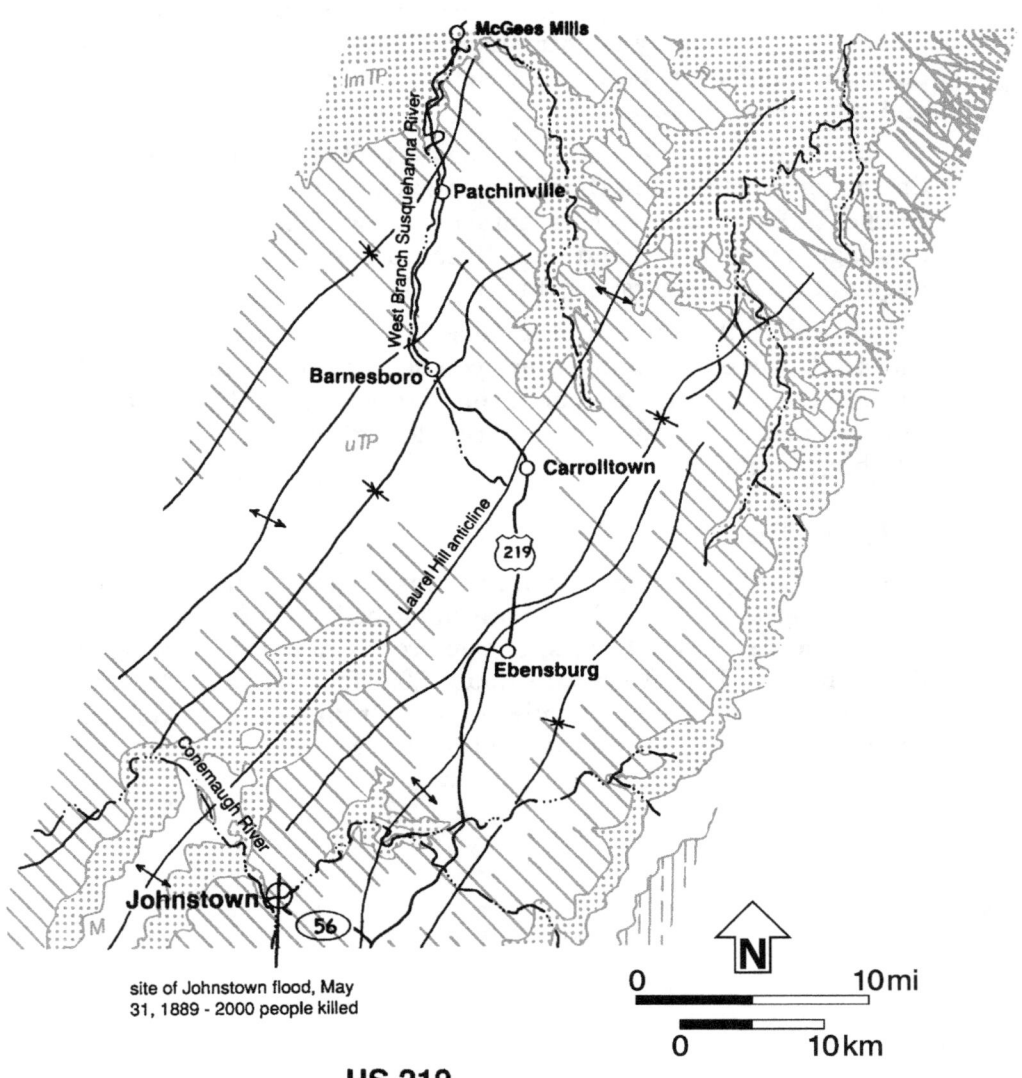

US 219
McGees Mills—Maryland Border
north half

US 219
McGees Mills—Maryland Border
111 mi./179 km.

In the 24 miles between McGees Mills and Carrolltown, you follow the west branch of the Susquehanna River to its headwaters along a peculiar path through the Valley and Ridge province. Between McGees Mills and North Bend, the river flows generally north or northeast, parallel to the trend of the plateau folding. Then it turns sharply southeast and spills off the Allegheny Front to Lock Haven, where it turns northeast again and follows Bald Eagle Mountain. Passing around the nose of Bald Eagle Mountain east of Williamsport, it turns sharply south and cuts across the ridges, joining the main branch at Northumberland.

All of the bedrock exposed along this section of highway belongs to the Pennsylvanian Allegheny and Conemaugh groups. Both are important sources of coal, so you see numerous strip mines. Brownish gray shale, siltstone, and sandstone are layered with thin coal seams. The route also crosses the edge of the region of shallow gas fields, and a few wells are visible — at Patchinville, for example.

Pennsylvanian Conemaugh sandstone, above, and shale, below, on U.S. 219 near Ebensburg, with a cut-and-fill stream channel at the lower left. Compare this channel with the larger one north of Exit 22 on I-79 shown in Road Log I-79: West Virginia Border—Zelienople.

Rubble at the base of shale outcrops often looks like this, in Conemaugh beds in DuBois.

Along the 14 miles of the Johnstown Expressway between Ebensburg and the junction with Pennsylvania 56, you see several cuts in the Conemaugh formation, some with coal seams. The high bridge over the south branch of the Little Conemaugh River has a historical marker at one end commemorating the Johnstown Flood. Near this point are the remains of the earthen dam which, on May 31, 1889, burst after several days of rain. Millions of tons of water flooded Johnstown, ten miles downstream on the Conemaugh River, and killed 2200 people.

Northwest of Johnstown, the Conemaugh River slices through the prominent ridge of the Laurel Hill anticline, exposing the upper Devonian Catskill beds of its core. Farther west, the river cuts through the next high ridge of the Chestnut Ridge anticline. These are the most distinct fold mountains of the plateau, although not the highest. Large expanses of Mississippian rocks crop out along their crests.

Between the Johnstown Expressway and I-76 junction near Somerset, 25 miles, the route parallels Laurel Hill. Nearly all of the bedrock, exposed in several good cuts, is Conemaugh formation, brownish gray shale, siltstone, and sandstone with some coal seams. The underlying Allegheny and Pottsville beds show up at two creek crossings. Somerset is on the northwestern flank, near the nose, of Negro Mountain, the most easterly prominent anticlinal ridge of this section of plateau.

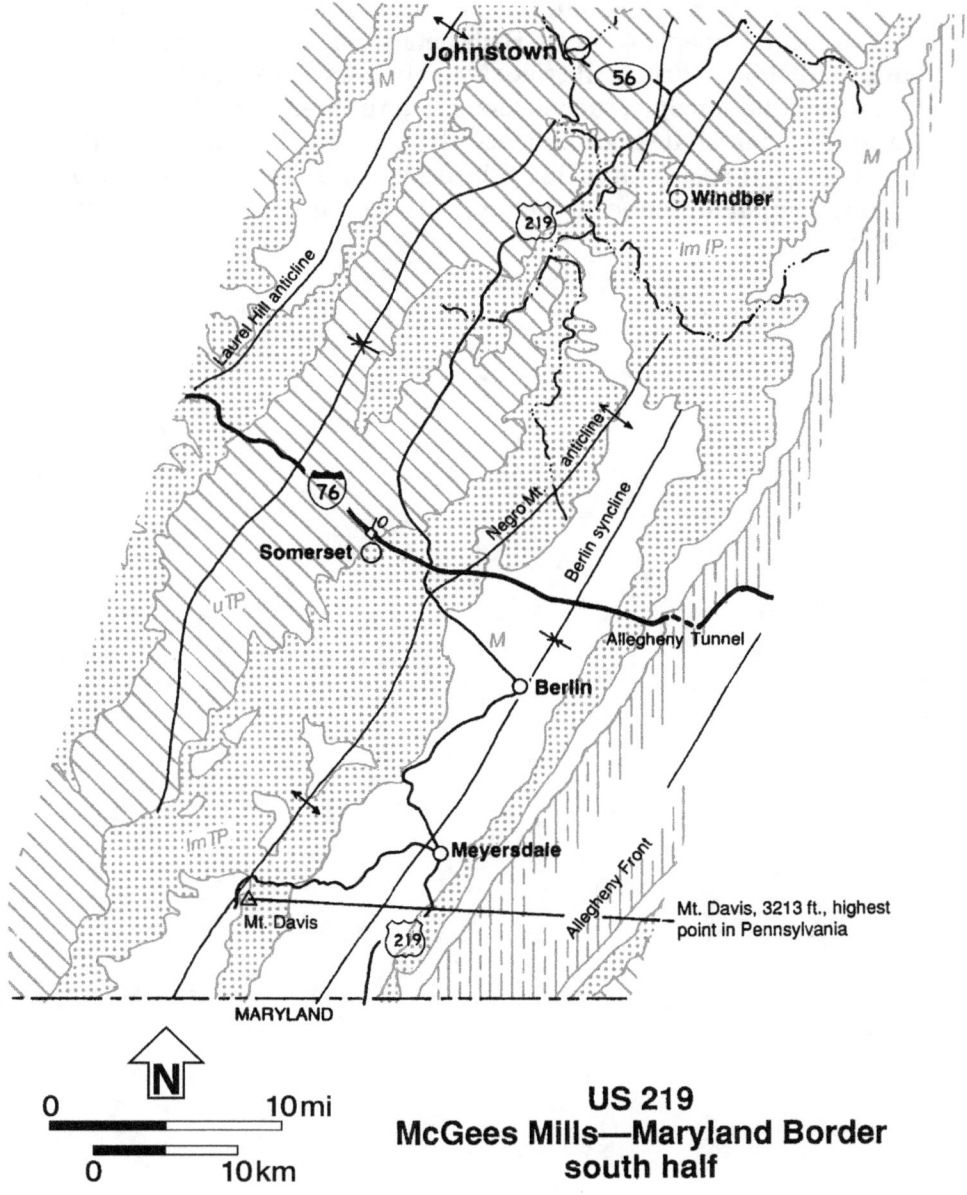

US 219
McGees Mills—Maryland Border
south half

Berlin stands on Pennsylvanian Monongahela beds in the trough of the Berlin syncline whose upturned southeastern limb forms the rim of the plateau and crest of the Allegheny Front. The fold formed during Alleghenian mountain building where the sole thrust that slices through weak Cambrian strata beneath the Valley and Ridge province curved steeply upward. Beds below the fault are almost flat. Near the Allegheny Front, the sole thrust breaks out of the Cambrian beds and ascends steeply, then levels off again at less than 10,000 feet in weak Silurian shale beds beneath the plateau. The rock package carried on the thrust fault is flexed into a syncline over the bottom of the ramp and an anticline at the top, with the Berlin syncline on its far side.

US 220
South Waverly—Williamsport
86 mi./139 km.

The contiguous towns of South Waverly, Sayre and Athens lie between the Chemung and Susquehanna rivers, which join immediately to the south at Tioga Point. Tioga Point is on a large flat gravel bar, the Queen Esther Flats, south of Athens.

US 220 crosses the Chemung River in the shadow of a 700-foot high scarp carved from upper Devonian Lock Haven beds. Round Top Recreation Area, atop the scarp, is an excellent place to see the joining of the rivers. Reach it by way of Station Road, which crosses the highway about a mile south of the Chemung bridge. The view reveals a very wide, flat valley with remnants of terraces along its sides, and a river that alternately banks against steep bedrock scarps on opposite sides of the valley.

As in the case of some other large Pennsylvania rivers, these features result from glaciation, but the exceptional breadth and development of the flood plain and terraces demand further explanation. The river courses were undoubtedly established long before the ice age, but glaciation modified them. The ice first scraped and gouged the valley deeper and wider. Later, as the last glacier receded, the Susquehanna River carried all of the drainage of the Finger Lakes region of New York for hundreds of years during which ice blocked the Mohawk Valley. The meltwaters deposited sediments in these valleys to considerable depths. When the ice receded farther

The low hill with no trees is a terrace of the Susquehanna River, near U.S. 220 south of Athens.

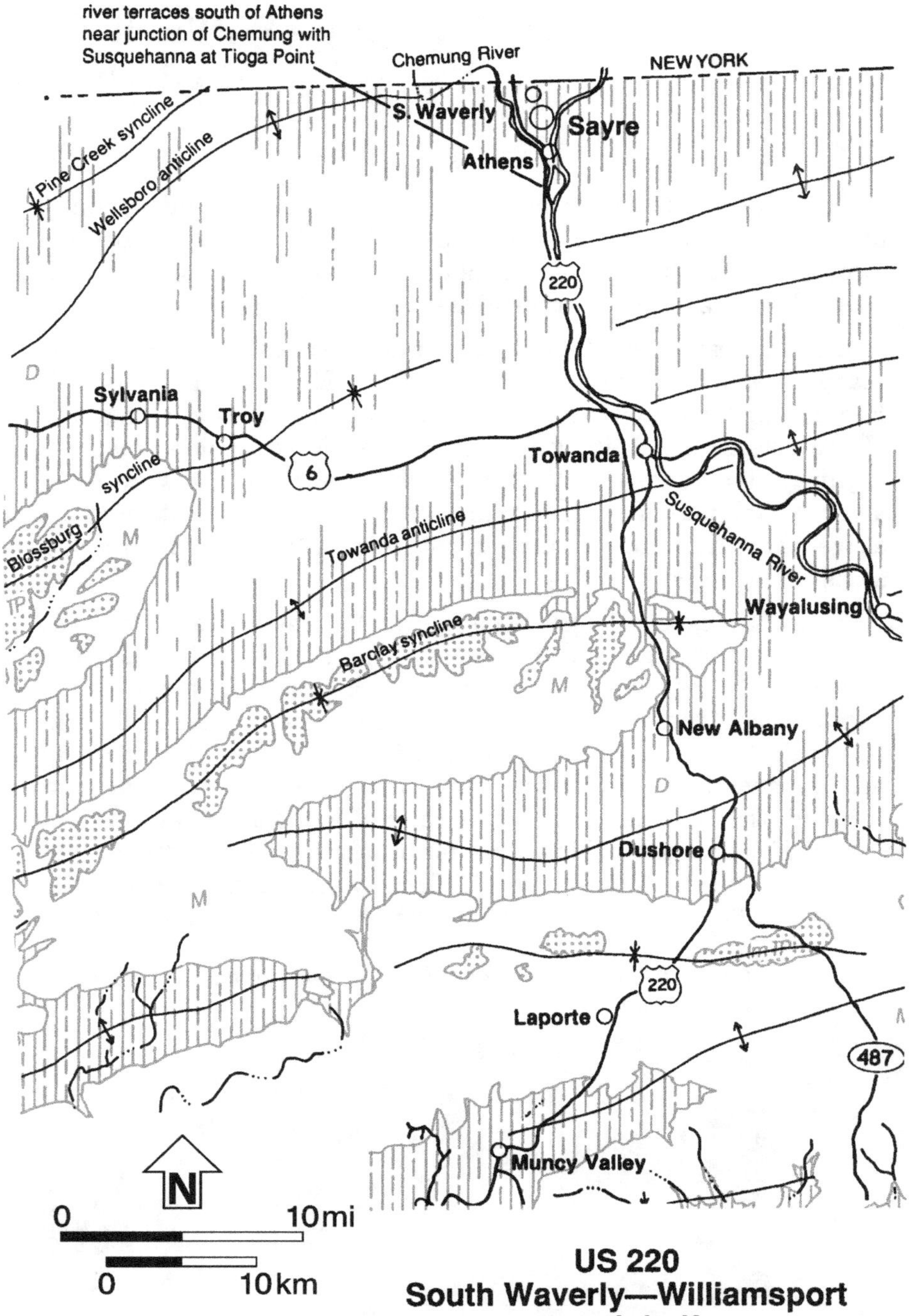

US 220
South Waverly—Williamsport
north half

north, waters from the Finger Lakes, as well as the forming Great Lakes began to flow to the Atlantic via the Mohawk and Hudson valleys, leaving the Susquehanna River much diminished. That is why it now appears small in comparison to the size of its valley. The valley is especially oversize here, because this was the initial dumping place for large quantities of glacial outwash. The route of the North Branch of the Pennsylvania Canal, now abandoned, lies on the floodplain just east of the highway between Athens and Milan.

The route follows the Susquehanna River between Milan and the US 6 intersection near Towanda, nine miles. Roadcuts are poor and far between, but the bedrock throughout is upper Devonian Lock

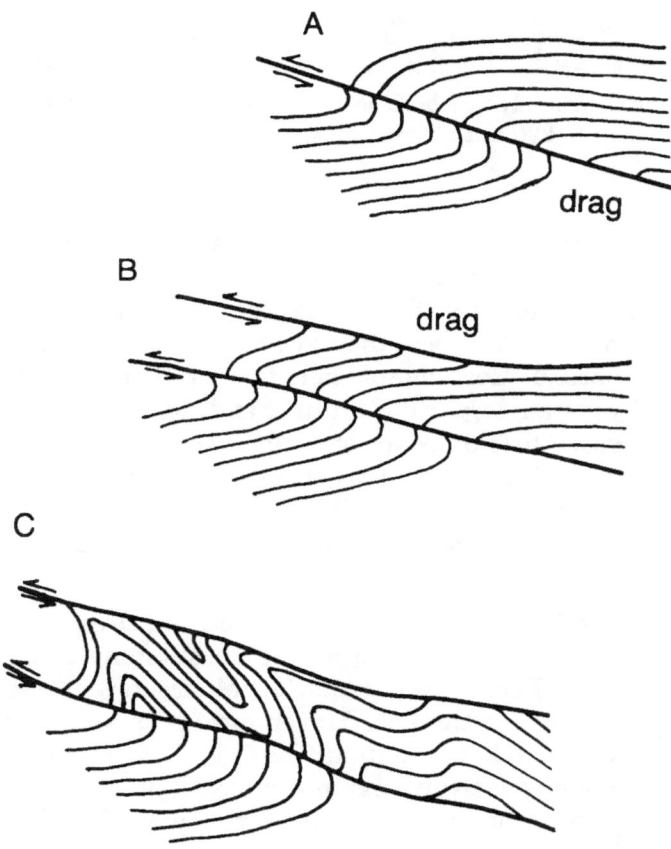

Hypothetical sequence of deformation in the Bridge Street thrust fault zone at Towanda. In A, beds are drag-folded by initial movement along lower thrust. In B, the upper fault moves, dragging the beds above and below it. C represents the present condition after much more movement, with severe crumpling of the beds sandwiched between the two faults. —Adapted from Pohn and Purdy (1981, p. 49)

Folded sandstone and shale beds of the upper Devonian Lock Haven formation on U.S. 220 near Monroetown.

Haven formation, an early deposit of the Catskill Delta. An excellent cut in this unit is on Bridge Street in Towanda, one-tenth mile west of the US 220 underpass. The cut exposes the Bridge Street thrust fault zone in one of the best displays of faulting of Pennsylvania's Allegheny Plateau. Most notable here is a fault slice, bounded above and below by low angle, parallel faults that dragged shales, siltstones, and sandstones into folds. Many other fault zones like this exist in the plateau, despite the apparently gentle nature of deformation. The deformation happened during Alleghenian mountain building, when North America and Africa collided during Pennsylvanian and Permian time.

Between the US 6 junction and Monroetown, seven miles, in one very large cut, you see dark olive gray shale, siltstone, and sandstone that belong to the Lock Haven formation. Monroetown is just north of the boundary between the Lock Haven formation and the overlying upper Devonian Catskill formation. The Catskill formation comprises the main body of relatively coarse sediments of the Catskill Delta derived from the Acadian Mountains during and after the climax of the uplift. Relics of the deeply eroded core of the mountain range probably lie deeply buried under the sediments of the Atlantic coastal plain.

In the nine miles between Monroetown and New Albany, you follow the south branch of Towanda Creek through a ridge carved from the Barclay syncline. You see Catskill redbeds; their contact with the overlying upper Devonian and lower Mississippian Huntley Mountain formation is on the slopes above each side of the road. The Huntley Mountain formation is a product of stream deposition transitional between the Catskill formation and the overlying, buff Pocono formation. It contains the Devonian-Mississippian time boundary.

Between New Albany and Dushore, six miles, you cross the Wilmot anticline over rolling farmland developed on Catskill redbeds.

Ricketts Glen State Park

This park is about 20 miles from US 220, either by way of Pennsylvania 487 south from Dushore or Pennsylvania 42, 239, and 118 east from Beech Glen, one mile south of Muncy Valley. The park straddles the Allegheny Front, the steep erosional scarp that marks the structural transition between the Allegheny Plateau and the Valley and Ridge province.

The name Ricketts Glen derives from the deep picturesque gorge that Kitchen Creek cut into the scarp. The glen undoubtedly achieved its present rugged configuration and great depth during glacial recession near the end of the last ice age. At glacial maximum, ice covered this section of the scarp. When the ice margin melted back onto the plateau, this and other glens along the scarp received torrents of sediment-choked meltwaters that carved them deeper into the edge of the plateau.

The Catskill beds exposed in the lower part of the glen are interbedded gray and red sandstone and siltstone, and red shale. In the upper slopes are outcrops of greenish gray, flaggy, cross-bedded sandstone and some interbedded red shale of the Huntley Mountain formation. Finally, on top of the plateau within the park are whitish, cross-bedded sandstone and conglomerate of the Pocono formation. This unit stands in relief, in places forming a south-facing step that tends to break up into large, joint-bounded blocks.

Grand View overlook provides an excellent view of the regional geologic setting. The viewpoint is at the rim of the Allegheny Front on Red Rock Mountain, nearly one mile west of Ricketts Glen. The sweeping view to east or west profiles the high plateau and the steep scarp of the Allegheny Front which, at Grand View, rises 1200 feet. The low hills south of the scarp are developed on the more highly folded, but relatively weak Catskill and Trimmers Rock beds that are within the Valley and Ridge province. The scarp here is held up by resistant sandstones of the Huntley Mountain and Pocono formations. Barely perceptible steps on the slope mark the positions of sandstone beds within the Huntley Mountain formation.

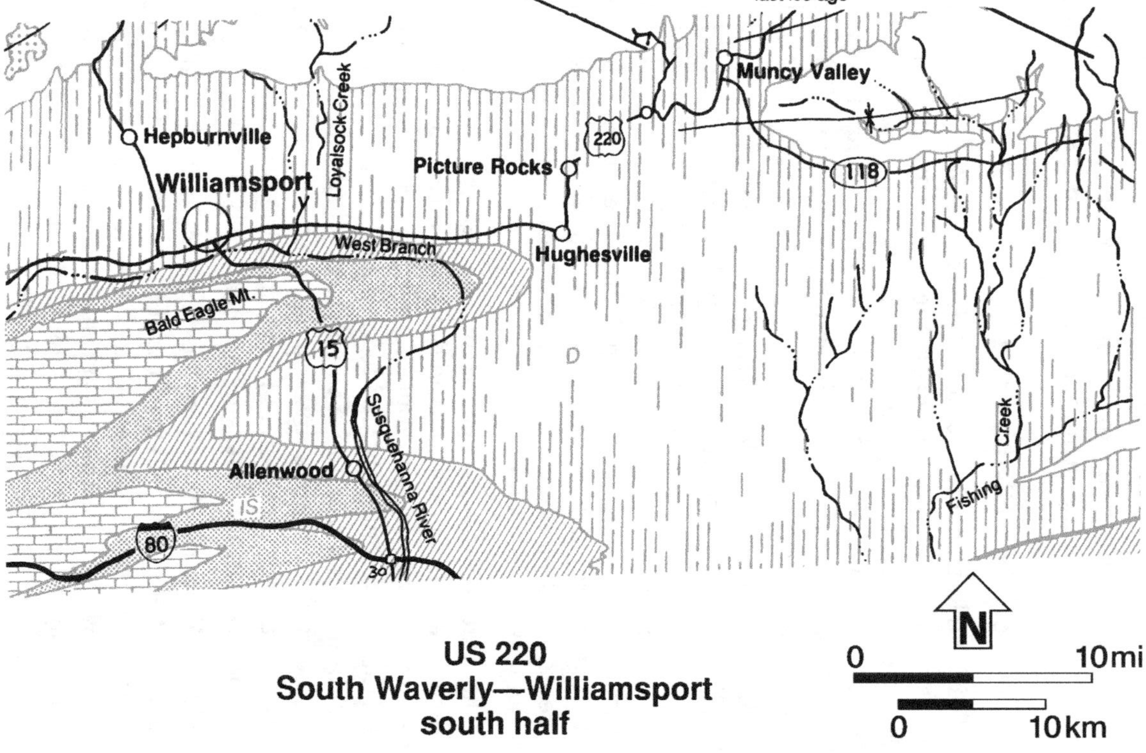

US 220
South Waverly—Williamsport
south half

- Between Dushore and Muncy Valley, 18 miles, you cross hilly country over Barbours syncline near the Allegheny Front. Ringdale lies on the fold axis, which can be traced for several miles east and west along Loyalsock Creek. The road climbs the stratigraphic section to there from either north or south, first crossing Catskill formation, then the Huntley Mountain and Pocono formations. Look for a very long cut in the Huntley Mountain formation between Laporte and Muncy Valley. Watch for the greenish gray, flaggy sandstones with a lot of cross-bedding and a few red shaley interbeds.

The route follows lovely Muncy Creek valley for 14 miles between Muncy Valley and Hughesville, across the Allegheny Front. The scarp here is unremarkable, in contrast to its steepness and abruptness at Ricketts Glen State Park. The red cliffs and roadcuts along the way are all Catskill sandstone. One cliff is at Glen Mawr, where you may want to take a side trip to Ticklish Rock. Another is at Tivoli, near the boundary with the underlying Trimmers Rock formation.

Ticklish Rock, an unusual balanced rock composed of thinly interbedded sandstone and shale of the upper Devonian Catskill formation. Note how the less resistant layers are etched out.

Ticklish Rock

Ticklish Rock is an unusual balanced rock on the rim of the Allegheny Front about four miles north of Glen Mawr. The rock is sculptured from a pinnacle that separated from a Catskill sandstone ledge in the same way that rock cities form in more massive sandstones and conglomerates. Vertical joints tend to open up where beds of hard rock project over weaker underpinnings on a hillslope. The process involves the erosional undercutting of the hard ledge to the point that it overhangs. Unsupported, the hard rock breaks along joints. Water that seeps into the joints and freezes, and tree roots that penetrate them help to rotate the blocks away from the cliff, so they eventually tumble. The freeze-thaw movement of the soil often aids in carrying the blocks slowly downslope.

Deeply weathered, undercut, and jointed upper Devonian Catskill cliff by Ticklish Rock, north of Glen Mawr (U.S. 220).

Really good rock cities form in very thick-bedded rocks of markedly greater resistance than those beneath. Such rocks tend to break along widely spaced, vertical joints, yielding huge, rectangular blocks that resemble buildings of a city. The Catskill sandstone here is, instead, thin-bedded, and appears to be somewhat tougher at the top of the cliff than at the bottom. Ticklish Rock is a pinnacle that somehow remained standing as it separated from the cliff, and since has been deeply weathered on all sides. The weaker layers have worn away to a slim pedestal, supporting a large, delicately balanced block of sandstone. Many balanced rocks and pinnacles may have formed along this rimrock in the geologic past. If so, all have suffered the eventual fate of this one — they tumbled.

• Hughesville is at the base of the Allegheny Front in middle Devonian Hamilton shales that wrap around the nose of Bald Eagle Mountain anticline. Between Hughesville and the US 15 junction at Willliamsport, 18 miles, the route enters the west branch of the Susquehanna valley that separates Bald Eagle Mountain from the Allegheny Front.

The Spring Mountain cut seen from the northbound lane of Interstate 81. The upthrown block right of the thrust fault exerts obvious frictional drag against Pottsville beds left of the fault, and rocks along the fault are crushed as a result of the movement.

II
Valley and Ridge Province

The Valley and Ridge province, proper, is a fold-and-thrust belt that extends the length of the Appalachian Mountains and forms the backbone of the range. Pennsylvania's small part of it is surely the most striking geologic feature of the entire range. In the satellite image on page 46, the region appears sharply confined between the Allegheny Plateau on the west and the Great Valley on the east. In between is an incredible array of pencil-sharp ridges, bunched tightly together in parallel alignment. Collectively, they trace a graceful curving path across the state that, in central Pennsylvania, is strongly convex to the northwest. This unique bend is most likely an imprint of the bulging rim of Africa, made when the continent collided with North America during the Pennsylvanian and Permian periods, generating the Alleghenian orogeny. This was the final major event in the assemblage of the supercontinent of Pangaea.

In the satellite image, the ridges and intervening valleys resemble folds in a carpet that has been shoved across a floor against a line of resistance — where, for example, furniture rests on top of the carpet. That, in fact, appears to be the way it happened. The carpet is the whole package of Paleozoic sedimentary strata that once lay flat over the whole state, or nearly the whole state. The force was provided by the approach of Africa. The oldest rocks in the package are Cambrian, the youngest at the start were Mississippian. The following is a hypothetical sequence of events that led to the present geological configuration of the Valley and Ridge and Allegheny Plateau provinces.

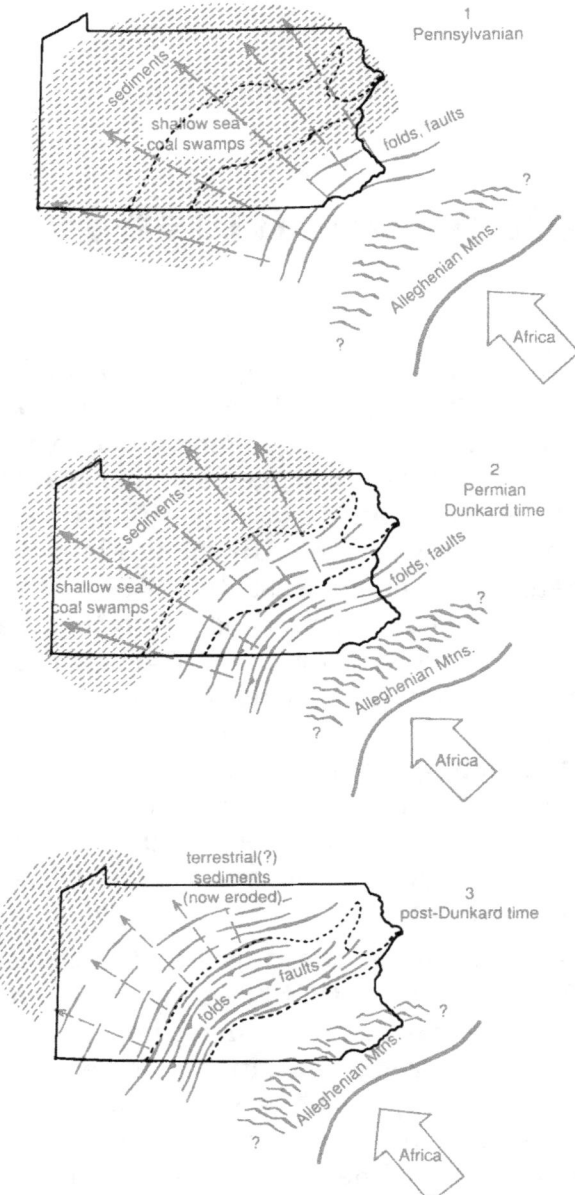

Series of schematic maps showing compression of Pennsylvania's Paleozoic strata during the Alleghenian orogeny, resulting in the fault and fold deformation of the Valley and Ridge and Allegheny Plateau provinces.

The earliest pulses of Alleghenian orogeny produced mountains that probably lay far east of the present state boundary. The rising mountains shed sediments westward, adding new Pennsylvanian strata to the carpet. Later pulses produced similar results; the carpet continued to thicken while the mountains and the deformation associated with them progressed steadily inland. And so it went into Permian Dunkard group time.

At some point, the Paleozoic sedimentary package caught in the squeeze began to react to the compression. The layers were not quite flat, but sloped gently southeastward. The whole pile began to move, en masse, shearing from the underlying bedrock along a sole thrust that sliced through a bed of weak Cambrian shale. The sole thrust extended its reach farther and farther northwestward, while numerous smaller faults branched from it and pushed steeply toward the surface, arching the overlying strata into anticlines. Thus, the Valley and Ridge folds and faults were generated. Along the line of the present Allegheny Front, the sole thrust broke free of its Cambrian sandwich and cut steeply upward through the overlying layers along a ramp. It leveled off again — under the Allegheny Plateau — in upper Silurian Wills Creek shales, and continued to extend its reach

Schematic sequence showing fault and fold deformation of Pennsylvania's Valley and Ridge and Allegheny Plateau provinces.

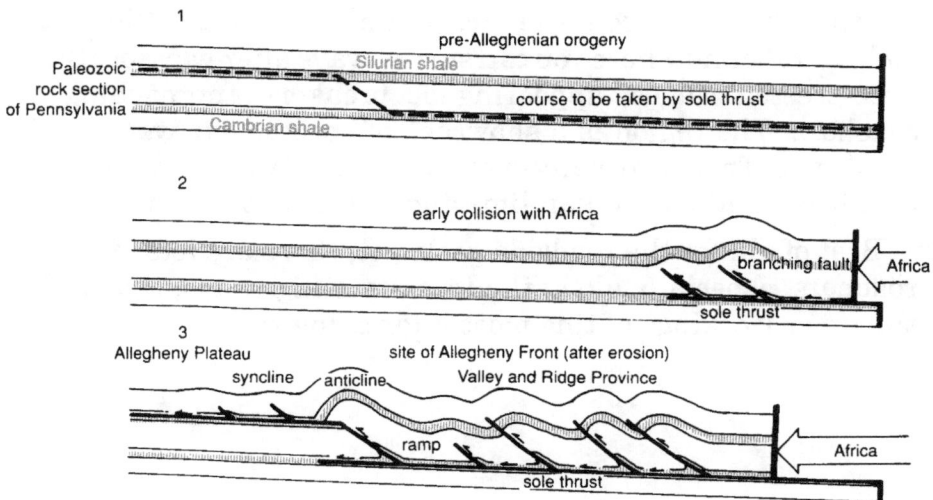

northwestward along those beds. Branching faults there also folded the overlying rocks but they are fewer in number and more widely separated than in the Valley and Ridge province.

The mechanical effect of the ramp on deformation in the Valley and Ridge province is uncertain. In one way, it may affect a release of pressure. In another way, it may, because of its steepness, represent the other side of the vice, a barrier to northwest transport of the whole rock package riding on top of the sole thrust. In the latter case, the ramp would enhance the severity of Valley and Ridge deformation.

The ramp produced a different brand of folds. Beds cut by the ramp were transported up its slope. Reaching the top, they bent over it into anticlines, with synclines immediately ahead of them. The present Allegheny Front is carved from the leading limbs of these anticlines, which are the same as the trailing limbs of the adjacent synclines.

Valley and Ridge erosion appears to be far more profound than that of the Allegheny Plateau. The anthracite-bearing Pennsylvanian rocks near its eastern boundary surely must have been part of the same coal-bearing sequences as those of western Pennsylvania. Only small pockets of those once continuous layers remain, preserved where they have been downfolded and thus, protected from erosion.

On the ground, the Valley and Ridge province is even more exciting than it is from the air. It is a visual feast. Ridgetop lookouts abound where you can survey wave after wave of other even-crested ridges, appearing as breakers approaching a seashore. The region is a showcase of spectacular water and wind gaps. There are many caves, particularly in the abundant Cambrian and Ordovician limestones and dolostones.

Best of all, for the roadside geologist, there are lots of large roadcuts, especially along the interstate highways, that bare the inner workings of this most interesting region.

I-70
Breezewood—Warfordsburg
21 mi./34 km.

This is a short, but geologically and scenically interesting route. In the four miles between Exit 28 and Exit 30, the route goes north-south near the base of Rays Hill, in view of its even crest. Upper Devonian Catskill redbeds crop out alongside the ridge. The Catskill redbeds were laid down during a period of rapid erosion during and after climax of the Acadian orogeny; the mountains rose east of the present state line.

North of Exit 30, the road goes through the ridge along Brush Creek Gap where you will see southeast-dipping, buff-colored, cross-bedded sandstones of the lower Mississippian Pocono formation. Those resistant rocks maintain the ridge. These sediments were eroded some 360 million years ago from the Acadian Mountains and carried west by rivers.

Rays Hill is carved from the west limb of a synclinal "canoe-shaped" structure with a pointed bow that rises 11 miles south-southwest of Exit 30. Rocks exposed beneath the Pocono formation in Brush Creek Gap are chocolate-brown shales and buff-colored sandstones of the Rockwell formation, a unit transitional to the underlying maroon to red Catskill beds. The Devonian-Mississippian time boundary probably lies within the Rockwell formation.

The route crosses diagonally over the axis of the syncline between Exits 30 and 31, where you will see Mauch Chunk redbeds that overlie the Pocono formation. The Mauch Chunk redbeds signal the onset of Alleghenian uplift and mark the end of the tectonic tranquility that succeeded the Acadian orogeny. Africa and North America were closing when they were accumulating, and would soon collide, causing the thrust faulting and wave-like folding seen here and throughout the folded Appalachians.

The northern four miles of the highway between Exits 31 and 32 is a long sloping traverse of the steep east flank of Town Hill, with sweeping views of the broad valley between Town and Sideling hills. Note again the Rockwell beds near the top of the hill and the Catskill formation below. The valley is all Catskill formation, with colorful red to greenish-gray to buff sandstones, siltstones, and shales crumpled into small anticlines and synclines.

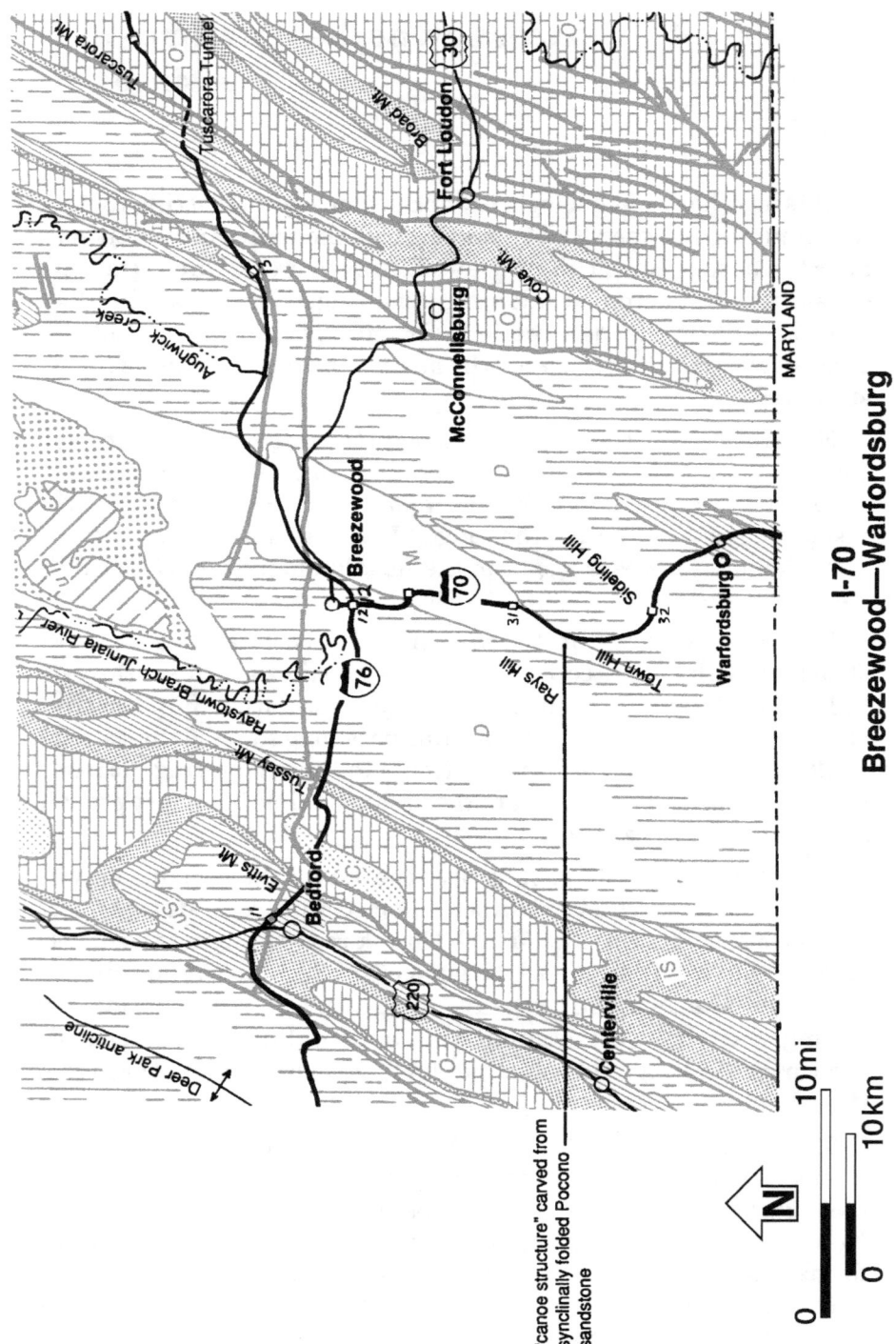

Exit 32 is at the eastern gate of Deneen Gap where Little Tonoloway Creek cuts through Sideling Hill. Resistant Rockwell sandstones hold up the ridge. The beds form the core of a syncline that gets deeper northward, carrying the Rockwell sandstone underground and exposing the overlying Pocono formation. In the gap, you'll see more Catskill redbeds beneath the Pocono formation. The southern three miles of this route descends into the core of a north-pointing anticline through increasingly older rocks.

Near Exit 33 to Warfordsburg, look for upper Silurian Wills Creek dark shales. The exit is in the middle of Pigeon Cove; Tonoloway Ridge to the west and Limestone Ridge to the east are both sustained by upper Silurian Keyser and Tonoloway limestones that overlie the Wills Creek. The limestones were deposited during a period of tectonic quiet like that of Cambrian and early Ordovician time. The Taconian Mountains were worn down then, and no longer supplied copious quantities of sand, silt, and clay to the depositional site; a marine environment developed, in which carbonates dominated. The stage was set for the Acadian orogeny.

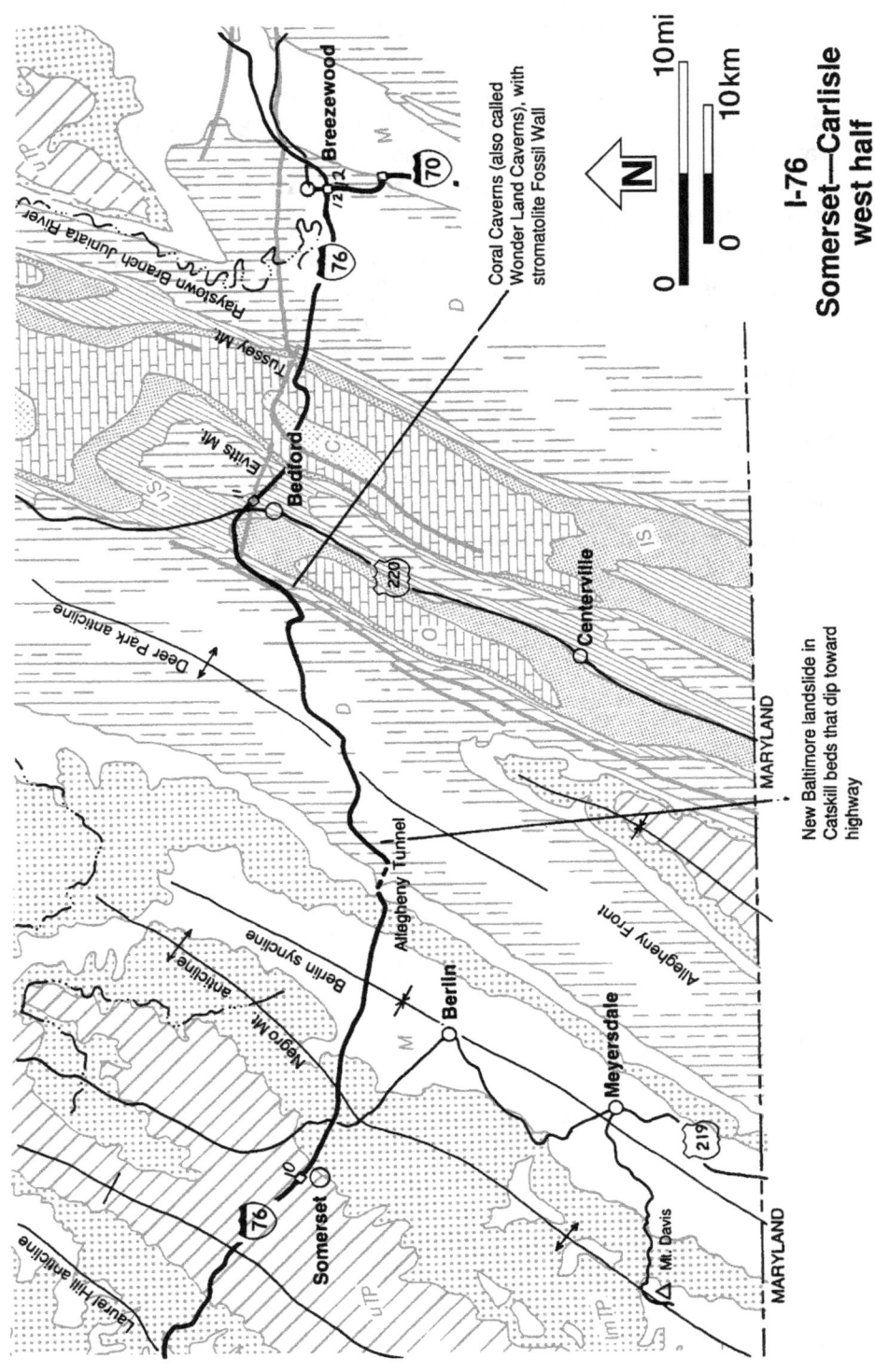

I-76
Somerset—Carlisle
115 mi./185 km.

The Pennsylvania Turnpike

The first 30 miles or so of this route east of Somerset are on the Allegheny Plateau; the turnpike crosses the Valley and Ridge province between New Baltimore and Carlisle.

Between Somerset and mile 117, you cross the Negro Mountain anticline with Allegheny beds in its core; the fold axis lies near the Somerset gas station. It would be easy to map the limits of the Allegheny beds from the air just because they contain many more coal mines than the outlying Conemaugh group. Watch for several beds of upper Kittanning and lower Freeport coals by the road. The large Negro Mountain cut at mile 116 replaces an old, partially driven tunnel just north of the highway. Mt. Davis, 3213 feet, is 15 miles southwest. It is the highest point on Negro Mountain, the highest in Pennsylvania.

Between mile 117 and the west portal of the Allegheny tunnel, you cross the Berlin coal basin. This is the easternmost major synclinal fold in this part of the plateau. The upturned Pottsville sandstones and conglomerates on the southeastern limb form the backbone of Allegheny Mountain and the crest of the Allegheny Front. This is also the northwestern limb of the deeply eroded anticline that formed where the sole thrust reached the Allegheny Front and leveled off in the upper Silurian Wills Creek shale. Considerable amounts of coal have come from the Allegheny formation on the gentle northwestern slope of the mountain.

The Allegheny tunnel is a little more than a mile long. The western portal is in Allegheny beds, and most of the tunnel passes through Mauch Chunk formation. The rock at road level by the eastern portal is buff, Burgoon sandstone full of cross-beds. Above this are the distinctively red Mauch Chunk shales, and above them at the crest of the mountain are cliffs of Pottsville sandstone. The Mauch Chunk shales were laid down by streams that drained from the Alleghenian Mountains, then rising in the present position of the Atlantic Coastal Plain. The overlying Pottsville sandstone records the transition that eventually led to widespread coal swamps. This part of North America was then on the equator.

The scarp face above the eastern portal of the tunnel is about 500 feet high; total relief of this part of the Allegheny Front is more than 1300 feet. Along most of the way between the tunnel and New Baltimore at mile 129, the road follows the narrow, scenic valleys of the Raystown branch of the Juniata River and Wambaugh Run.

You pass downward through the Mississippian-Devonian time boundary in the Rockwell formation near mile 124. Between there and New Baltimore, nearly all of the reddish-brown to greenish-gray sandstones and shales are in the Catskill formation.

The New Baltimore landslide is east of the highway. Numerous slides have required removal of thousands of cubic yards of rock since the road was built. Sliding is always a threat where a road slices through beds that dip towards it as these do, especially in formations that contain weak shale interbeds. Roadcuts remove support from the base of the dipslope permitting slippage along bedding planes, particularly when water reduces friction. The pronounced folding of the Valley and Ridge province presents many landslide-prone slopes, trouble for the highway department.

Between miles 128 and 139, you cross the Deer Park anticline. Watch for olive-gray shales and sandstones between the Catskill redbeds and Helderberg group limestones. Some people consider them the lower part of the Catskill formation. Sandstones in these flaggy beds commonly project as ledges because they better resist erosion than the intervening shales.

Upper Devonian Trimmers Rock sandstones and shales at mile 138 on I-76.

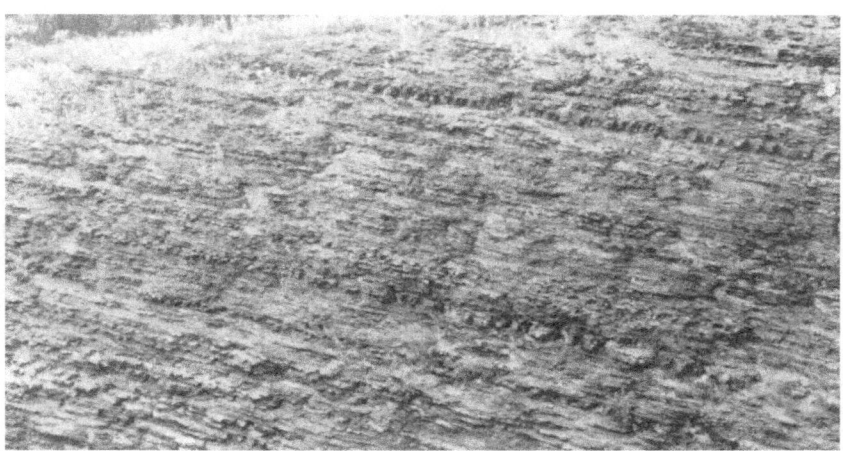

Thin-bedded, olive gray siltstone and shale of the Trimmers Rock formation at mile 138 of I-76, formed from sediments deposited in a continental sea in late Devonian time about 385 million years ago. Original level bedding was tilted during the Alleghenian orogeny.

An 8-foot bed of Tully limestone shows up on both sides of the highway near mile 141, dipping 87 degrees to the southeast. It is a persistent thin unit at the top of the Mahantango formation, an easily recognized bed helpful for geologic mapping in New York and Pennsylvania. In some areas it is more of a limey shale than a true limestone. A 60-foot high cut near mile 142 exposes Keyser limestone.

This region contains a number of caves developed along steeply tilted bedding planes in the Keyser limestone, some commercialized Coral Caverns near the village of Mann's Choice, about five miles southwest of Exit 11, features a "Fossil Wall" with stromatolites and other fossils, a large cathedral room, stalactites and stalagmites.

Stromatolites

Stromatolites are thinly layered limestone structures built by blue-green algae growing in shallow marine environments, particularly in tidal flats where they alternately submerge and emerge. They occur in a variety of forms: flat, wavy, columnar, or in mounds that vary widely in size. The same rock may also contain fossils of other marine organisms. The oldest-known fossils are stromatolites in Precambrian rocks more than 2000 million years old. Yet modern stromatolites growing today are identical to those of the geologic past.

Stromatolites in the Keyser limestone, near the base of the Devonian section, are about 400 million years old, showing that this area was under shallow water then.

Stromatolites form as the sediment surface, whatever its shape, is first coated with a layer of cells of blue-green algae organized into thread-like filaments, an algal mat that binds sand and silt particles into a coherent layer. As waves or currents bring in more sediment particles, the filaments trap them, covering the algal mat. Within hours, algal filaments grow up through the new layer of sediment and incorporate it into the mat while repopulating the new top surface. Repetitions of this process produce fine laminations. The stromatolites are not actually fossils of the algae, but are instead a kind of trace fossil that preserve their form.

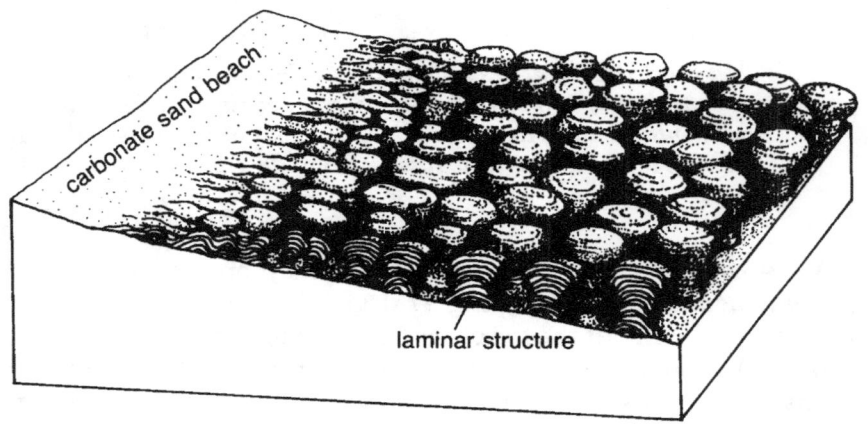

Modern stromatolites, like these at Sharks Bay, Australia, build in the intertidal zone, where they are subject to wave action and alternating emergence and submergence. —Adapted from a drawing by Harlan J. Johnson, 1961, Limestone-building algae and Algal Limestones, Johnson Publishing Company.

- Exit 11 at mile 145, is on the nose of the Wills Mountain anticline. A few miles to the southwest, the mountain forks, with Buffalo Mountain carved from the northwestern limb and Wills Mountain from the southeastern limb of the anticline. Ordovician rocks are exposed between the ridges, in the core of the fold. The rock that makes the ridges is lower Silurian Tuscarora quartzite, at the top of the Queenston Delta. It records the demise of the Taconian Mountains and the beginning of a transition from terrestrial to shallow marine environment in eastern North America — about 425 million years ago.

West of Exit 11, you will see Clinton group redbeds like those that have produced so much iron ore in parts of the Appalachians. Clinton

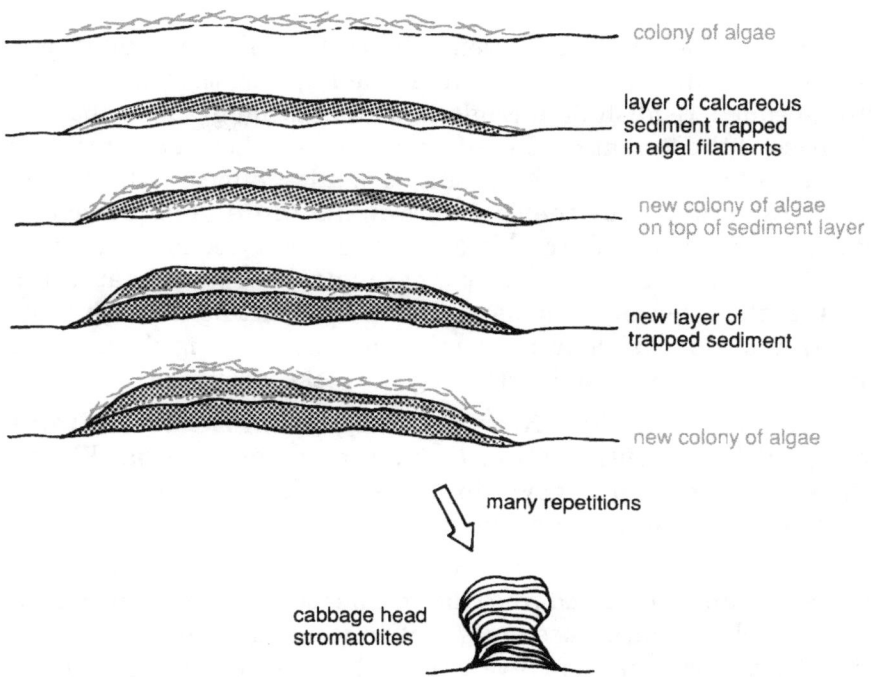

Step-by-step formation of stromatolites by blue-green algae.

ores were the foundation of the steel industry in Birmingham, Alabama. Old colonial iron workings in these beds survive near Bedford, Pennsylvania.

At the Bedford exit are grayish, calcareous shales and sandstones of the upper Silurian Wills Creek formation with some cherty Tonoloway limestone near the top. The Tonoloway limestone records a return to tectonic quiet and shallow marine carbonate conditions like those that preceded the Taconian orogeny.

You cross the Bedford syncline between miles 145 and 148 with Mahantango shales exposed in its core. These are flysch deposits, dark, fine-grained sediments laid down in the deep, poorly oxygenated waters of a foredeep; North America and Europe were beginning to collide, and the crust buckled down at the start of the Acadian orogeny.

Between miles 148 and 149 you pass through Bedford Narrows, a water gap where the Raystown branch of the Juniata River cuts through Evitts Mountain. The mountain is a ridge of white Tuscarora quartzite on the western limb of the Friends Cove anticline. Watch for the nearly vertical beds of quartzite at the northern end of the half-mile long cut beside the east-bound lane. South of the Tuscarora quartzite exposures, you descend the section, first through upper Ordovician Juniata sandstone and shale redbeds, then through the cross-bedded, reddish Bald Eagle sandstones that grade to shales at the base of the formation. Finally, near the southern end of the cut, grayish siltstones of the Reedsville formation appear. All the beds stratigraphically below the Tuscarora quartzite are late Ordovician in age. Several small faults cut through the exposed section. One or two smooth subvertical faces on the cut parallel to the road are faults that parallel the river and may have guided it in cutting through Evitts Mountain. Also visible are several small east-dipping thrust faults that shorten the section by stacking the rocks westward.

Just east of the Bedford Narrows, you cross the Friends Cove thrust fault, which dips gently eastward and moved upper Cambrian Warrior limestone over lower Ordovician Bellefonte dolostone. The extent of fault displacement is unknown; it may be several miles.

Between miles 153 and 154, you pass through Aliquippa Gap in Tussey Mountain, carved from the southeastern limb of the Friends Cove anticline. The exposed rock section is virtually identical to that of Bedford Narrows, but in reverse order because it is on the other side of the fold. Bald Eagle formation appears at the western end, Tuscarora quartzite at the eastern end.

Between mile 154 and the Breezewood exit, you cross a broad syncline with Catskill redbeds in its core. They are part of the Catskill Delta, terrestrial sediments shed from the rising Acadian Mountains as Europe and North America collided, and the Proto-Atlantic Ocean clamped shut.

Vertical Juniata sandstone beds with red shale partings in Bedford Narrows, at mile 148 of I-76.

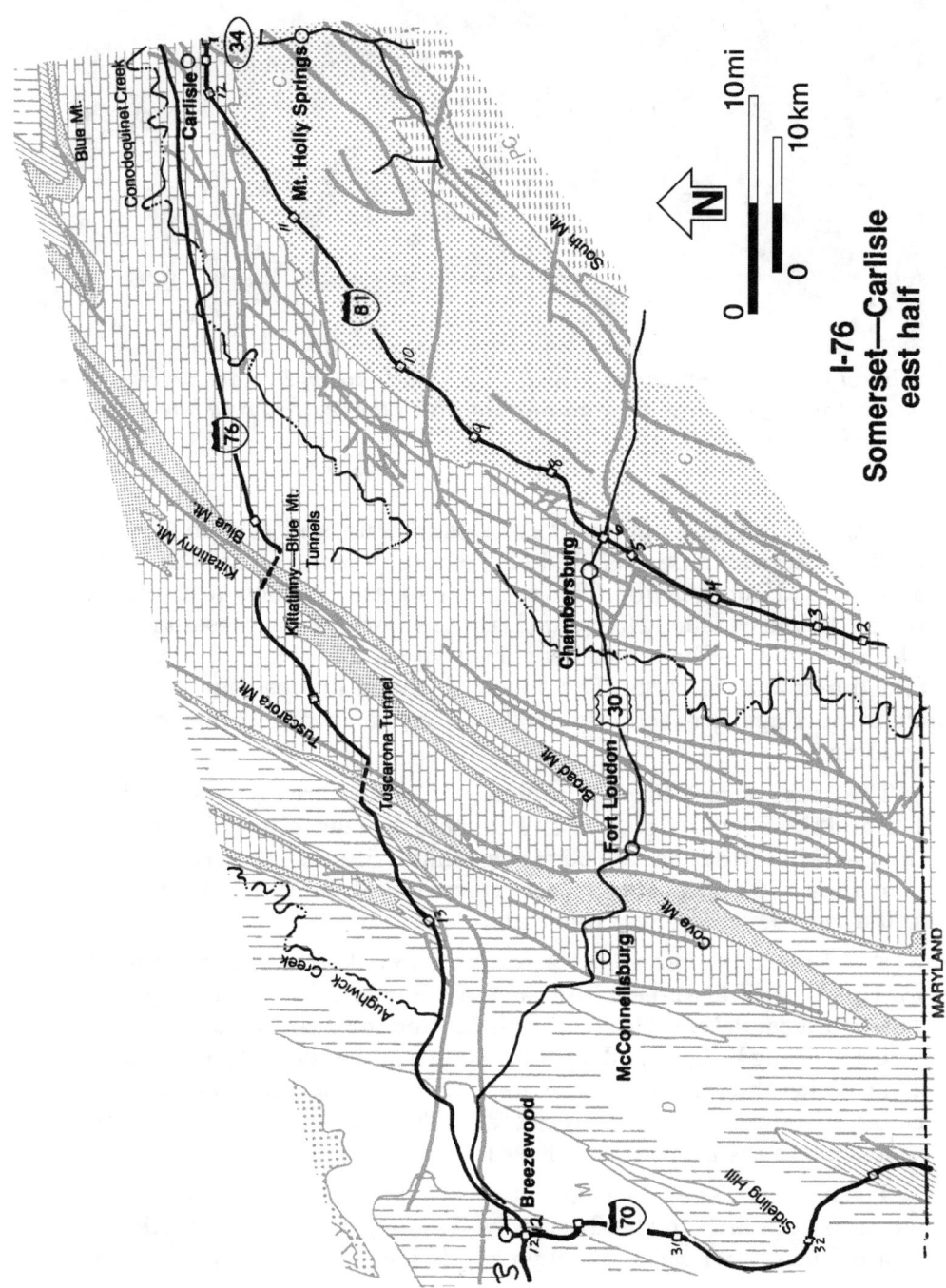

Almost horizontal Pocono sandstone beds at mile 168 of I-76 are cut by several steeply-dipping shear zones from which the crushed rock has eroded.

Watch for Mississippian Pocono sandstone and Mauch Chunk redbeds between miles 165 and 172, where the road crosses the core of the Broad Top syncline. This section bypasses the now-abandoned Rays Hill and Sideling Hill tunnels that lie north of the present route. The new route affords excellent open views and long exposures of folded and faulted rocks. The steeply-inclined planes that look like shale partings are actually eroded cleavage zones transverse to bedding.

The route passes through a highly deformed area between miles 172 and the Fort Littleton exit. The redbeds near mile 174 belong to the Catskill formation. the olive gray shale, siltstone, and sandstone near mile 176 is probably middle Devonian Mahantango formation and upper Devonian Foreknobs and Scherr formations.

Exit 13 is at the nose of a narrow anticline delineated on the west by Blacklog Hill and on the east by Shade Hill, both ridges of Tuscarora sandstone. Rocks at Exit 13, and for some distance either side of it along I-76, were deformed along a major, near-vertical, fault zone, which the road follows between miles 178 and 182, as its trend changes from east-west to northeast. The faults bring Ordovician limestone to the surface near Fort Littleton.

The brick red shales visible between miles 180 and 187, along the western approach to the Tuscarora Tunnel, belong to the Clinton group. The tunnel passes through a mile-long, complexly folded rock section that includes Wills Creek shale and Juniata and Bald Eagle formations. The eastern portal is in the much older, upper Ordovician Martinsburg shale which has been thrust faulted upward against the Bald Eagle formation. The fault is one of the many that branch upward from a deep-seated, flat-lying sole thrust under the Valley and Ridge province. Where the branching faults do not break the surface, anticlinal folds reveal their presence at depth. On a regional scale, the faults stacked westward with eastern slices over western, while the rock beds were tightly corrugated to accommodate the vice-like compression of colliding continents.

Light grayish lower-to-middle Ordovician Bellefonte limestone appears in several places east of the eastern portal. It was thrust up along another fault and shoved westward into its position against Martinsburg shales. The Bellefonte limestone is at the top of the Cambro-Ordovician carbonate sequence deposited in shallow water on the continental shelf during the long period of tectonic quiet that preceded the Taconian orogeny. By contrast, the dark Martinsburg shales were deposited in the depths of a basin during the earliest phase of the uplift.

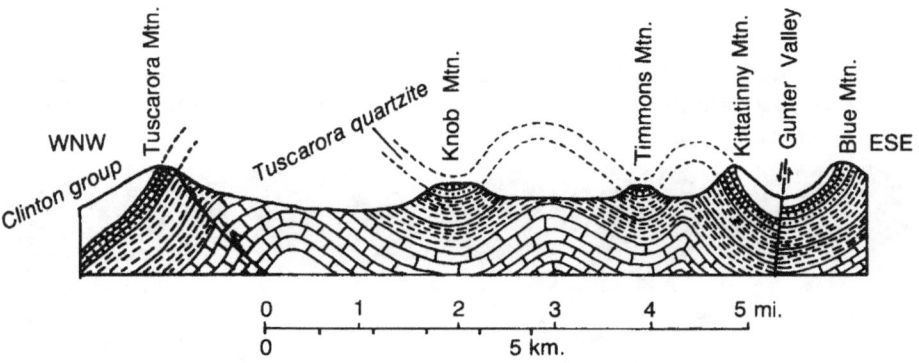

All anticlines are eroded to valleys, and synclinal folds underlie Knob and Timmons mountains. Only Gunter Valley is carved from a syncline, where fault-crushed rock was easily eroded. —Adapted from Cleaves and Stepherison (1949, p. 24)

More Martinsburg formation appears between miles 191 and 197 in the form of interbedded, olive gray shales and siltstones. This open traverse provides excellent views across the west branch of Conococheague Creek valley. The even-topped ridge of Tuscarora Mountain lies to the west, Kittatinny Mountain to the east, with two

175

truncated ridges called Knob and Timmons Mountains in between, both eroded from narrow synclinal folds. All four ridges and also Blue Mountain east of Kittatinny Mountain are ridges of Tuscarora quartzite.

The combined length of the double tunnels plus the short stretch of open highway in Gunter Valley between them is about two miles. Dark Martinsburg shales appear at either end of the tunnels. Redbeds of the upper Silurian Bloomsburg formation and Clinton group lie beneath Gunter Valley. The central part of each tunnel is cut through the intervening section consisting of Tuscarora, Juniata, and Bald Eagle formations. The eastern portal of the Blue Mountain tunnel is at the edge of Cumberland Valley. This is the southern segment in Pennsylvania of the Great Valley physiographic province that extends almost continuously from New York to Georgia.

Cumberland Valley is topographically and structurally distinct from the Valley and Ridge province. It is low because the rocks are weak. The Cambrian and Ordovician rocks are almost entirely sandstones and carbonates deformed during Taconian mountain building. The much younger Alleghenian folding and thrust faulting in the Valley and Ridge province produced folds that generally stand

Where cleavage cross-cuts bedding, as in the case of the middle Devonian Martinsburg formation at mile 225 of I-76, slate rubble may take the form of long pencil-fragments.

upright. The Taconian deformation in the Cumberland Valley was extremely intense and produced complex folds of enormous size that lean to the northwest. The rocks folded almost as though they were fluid because they were close to the heated, metamorphosing core of the rising Taconian Mountains. The deformation resembles that of many parts of the Alps of Europe.

Here, the Great Valley is divided into two, subtly differing parts, the slightly hillier western part developed over Martinsburg shales and the lower, flatter eastern part over weaker soluble Cambrian and Ordovician limestone and dolostones adjacent to South Mountain.

You cross diagonally over the western segment of the valley between the Blue Mountain tunnel and mile 215. Dark Martinsburg shales display small folds and shingled slopes.

Between miles 215 and the Carlisle exit, nearly all exposed bedrock is carbonate. Roadcuts are scarce, but numerous pinnacles of light gray rock project through the turf along the highway. This is a limestone topography developed by solution, which formed sinkholes and caves. Streams disappear underground and then reappear elsewhere as large springs. At Carlisle are large cuts in middle Ordovician St. Paul limestone.

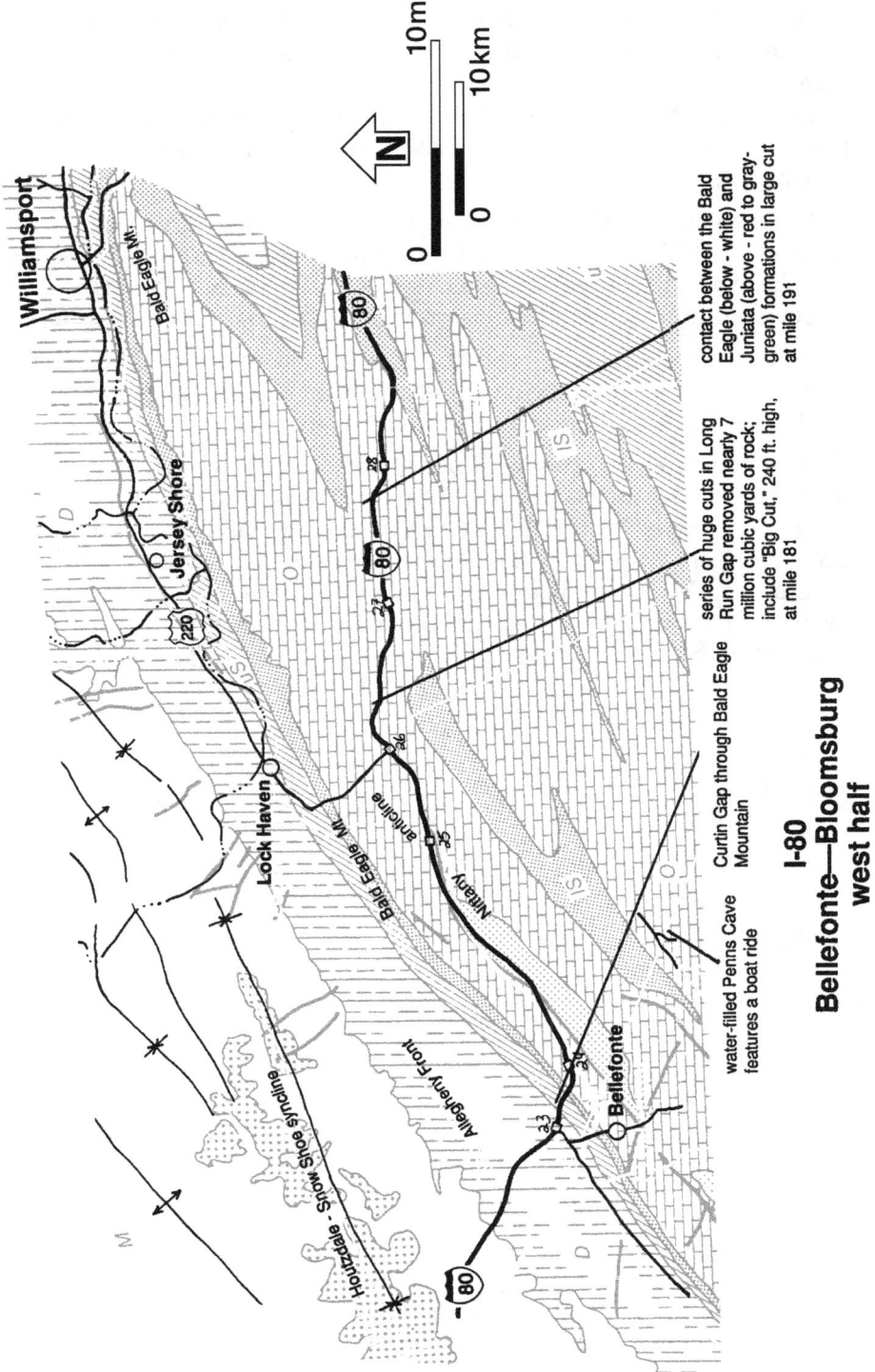

I-80 Bellefonte—Bloomsburg west half

I-80
Bellefonte—Bloomsburg
72 mi./116 km.

The highway obliquely crosses a succession of anticlinal and synclinal folds, through Cambrian to Devonian beds, with many large roadcuts. Between Exits 28 and 34, the route crosses the largest preserved tract of older glacial deposits in the state, a region where the early ice sheet reached much farther south than that of the last ice age. The deposits were not overridden, buried, or bulldozed by the later ice advance.

Between miles 160 and 178, you go northeast-southwest along Nittany Valley near Bald Eagle Mountain. Like so many of the hard-capped ridges of the region, this one has a remarkably even summit, but it is creased by water and wind gaps. The dry wind gaps were apparently carved by streams that later abandoned them. The flanking ridge to the south between miles 160 and 173 that separates the valley into two parts, is the upper Cambrian Gatesburg formation of interbedded sandstone, limestone, and dolostone. Most of the rocks under the valley are Ordovician carbonates.

The big gap four miles northwest of the Lock Haven Exit 26 is where Fishing Creek, which drains Nittany Valley, cuts through Bald Eagle Mountain, joining Bald Eagle Creek on the other side. Bald Eagle Creek, in turn, joins the west branch of the Susquehanna River a few

This 240-foot high "Big Cut" at mile 180 of I-80, the largest roadcut in Pennsylvania, exposes dipping, varicolored beds of the upper Ordovician Juniata formation.

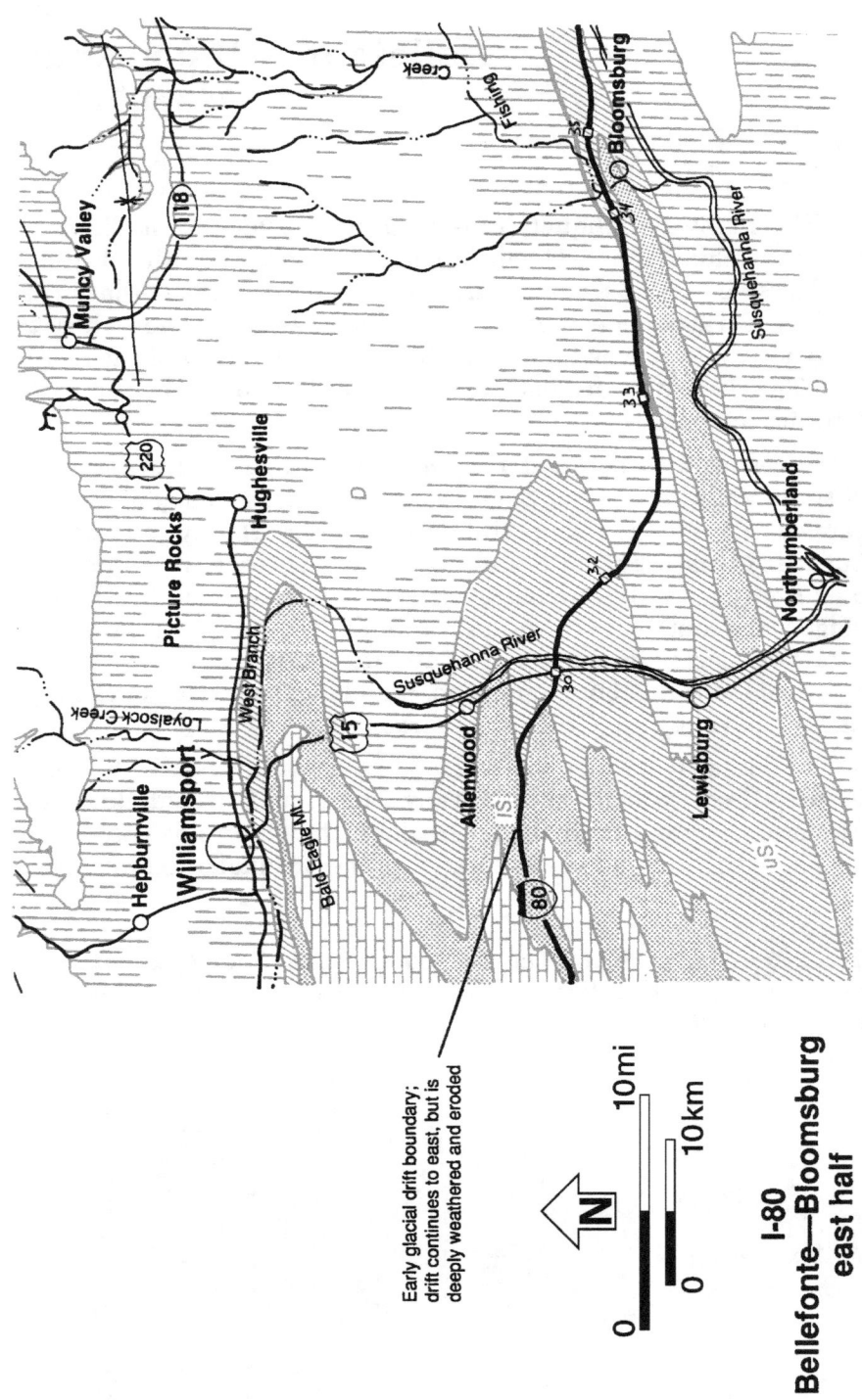

I-80 Bellefonte—Bloomsburg east half

miles farther northeast at Lock Haven. At mile 180 is Long Run Gap through the Bald Eagle conglomerate of Big Mountain. Just east of this, you see the first of a series of spectacular large cuts in the upper Ordovician Juniata formation. Big Cut, at mile 181, is 240 feet high, the largest in Pennsylvania. In all, an astounding seven million cubic yards of rock were removed from these several cuts. They are strikingly colored with purplish-red to gray banding of the interbedded sandstone, siltstone, and shale. In the Big Cut, beds appear to dip about 30 degrees east; in the fourth and fifth cuts to the east, beds are horizontal. The sixth and seventh cuts expose broad anticlines. You see more of the Juniata formation at mile 185, and near miles 186, 187, and 191.

At mile 191, look for the gradational boundary between the Juniata sandstone and underlying Bald Eagle conglomerate. Bedding dips rather steeply northward away from the axis of a large east-northeast-trending anticline. The Bald Eagle formation appears as thick, white conglomerate interbeds that contrast markedly with the red to greenish gray, thin-bedded siltstones and sandstones near the base of the Juniata formation. Conglomerate dominates in the lower part of the section. The Bald Eagle formation rims the anticline; older Bellefonte dolostone occupies the fold core a few miles farther west. The route crosses to the southern limb of the fold between miles 191 and 193.

You pass through narrow, woodsy, White Deer Creek Valley, also known as Sugar Valley Narrows, between miles 193 and 210. The pale boulders and talus on the flanking slopes are Tuscarora quartzite derived from the caprock.

The boundary of the older glacial deposits is about at mile 200 near the western end of Sugar Valley Narrows. This is the first contact, traveling eastward, with the glaciated lands of northeastern Pennsylvania.

You cross the west branch of the Susquehanna River at mile 211. Upper Silurian Bloomsburg red shales underlie this section. Between miles 207 and 232, you cross a rolling lowland so large that you might wonder how it can possibly be part of the Valley and Ridge province. It is low because the underlying Silurian and Devonian rocks, including Wills Creek shale, Keyser limestone, Onondaga limey shale, Mahantango shale, and Trimmers Rock shale and siltstone offer little resistance to erosion. Between Exits 33 and 34, you cross a fault that places Wills Creek shale against Mahantango shale, completely eliminating the intervening Keyser and Onondaga formations.

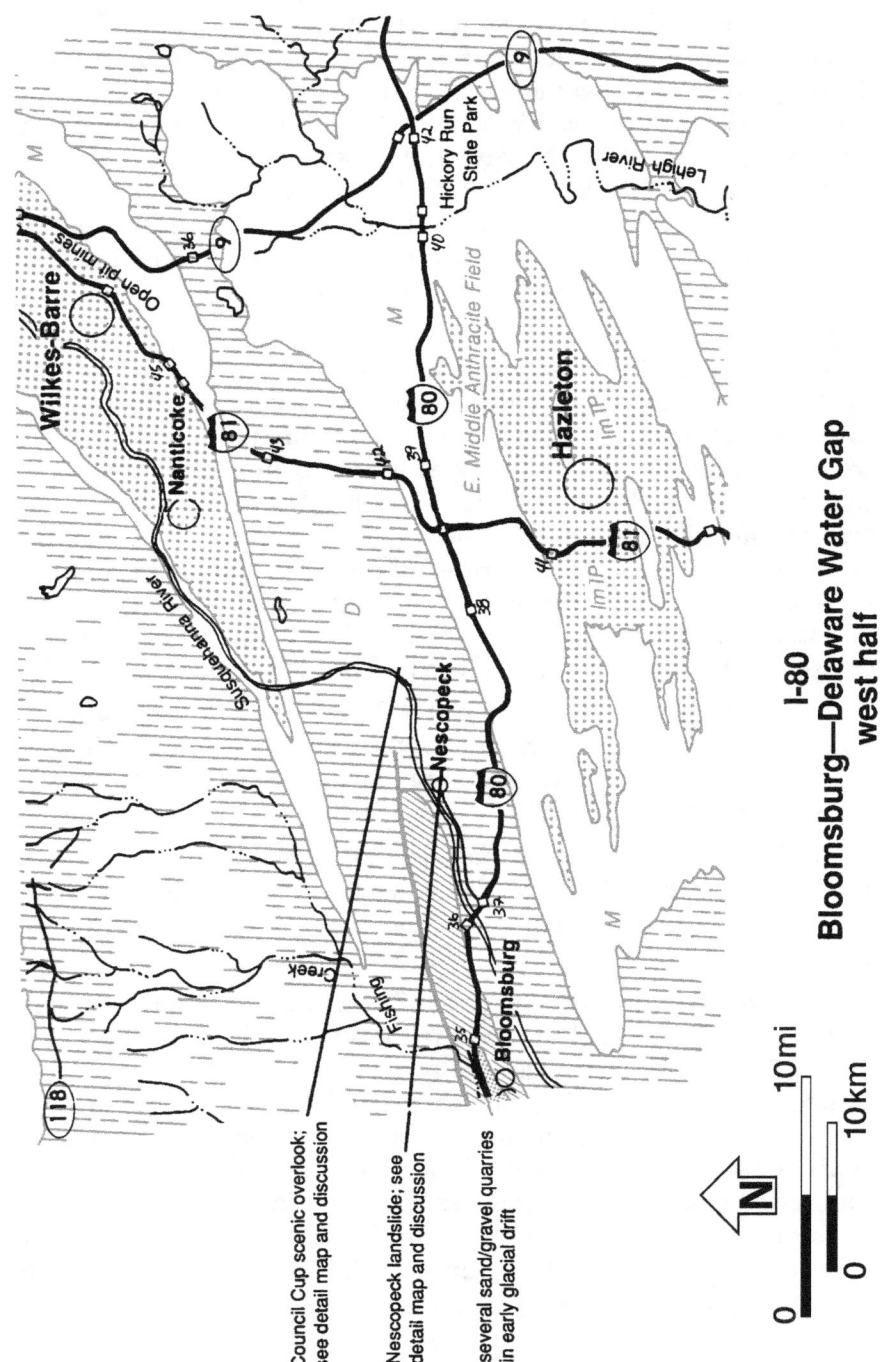

I-80 Bloomsburg—Delaware Water Gap west half

I-80
Bloomsburg—Delaware Water Gap
81 mi./131 km.

Between miles 232 and 241, you follow a broad stretch of Susquehanna Valley flanked by Huntington Mountain on the north and Nescopeck Mountain on the south. In so doing, you cross the axis of a large east-northeast-trending anticline through weak shales of the Silurian Wills Creek, Bloomsburg and Mifflintown formations and of the Clinton group, into which the river has excavated its valley. Farther west, the topographic expression of the anticline becomes a ridge instead of a valley where the tough Tuscarora quartzite in its core rises to the surface in Montour Ridge, nearly 1000 feet above the valley floor. Sand and gravel pits near mile 235 are in older glacial deposits.

You cross the Susquehanna River between the Berwick and Mifflinville exits. The broad, flat plain at Mifflinville that extends south through Catawissa and Nescopeck mountains appears to be an abandoned old channel of the river.

Nescopeck Landslide

Nescopeck is on the south bank of the Susquehanna River five miles east of Mifflinville along Pennsylvania 339. The landslide, one-half mile southeast of the village, is on River Hill, a steep, 300-foot-high rocky scarp carved in upper Devonian Trimmers Rock sandstones,

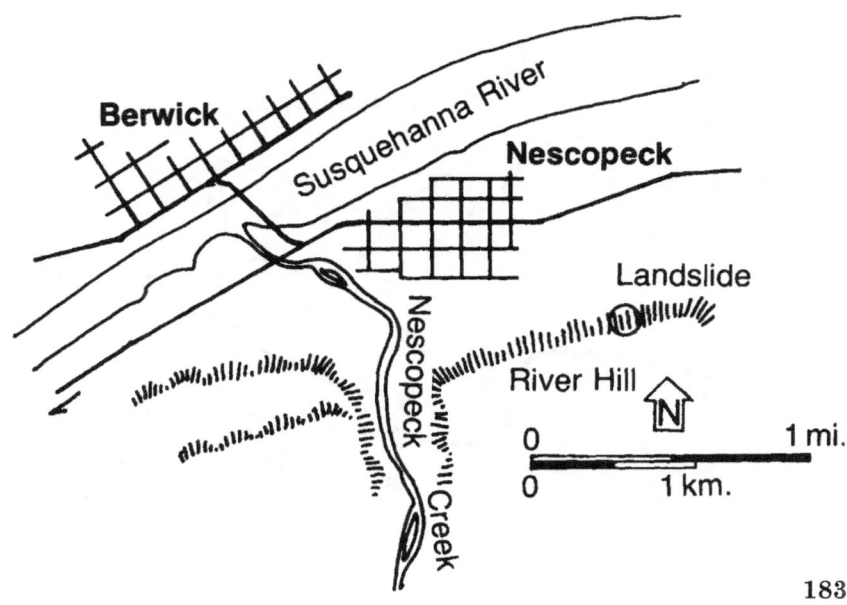

siltstones, and shales. Sidecutting by the river and glacial ice carved the scarp. The thin-skinned slide appears as a long narrow scar extending from near the top of the hill to a fan of debris at the bottom. Sandstone ledges appear in three places within the scar.

The landslide moved in June, 1972 after heavy rains from tropical storm Agnes saturated the soil. Sandstone cliffs flanking the top of the slide may have contributed by channeling excess runoff to the soil-covered area between them. Saturation promotes landsliding principally by adding weight to the slide mass and reducing friction along the slide surface. The straight, narrow track of this slide suggests that it moved very quickly. Trailer homes near the base of the scarp may be in danger from future slides.

• Between miles 242 and 256 you pass through Nescopeck Mountain along Nescopeck Creek; Conyngham Valley is on the south side. On the north side of the gap are steeply dipping red shales and sandstones of the upper Devonian Catskill formation. Those of the south side are grayish sandstone and conglomerate of the lower Mississippian Pocono formation and upper Mississippian Mauch Chunk redbeds.

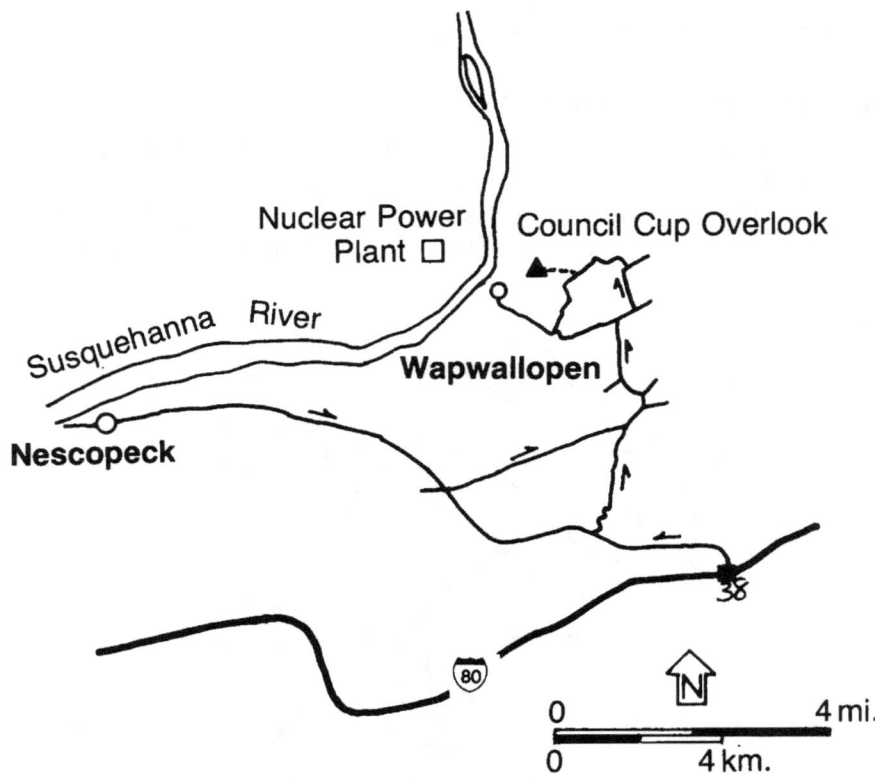

Mississippian Mauch Chunk redbeds at mile 251 of I-80 erode in steps that reflect the greater resistance of sandstone beds relative to interbedded shales.

Council Cup Scenic Overlook

Find this fine overlook on the west rim of a steep, rocky scarp known as Hess Mountain, 600 feet above the Susquehanna River. It is about seven miles northwest of Exit 38, near Wapwallopen. The accompanying map gives details. The view encompasses a broad expanse of valley with conspicuous terrace remnants visible in several places on either side of the river. The terraces, as well as the nicely displayed water gap through Lee-Penobscot Mountain to the north, record the downcutting progress of the river, which continues.

• Between miles 251 and 276, you cross Mauch Chunk redbeds on the north side of the Eastern Middle Anthracite Field. The more massive beds of this unit, mainly siltstone and fine sandstone, are prone to weather spheroidally. Watch for a good example near mile 257, south of the highway.

Spheroidal Weathering

Weathering is the chemical decomposition and mechanical breakdown of rocks and minerals to soil. Spheroidal weathering rounds angular blocks of rock by gradually breaking them down from the outside in. It works best in massive, homogeneous rocks like some sandstones and siltstones, and most basalts and granites. Criss-crossing fractures and bedding first separate angular blocks, exposing

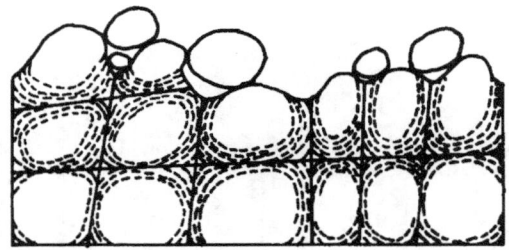

Spheroidal weathering, common in some massive rocks, begins along joints, bedding, and other passageways penetrated by air and water, and works inward. Each new weathered shell peels away like the layer of an onion, leaving an increasingly spherical core rock.

all sides to weathering processes along the openings. Corners are attacked from all sides, and thus more vigorously than faces, which are only attacked from one side. As the corner rock crumbles, the remaining fresh block becomes rounded. The core rock becomes more spherical as it sheds layers of decomposed rock.

Hickory Run Boulder Field

The boulder field at Hickory Run State Park is one of the most remarkable relics of the ice ages in Pennsylvania. Take Exit 41 to the park entrance.

The boulder field covers a flat segment of Hickory Run valley, about 1600 feet long east-west and 400 feet wide north-south; the layer of boulders is probably at least 12 feet deep. East-west-trending bedrock ridges rise about 200 feet above the field. A shallow, marshy drainage

Spheroidal weathering in Pocono sandstone at Exit 42 of I-80 — the Pennsylvania 9 interchange.

186

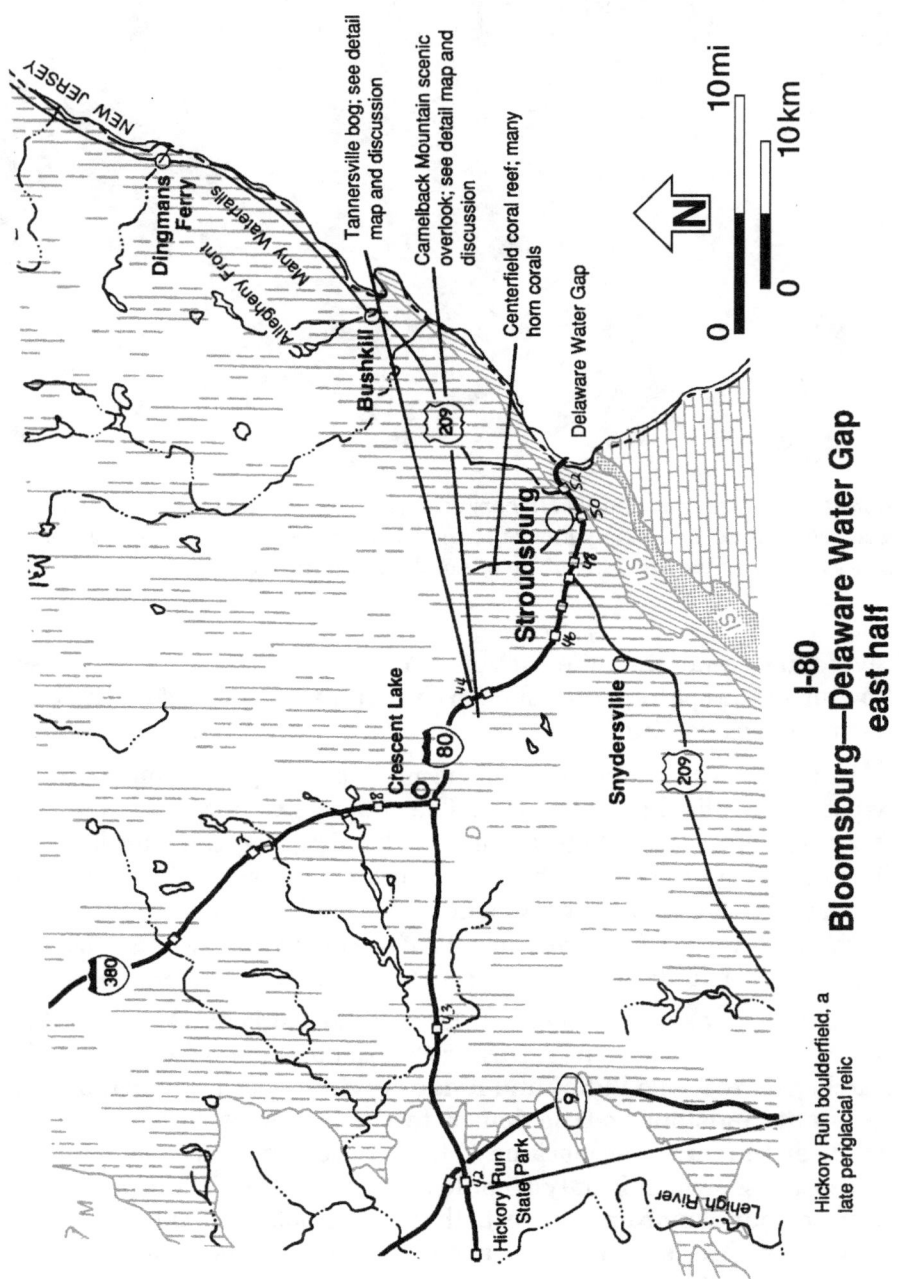

I-80
Bloomsburg—Delaware Water Gap
east half

Boulders in this remarkable Hickory Run field began as talus at the base of a nearby scarp, and slowly moved to their present position by a freeze-thaw process during the last glaciation of northern Pennsylvania.

divide lies about one mile east of it. You can sometimes hear water flowing in the open spaces between the boulders of the field on its way down to Hickory Run.

In addition to its large size, the most striking and unexplained aspect of this field is its flatness. The boulders vary widely in size; a few of the largest are 25 feet long, yet the view over the length of the field reveals an exceptionally level, boulder-top horizon. The rocks are almost entirely reddish-gray sandstones and conglomerates of the Catskill formation from the cap rock of the nearby ridges. How do you suppose the boulders got down from the ridges?

Let's look at several other similar boulder fields in the state: Blue Rocks, Devil's Race Course, Devil's Potato Patch, and Devil's Turnip Patch. All are in the Valley and Ridge province near ridges capped by resistant sandstone and conglomerate, the sources of the boulders. all are near the terminal moraine of the last ice age, except for Devils Race Course, which is only 15 miles southeast of it. Their proximity to the terminal moraine suggests that all the boulder fields formed at the climax of the last glaciation, about 20,000 years ago. Temperatures then were colder than they are now. A lot of meltwater was issuing from the glacier, and frost heave was very active. Talus that accumulated on the sides of the ridges migrated downslope to fill the valley bottoms.

Each freeze-thaw cycle carried the talus a little farther downslope. Judging from the large size of the moraine, the ice front remained stationary for a long time, maintaining the cold climate necessary for the boulder field to form. Perhaps boulder fields are poorly developed within the glaciated region north of the moraine because the receding glacier did not stabilize in any one position long enough.

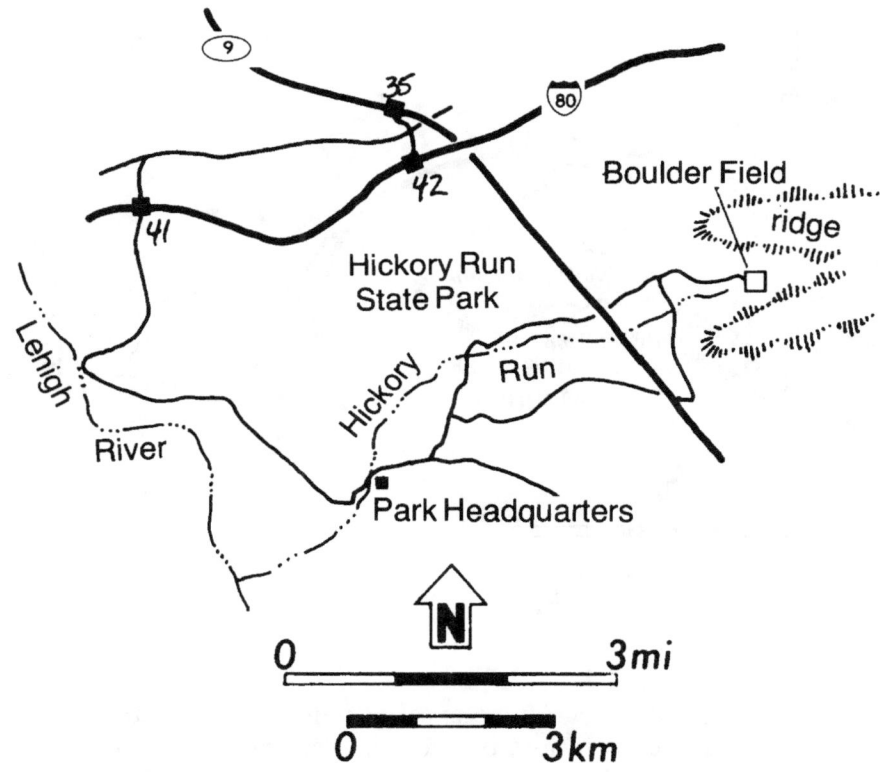

- You cross the Pocono Plateau between miles 274 and the I-380 interchange, passing over or near the terminal moraine of the last ice age. The plateau is a region of abundant lakes, glacial drift, and bogs. Note the bogs near miles 287, 290, and 292, and the gravel pit near mile 202. Bedrock in this section is largely concealed. Pale conglomerate and sandstone boulders near mile 275 derive from the underlying Pocono formation at the synclinal nose of the Eastern Middle Anthracite Field. The contact between the Spechty Kopf formation at the base of the Mississippian section and the underlying Catskill formation lies concealed near mile 280.

Camelback Mountain Overlooks

Camelback Mountain, in Big Pocono State Park, is a 1000-foot high, east-pointing finger of the Allegheny Front at the rim of the Pocono Plateau. It juts from the main scarp where Pocono Creek excavated the crushed rock of an east-west fault zone on its north side. The east end of the mountain is strikingly profiled as you travel east between the rest area at mile 289.5 and Exit 45.

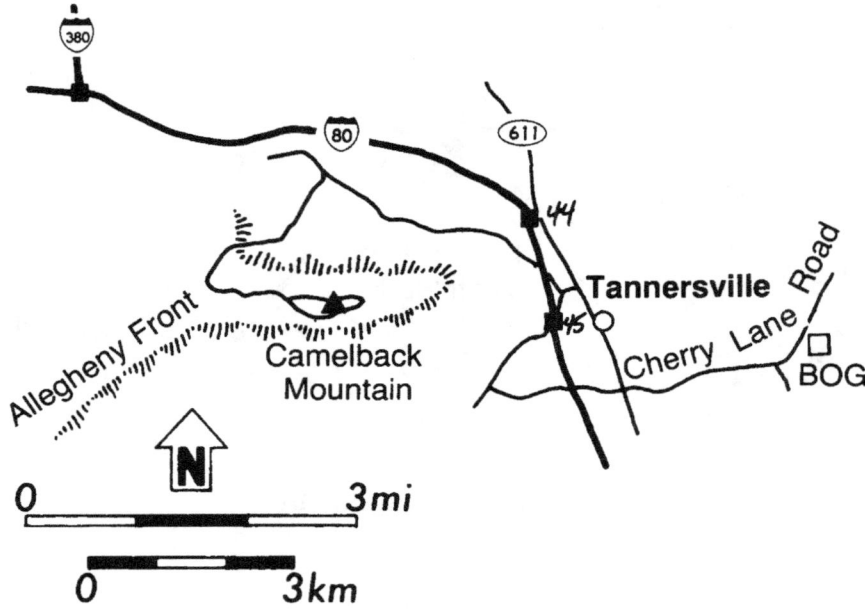

From the summit, you have excellent views of the surrounding countryside. Reach it in just fifteen minutes from the Tannersville Exit 45, by way of the Big Pocono Ski Area road on the north slope. The most dramatic view is the profile of the Allegheny Front to west-southwest; watch back to the right as you approach the the fire tower. The distant scarp is continuous with Camelback Mountain and sustained by the same tough conglomeratic caprock of the Catskill formation that appears in broken confusion at the tower base. Walk to the east end of the summit loop drive for another spectacular view, a 130-degree panoramic profile of Blue Mountain ridge with its several gaps and otherwise even crestline.

Broken upper Devonian Catskill conglomerate bed exposed by the Camelback Mountain fire tower forms part of the rim of the Pocono Plateau.

Camelback Mountain stands in the terminal moraine of the last ice age. Only a few small patches of glacial drift are on the summit, but moraine covers the lower north and south slopes and lies around the base. The mountain apparently exerted considerable influence on the southwestward advance of the ice, cleaving it into two separate lobes. To the south, a giant lobe continued to Brodheadsville and halted, leaving a horseshoe-shaped moraine ridge at its final position, with one end at Camelback and the other at the Delaware Water Gap. From the Camelback end, the moraine makes a hairpin turn and continues straight west across the Pocono Plateau.

Tannersville Peat Bog

Reach this bog in a few minutes drive east of Exit 45. The bog occupies about 300 acres in slack headwater sections of two small creeks, including Cranberry Creek. The moist perimeter is lush with rhododendron, marsh and cinnamon ferns, burreed, poison sumac,

Profile of Blue Mountain in 130° panorama from Camelback Mountain, showing even crestline and four gaps. —From Frakes and others (1963, p. 13)

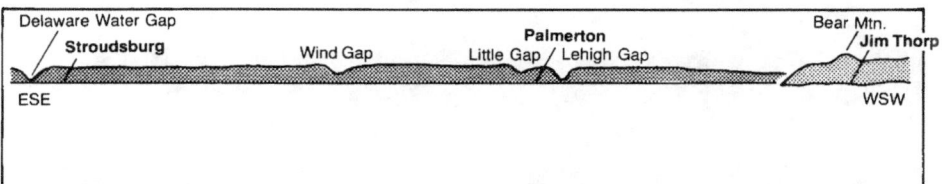

and maleberry. In the surrounding woods on peat soil grow black spruce, red maple, yellow birch, rhododendron, and poison sumac over a rich undergrowth.

Peat bogs are numerous in the glaciated regions of North America, including Pennsylvania. Glacial erosion and deposition left many water-filled depressions independent of local stream drainages. Stagnant water is poorly oxygenated, so vegetation that accumulates in it decays at only a small fraction of the rate it would in air. Partially decayed vegetation accumulates to form peat.

The northern climate promotes growth of peat-forming plants in these wet environments, mainly trees, heath shrubs, sedges, grasses, mosses, pondweeds, waterlilies, reeds, cattails, algae, and ferns. Sphagnum moss commonly grows out from the shore to form a floating mat in which other plants root, including even pitcher plants and other insectivores, and provides a plentiful supply of peat materials to the lake bottom. Walking on one of these mats is like walking on a water bed.

Many more active bogs existed in Pennsylvania at the close of the ice age than now. Recognize active bogs, those that have completely filled, as basins covered with wetlands vegetation. The state of Pennsylvania has long been interested in exploiting its peat resources. Although inferior to coal, they do contain a sizeable alternate source of energy. The U.S. Geological Survey estimates that peat reserves in northeastern Pennsylvania exceed 10 million short tons. Some deposits that have been forming since the last glaciers receded are 18 feet thick; their rate of formation must average about one or two inches per century during the last ten to twenty thousand years.

The highly fossiliferous Centerfield coral reef was once a thriving community of corals and other marine animals in a shallow sea that covered much of Pennsylvania about 390 million years ago, duirng middle Devonian time.

Fossils, especially horn corals, in the Centerfield coral reef structure. Note the curved horn coral immediately left of the lens cap and the oval cross-sections elsewhere.

- Between miles 292 and 307, you cross the lower part of the Catskill formation and the Devonian section beneath it. Note Trimmers Rock olive-gray siltstones and shales at mile 300, Mahantango brownish shales at miles 302 and 303, and the underlying Marcellus black shales at mile 307. The cherty, streaky limestone also at mile 307 and between miles 309-310, belongs to the middle Devonian Buttermilk Falls formation that lies beneath the Marcellus shale.

These generally dark rocks record the filling of the Acadian foredeep basin before the main uplift. The basal chert, an aggregate of microscopic quartz grains, represents the earliest, deepest sedimentation. The Trimmers Rock formation at the top is a relatively shallow water deposit formed at the beginning of the main orogeny when the foredeep basin was nearly filled up. It is transitional to the overlying, largely terrestrial, Catskill formation.

Centerfield Coral Reef

This fossil reef in the Mahantango formation is exposed on Pennsylvania 191, four miles north of I-80 Exit 50, just south of the Pennsylvania 447 junction. It contains abundant corals, lacy bryozoans, brachiopods that look almost like clams, and one kind of trilobite. The reef grew nearly 390 million years ago, when this area lay submerged beneath a shallow inland sea that teemed with life. It is part of the widespread Centerfield fossil zone of the Mahantango formation exposed in several places in eastern Pennsylvania.

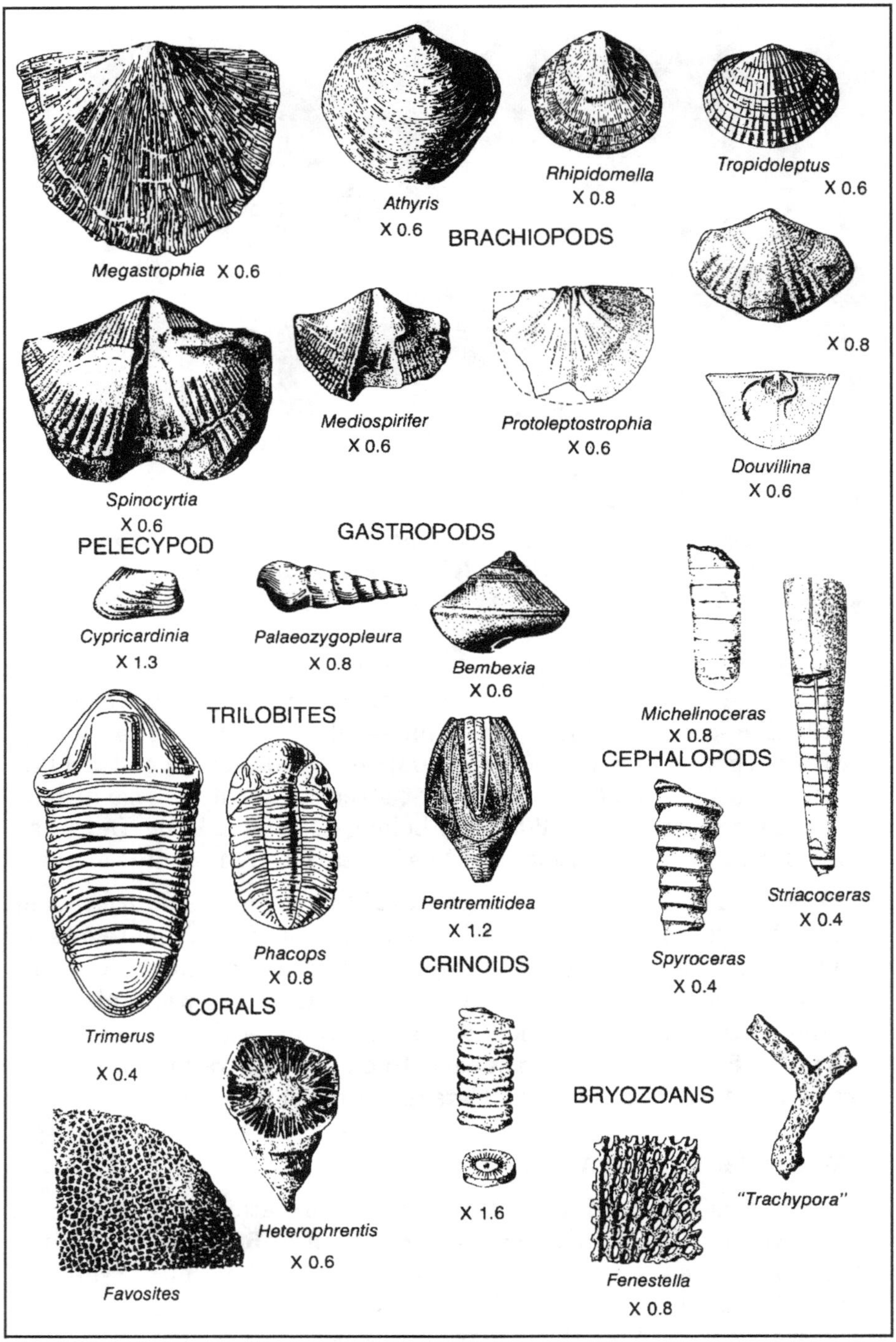

Some of the middle Devonian fossils found in the Centerfield fossil zone.
—Adapted from Hoskins and others (1983, p. 200-206)

- Near both ends of the Delaware River bridge are cuts in grayish calcareous shale and siltstone of the upper Silurian Wills Creek formation.

The Pennsylvania segment of I-80 ends at the Delaware Water Gap, perhaps the most striking of all the Appalachian water gaps. At the nearby National Park Service headquarters is a good view of Mt. Tammany on the northeast side of the gap, partly obscured by the lower slopes of Mt. Minsi on the other side. Both are part of Kittatinny Mountain, a prominent ridge capped by lower Silurian Shawangunk quartzite and conglomerate. The ridge continues northeast to Kingston, New York, where it forms the backbone of the Shawangunk Mountains with their magnificent white cliffs. It continues southwest almost all the way across Pennsylvania where the caprock is the Tuscarora quartzite, equivalent to the lower part of the Shawangunk conglomerate. Near Harrisburg, the ridge is called Blue Mountain. Its total length is more than 260 miles.

Bare cliffs visible from park headquarters are Shawangunk conglomerate beds that dip about 35 degrees to the northwest. An apparent concealed fault parallel to the bedding — a bedding fault — separates the Shawangunk formation from the underlying upper Ordovician Martinsburg shales, which lie beneath talus shed from the Shawangunk conglomerate. The overlying fault block moved relatively downward, most likely as a result of the original backtilting of all the beds. The back lower slope of the ridge is red sandstones, siltstones, and shales of the upper Silurian Bloomsburg formation, which lies above the Shawangunk conglomerate.

Hogbacks

If Kittatinny Mountain were in Colorado it would be called a hogback. It would keep company with other erosional forms common there — mesas and cuestas. All start with unfolded, or only slightly folded, interlayered hard and soft rocks. Everyone knows that a mesa is a flat tableland with a cliffy rim carved from its hard caprock layer. The cliff recedes, and the mesa gets smaller as the softer layers under

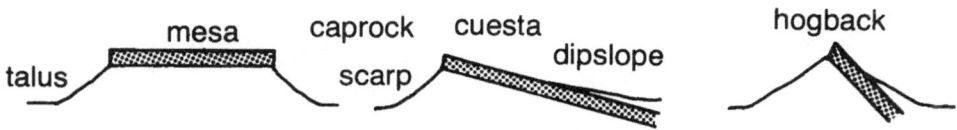

the caprock wear away, causing the caprock to overhang and break off. The same mechanism maintains the cliff of Horseshoe Falls at Niagara, and many other waterfalls. A cuesta is like a slightly tilted mesa, with a cliff only on the uptilted side. On the other side, the dipslope, streams tend to migrate down the surface of the caprock, stripping away the softer beds above it. Tilt the mesa even more, even to vertical, and you get a Kittatinny Mountain, a hogback, with the hard layer sticking up like the backbone of a hog.

• Terminal moraine of the last ice age blocks the valley behind the ridge eight miles southwest of the Delaware Water Gap. This was the site of short-lived glacial Lake Sciota, which achieved a maximum depth of about 200 feet and a length of eight miles. Like so many of the meltwater lakes formed during glacial retreat, this one banked against the moraine at one end and the ice at the other — but only while the gap was blocked. Once the ice uncovered the gap, Lake Sciota emptied.

I-81
Junction I-78—Wilkes-Barre
80 mi./129 km.

South of the I-78 junction, I-81 traces a curving path through the Cumberland Valley parallel to the regional structures, passing very few exposed rocks. By contrast, this section trends generally across the highly deformed rocks of the Valley and Ridge province, alternately passing through water gaps and following stretches of valley that separate the ridges. Numerous large cuts reveal a great deal of bedrock. The accompanying Geologic Time-Travel Map schematically illustrates the rock formations traversed by the highway.

In crossing Pennsylvania, I-81 cuts through more than 200 million years of geologic history, as shown by this time-travel map. —Adapted from Bolles, W.H. and Geyer, A.R., 1976, Geologic Guide of Pennsylvania Interstate 81, Pennsylvania Department of Education

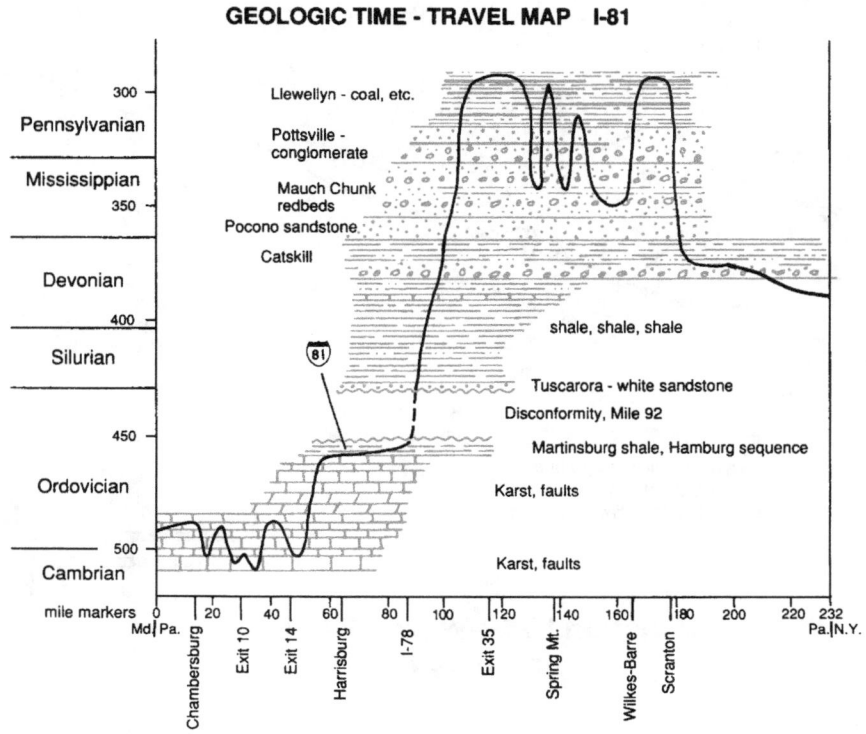

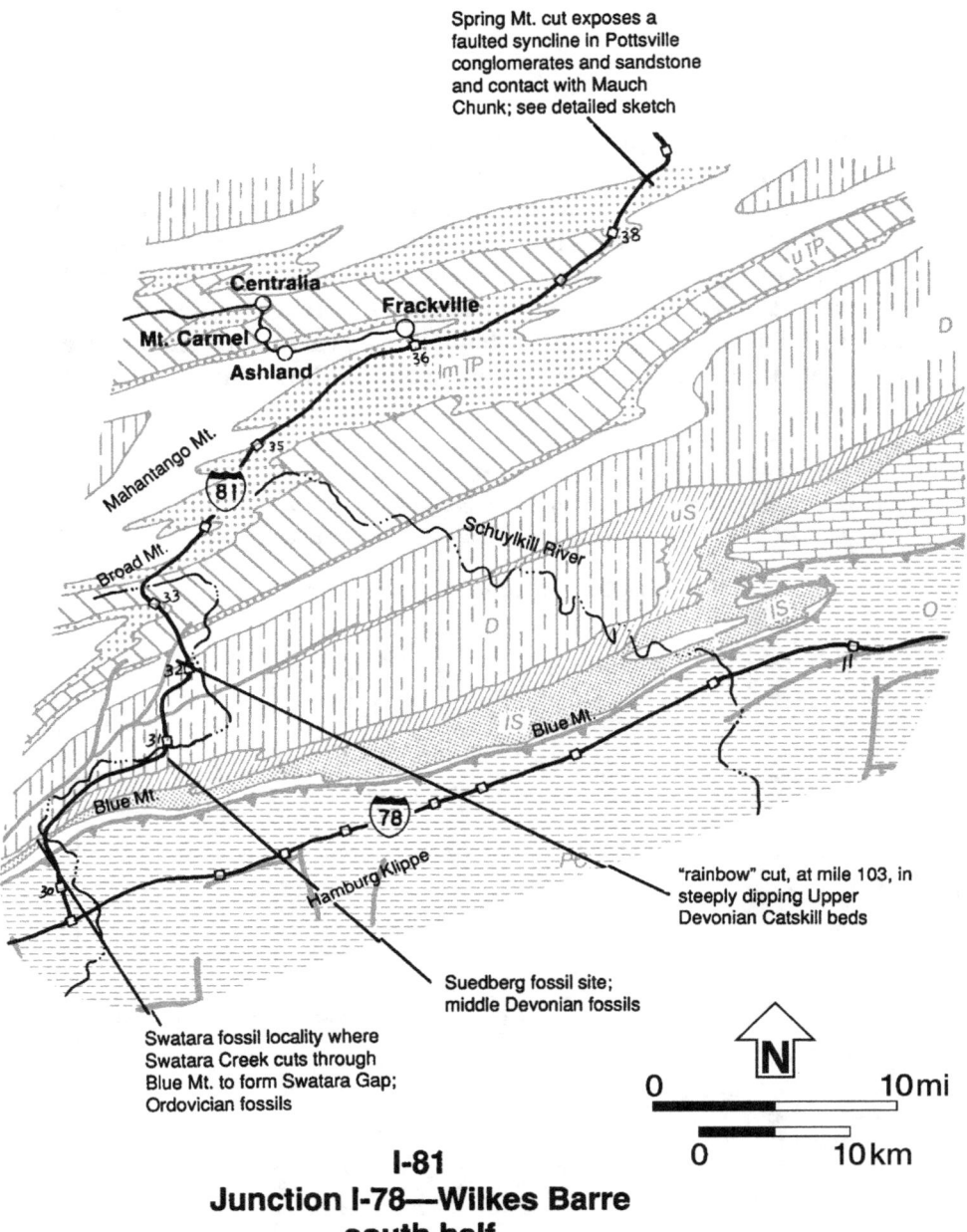

**I-81
Junction I-78—Wilkes Barre
south half**

Swatara Gap Fossil Locality

To reach this well-known collecting locality, take Exit 30, less than two miles north of the I-78 junction and two miles south of Swatara Gap, to Lickdale. From there, go north on Pennsylvania 72 for two miles to an abandoned quarry under the northbound I-81 bridge over Swatara Creek, at the entrance to Swatara Gap through Blue Mountain.

The bedrock here is the upper Ordovician Martinsburg formation: grayish to brownish shales and siltstones. This is one of only a few places where this formation yields fossils, all of which are either molds or casts. The original shells were buried, then dissolved by groundwater,

Upper Ordovician fossils from the top of the Martinsburg formation at Swatara Gap. —Adapted from Hoskins and others (1983, p. 197-198)

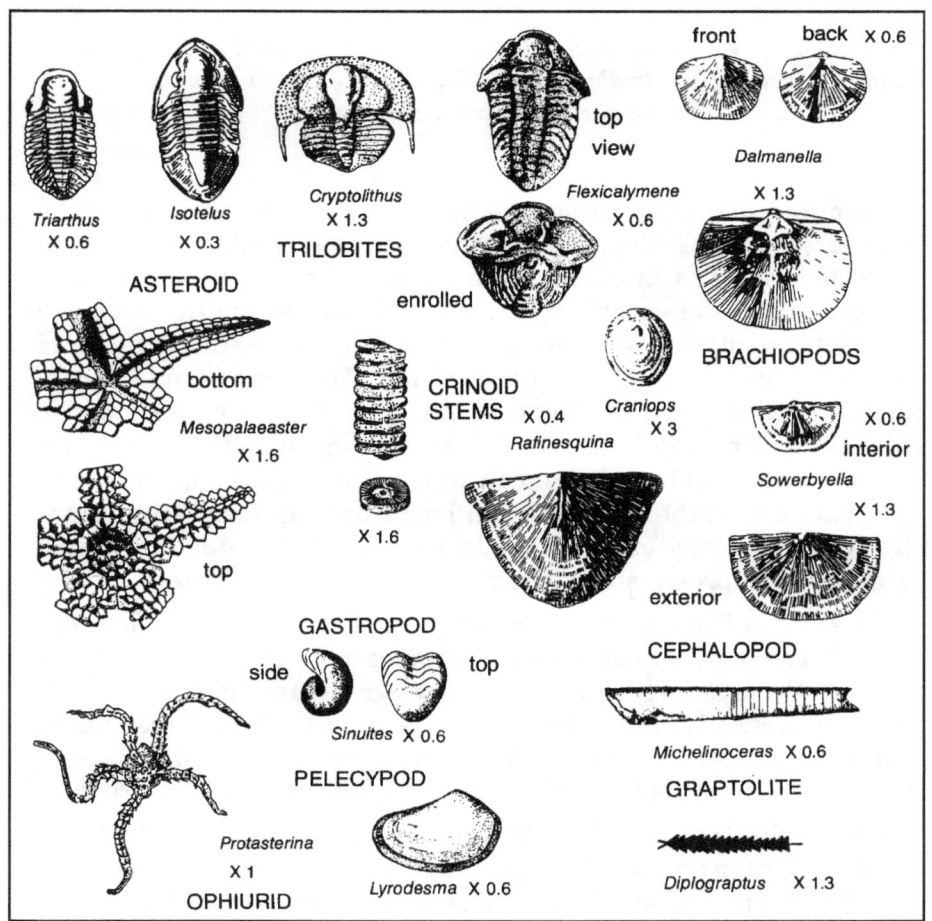

leaving only impressions; some impressions were filled with new deposits, creating casts. The accompanying drawings illustrate some of the more common fossils.

- Swatara Gap is much the same as the Susquehanna gaps, with even-crested ridges on either side truncated by the stream. Here the truncated ridges slope more gently to Swatara Creek.

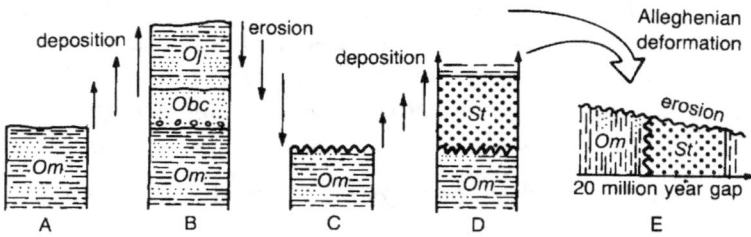

Geologic history recorded in Swatara Gap, mile 92, A-E. Om, Obe, Oj = upper Ordovician Martinsburg, Bald Eagle, and Juniata formations, respectively; St = lower Silurian Tuscarora quartzite.

At mile 92 in the gap you see the disconformity that is schematically illustrated on the Geologic Time-Travel Map. A disconformity is a type of unconformity, a buried erosion surface that separates parallel sequences of sedimentary layers. As in all unconformities, this one represents a missing rock record. Where exposed on the southbound lane, the disconformity is quite striking. The beds stand on end, having been rotated from their original horizontal position during the Alleghenian orogeny. The darker Martinsburg shales and siltstones on the southeast side make sharp contact with the pale quartzites and conglomerates of the lower Silurian Tuscarora quartzite. Missing are the whole sections of upper Ordovician Juniata and Bald Eagle formations, approximately 20 million years of Earth history.

This is the Taconic unconformity, a result of the late Ordovician Taconian orogeny. Farther east, it becomes an angular unconformity in which the erosion surface cuts at an angle across tilted Ordovician rocks. It means that the older rocks were first tilted, then eroded, before the Tuscarora sands were laid down on top of them. This area was farther from the locus of mountain uplift, and less disturbed.

Within Swatara Gap, the highway crosses red shales of the lower Silurian Clinton group and gray to grayish-red sandstones, siltstones, and shales of the upper Silurian Bloomsburg formation. Between mile 93 and the Suedberg Exit 31, the route goes northeast-southwest along Swatara Creek between Blue and Second mountains. Blue Mountain ridge on the south is Tuscarora quartzites and conglomerates; Second Mountain on the north is Pennsylvanian Pottsville sandstones and conglomerates at the edge of the Southern Anthracite Field. The valley is eroded principally in weak middle Devonian Hamilton group shales, with brownish Mahantango shale above and black Marcellus shale below. The Onondaga limestone lies beneath the Marcellus shale.

Suedberg Fossil Site

To reach this site, take Exit 31 to Suedberg, then turn south to cross Swatara Creek. A borrow pit along an unimproved road that parallels the south bank of the creek, about one-half mile from Suedberg, exposes fossil-bearing beds of the Mahantango formation. Most of the fossils are molds and casts, like those of Swatara Gap; but the animals existed 65 million years later, during middle Devonian time.

• Between Suedberg and the Dorrance Exit 42, 56 miles, you diagonally cross a huge, northeast-trending, complex syncline, the Anthracite synclinorium, that contains the Southern, Western Middle and Eastern Middle anthracite fields. Numerous wrinkles, smaller anticlines and synclines, are within the master fold; the Pennsylvanian coals are preserved only in the synclines, where they lie below the current level of erosion.

The deformation is Alleghenian, a product of collision between North America and Africa during Pennsylvanian and Permian time. Compression was relieved primarily by movement along a flat-lying sole thrust that sliced through weak middle Devonian shales 15,000 to 20,000 feet below the surface. Strata beneath the sole are undeformed. Rocks above it are much folded where numerous reverse faults branch from the thrust and ascend steeply toward the surface. Most of the faults do not reach ground level.

The cross-section helps to explain the synclinal form of the major fold. To the south, the same sole thrust lies more than 40,000 feet beneath the Lehigh Valley, where it cuts weak Cambrian strata. Near the Valley and Ridge margin, which is distinguished on the ground by Blue Mountain ridge, the sole thrust splits and the upper branch ascends along a ramp and then levels off at 15,000-20,000 feet. The bend at the top of the ramp arched the rocks into an anticline.

This is analogous to folding along the Allegheny Front, where an anticline also lies at the top of the ramp, next to a syncline. There the sole thrust levels off at a much shallower depth under the Allegheny Plateau, and plateau strata are only mildly folded. Here the more

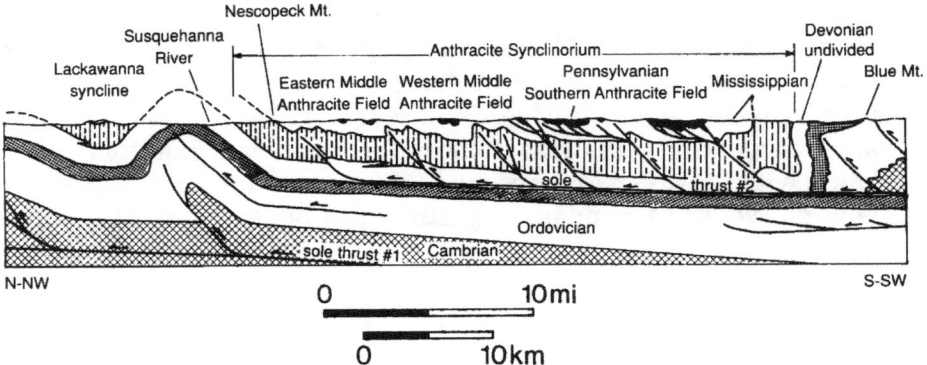

Geologic cross-section north-northwestward from Blue Mountain near Hamburg through the Anthracite synclinorium and Lackawanna syncline. The Synclinorium has an overall synclinal form, but contains many smaller synclines and anticlines. Coals are preserved in the synclines.
—Adapted from Socolow, A.A., 1980, Geologic Map of Pennsylvania, Pennsylvania Geological Survey

severe deformation reflects the greater resistance to compression offered by a thicker rock package, but there is more to the story.

Recall that the sole thrust split beneath the Lehigh Valley margin. The lower branch slices through Cambrian beds for about 50 miles, ascending gradually to a depth of about 30,000 feet. Finally it rises into weak Silurian beds about 10,000 feet beneath the Allegheny Front, where it levels off. The rock package between the upper and lower sole thrusts is virtually flat-lying and undeformed, except where steeply-dipping reverse faults split off and ascend from the lower thrust. The resulting Anthracite synclinorium resembles a cake pan in cross-section: shallow with sharply turned edges.

A very long cut at mile 103, where I-81 passes through Second Mountain, exposes a rainbow of steeply-dipping, multicolored sandstone and shale of the upper Devonian Catskill formation. The overlying lower Mississippian Pocono formation, brownish to yellowish-gray sandstone, crops out at the northern end of the cut.

Between miles 105 and 143, you cross the Southern and Western Middle Anthracite Fields, where Pennsylvanian Pottsville and Llewellyn coals have been mined. Visible throughout are immense rock waste piles and strip mines, unreclaimed testimony to the intense mining of the past. About seven billion tons of coal have been taken from the eastern Pennsylvania anthracite fields since 1769, and an estimated three times that amount remains in the ground. The industry is now at a standstill and may never revive. The remaining coal is deep underground and very expensive to extract. Subsurface coal mining is notorious for its accidents and health problems.

At mile 106 you see Pottsville sandstones, siltstones, and coals at the northeast-pointing nose of the Joliet anticline. To the southwest, over a distance of 25 miles to the Susquehanna River, the fold encloses Mississippian and Devonian rocks in its core. Rocks in the middle of eroded anticlines are invariably older than those on the flanks

On the southbound lane, at mile 107, is a beautiful small fold, the Tremont syncline, in Llewellyn sandstones. The northern end of the fold is cut by a small, steep fault on which the northern side appears to have moved up a few feet relative to the southern side. Elsewhere in the cut are several thin coal beds. Old strip mines nearby are in the Primrose coal seam of the Llewellyn formation.

A scenic viewpoint on the southbound lane at mile 109 overlooks Hegins Valley from the crest of Broad Mountain. The opposing ridge four miles to the north is Mahantango Mountain, capped by Mississippian Pocono sandstone, on the south limb of another northeast-pointing anticline similar to the Joliet anticline.

Between miles 108 and 140, you follow the northern edge of the Southern Anthracite Field and eastern end of the Western Middle

The Tremont syncline at mile 107 of I-81 in Pennsylvanian Llewellyn sandstones and shales, is cut by a small, steeply-dipping fault on the right limb.

Water-filled, abandoned strip mine at mile 120 of I-81, a common sight in the anthracite region.

Field, nearly parallel to the trend of the folds. You see many exposures of quartz pebble conglomerates of the Pottsville group, easily recognized by their speckled appearance. Numerous cuts also contain sandstone, shale, and thin coal interbeds.

Most of the abandoned coal mines visible from the highway are in the younger Llewellyn formation, the principal coal-producer of the anthracite region. At mile 120 is a good example of an unreclaimed, water-filled strip mine in the Birch Mountain #5 coal seam. The mine is in the New Boston syncline.

Most of the pebbles in this Pennsylvanian Pottsville conglomerate at mile 133 of I-81 are milky quartz.

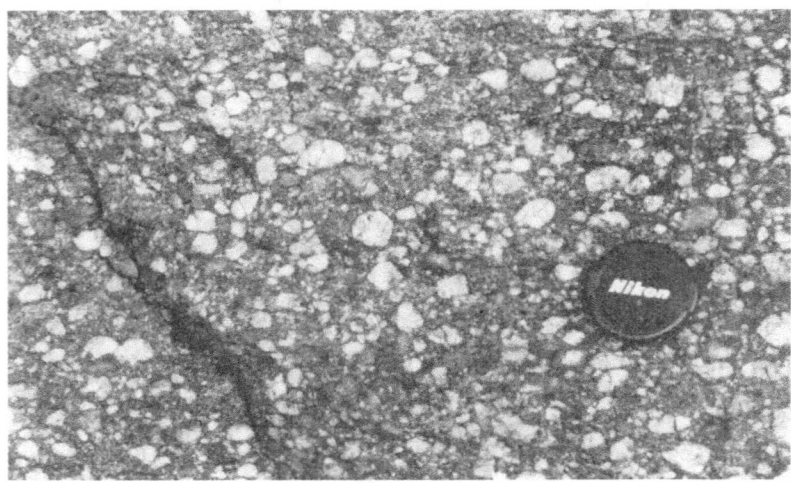

204

Acid mine drainage is a constant, nagging problem in Pennsylvania, principally from underground mines. Sulfuric acid forms as water and oxygen react with the mineral pyrite (iron sulfide) which is abundant in the coal and in shales associated with it. The red-orange color of dissolved iron sulfate announces the presence of acid water. Burning coal that contains pyrite creates sulfur dioxide, which pollutes the atmosphere.

A half-mile long roadcut at mile 134 exposes Pottsville conglomerates and, at the northern end, its basal contact with redbeds of the upper Mississippian Mauch Chunk formation.

The huge Spring Mountain roadcut at mile 138 is one of the most spectacular bedrock exposures in the state. On the north-bound side you see a shallow syncline in light brownish Pottsville sandstone and conglomerate beds, cut by one major south-dipping reverse fault on which the south side moved up and north perhaps 50 feet relative to

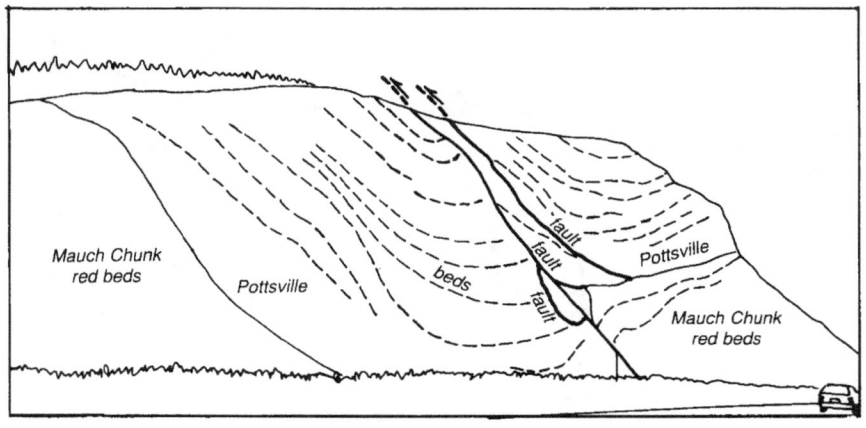

The enormous Spring Mountain cut at mile 138 of I-81 exposes a faulted syncline in Pennsylvanian Pottsville sandstones and conglomerates, above Mississippian Mauch Chunk redbeds.

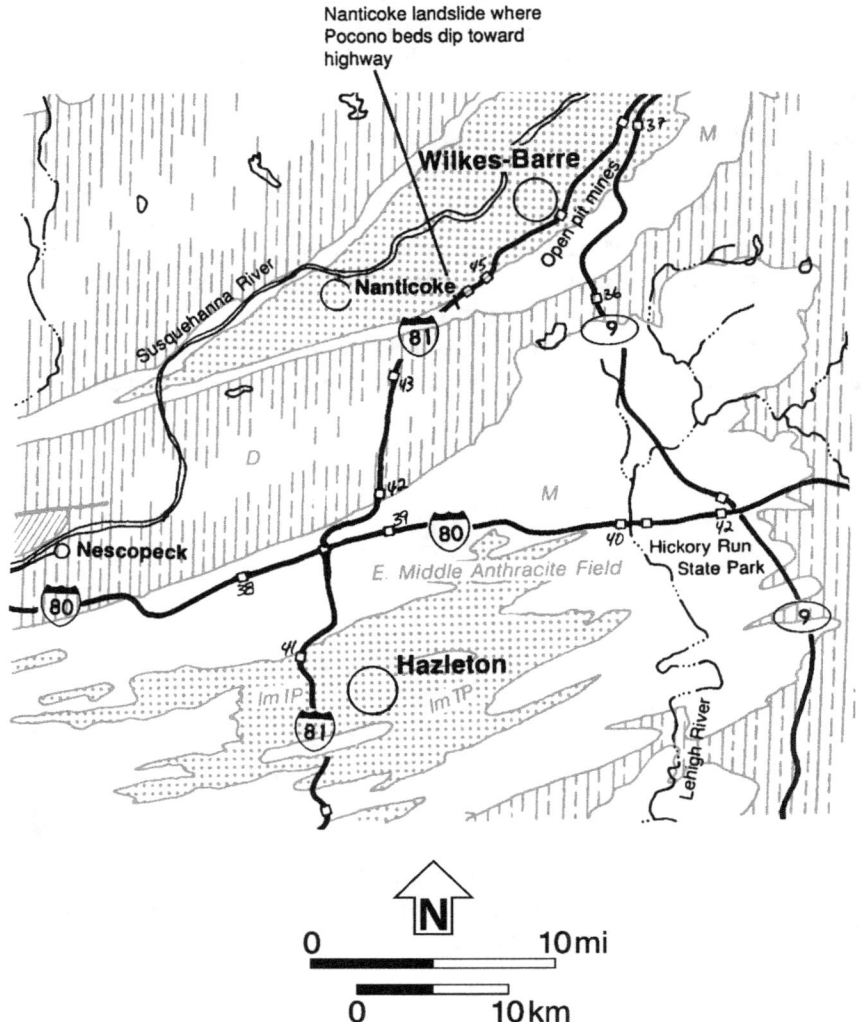

**I-81
Junction I-78—Wilkes-Barre
north half**

Typical speckled Pottsville conglomerate and sandstone at mile 133 of I-81.

the north side. In places the beds adjacent to the fault have clearly been dragged by the friction associated with movement of the blocks against each other. There are also several minor faults, as indicated on the accompanying diagram. At the northern and southern ends of the cut, the Pottsville sandstone is in contact with the reddish, brownish, and light grey shales and siltstones of the underlying Mauch Chunk formation.

Between the Spring Mountain cut and Hazelton Exit 41, you cross the Eastern Middle Anthracite Field. Pottsville and Llewellyn beds appear along the road except near mile 141, where Mauch Chunk formation is exposed.

Between mile 145 and the I-80 interchange you see mainly the red, maroon, and purple siltstones and sandstones of the Mauch Chunk formation. The rest stop at mile 147 offers a picturesque view of

Conyngham Valley from the crest of Butler Mountain. The lone, conical hill in the mid-distance, Sugarloaf Mountain, was carved from Mauch Chunk beds. Nescopeck Mountain in the distance is a ridge of sandstones and conglomerates of the Pocono formation beneath the Mauch Chunk formation.

Between the rest stop and mile 151, you cross Conyngham Valley and then pass diagonally over Nescopeck Mountain. The pale Pocono beds at mile 152 dip south toward the synclinal core. Between miles 153 and 161, you see reddish-brown to greenish-gray sandstones of the Catskill formation that underlie the Pocono formation and occupy the core of Berwick anticline.

Marshy lowlands visible from mile 155 signal the edge of the glaciated region of northeastern Pennsylvania. North of here, the landscape is imprinted with products of continental glaciation: wetlands, kettle holes and lakes, outwash and moraine deposits, erratic boulders and glacial striations.

Between miles 161 and 163, the north and south-bound lanes of I-81 separate widely along opposite sides of Penobscot Mountain, which consists of the Pocono formation. This is the northern limb of the Berwick anticline, which is also the southern limb of the Lackawanna syncline. The southbound lane traverses the dipslope of the ridge where Pocono beds slope 30-50 degrees toward the highway and are prone to sliding.

This is where the catastrophic Nanticoke Landslide of 1986 blocked all three lanes of the highway. The Pocono sandstone beds here are separated by thin, discontinuous shale partings and cut by steeply inclined joints oriented almost at right angles to the bedding planes and the road. The joints and bedding planes both facilitated infiltration of rainwater, which reduced friction between the beds, especially along shale partings, and allowed them to slide. The slide scar and debris at its base are still visible.

The north- and south-bound lanes rejoin just east of the slide, continue along the dipslope, then pass through the next lower ridge, called Wilkes-Barre Mountain, at mile 163, along Sugar Notch. Here you see red shales, sandstones, and siltstones of the Mauch Chunk formation and beyond the notch, Pottsville beds. All beds dip north into the Lackawanna Valley. Old strip mines abound. They worked coal seams in Llewellyn beds in the central valley and in the Pottsville formation up the sides of the valley. This segment of I-81 ends at Exit 45 on the south side of Wilkes-Barre, where the whole hillslope east of the highway was strip-mined.

Large cut near I-380 Exit 1 exposes white Pottsville conglomerate at the top that rests on an erosional unconformity over Mauch Chunk shales and sandstones. Pocono sandstones crop out beneath Mauch Chunk at the far end of the exposure.

I - 380
Scranton—Crescent Lake
Junction I-80
28 mi./45 km.

Between the I-81 intersection in Scranton and Exit 2, two miles, I-380 climbs quickly out of the city, following Roaring Brook Gap through Moosic Mountain. The city nestles in the trough of the Lackawanna syncline, which houses the Northern Anthracite Field. All rocks exposed in the trough belong to the middle to upper Pennsylvanian Llewellyn formation, the principal coal-producer of this part of the state.

A large cut near Exit 1 exposes some of the beds that lie stratigraphically below the Llewellyn formation and project up the side of the syncline, in part holding up Moosic Mountain. The cut contains three principal units. White conglomerate at the top is at the base of the lower Pennsylvanian Pottsville group. This rests on an erosional surface, an unconformity, on upper Mississippian Mauch Chunk beds, that consist mostly of colorfully-banded rusty to buff colored sandstone, along with some siltstone, and shale. The different

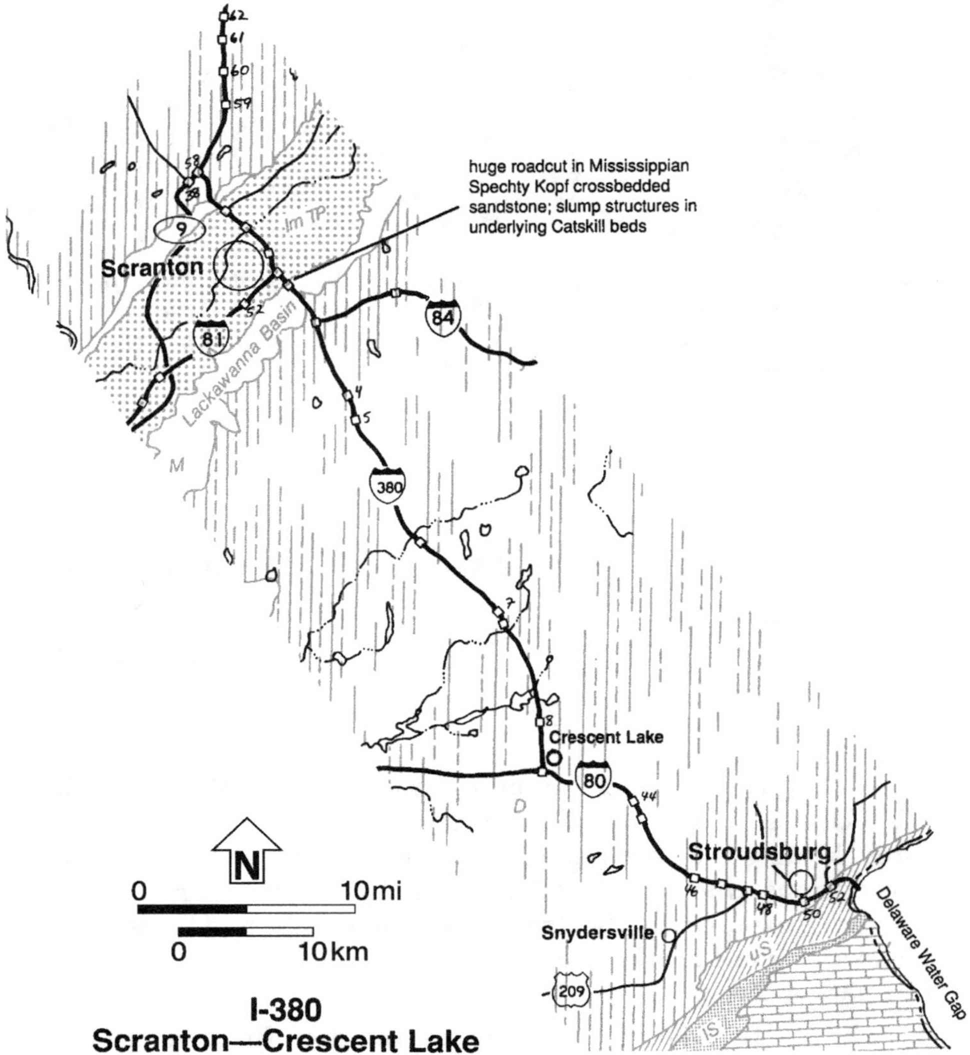
huge roadcut in Mississippian Spechty Kopf crossbedded sandstone; slump structures in underlying Catskill beds
Scranton
Lackawanna Basin
Crescent Lake
Stroudsburg
Snydersville
Delaware Water Gap
0
10mi
0
10km
N
380
81
84
80
209

I-380
Scranton—Crescent Lake

Huge cut in Spechty Kopf formation near Pennsylvania 435 interchange of I-380 exposes large-scale cross-beds in sandstone that supports Moosic Mountain. The base of the cut contains red-brown mudstone with considerable deformation resulting from slumping of the original unconsolidated muds.

rock types are easy to recognize because the sandstone is coherent and cliffy, whereas the shaley or silty beds break down to rubbly shelves. Underlying the Mauch Chunk formation in the lowest part of the cut are lower Mississippian Pocono sandstones. All of the beds dip gently northwest toward the axis of the syncline, and the dip steepens slightly toward the bottom of the outcrop.

An even larger and more striking cut appears a little farther east near the Pennsylvania 435 interchange. The entire exposure is Spechty Kopf formation, the lowest Mississippian unit in this part of the state. It lies unconformably below the Pocono formation. This is

Left end of same cut shown above, showing uniformly-dipping sandstone beds that overlie the cross-beds. These beds are on the upturned southeast limb of the Lackawanna syncline.

the rock that holds up the crest of Moosic Mountain. Most of the cliff consists of buff sandstone in well-defined beds that dip uniformly northwestward toward the synclinal axis at an angle of about 30 degrees. Near the southeastern end of the cut, planar beds give way to underlying giant crossbeds, which in turn give way to a transition zone in which sandstone and red-brown shale, or mudstone, are thinly interbedded. The base of the zone is sharply defined by a thin uniform sandstone bed, below which the mudstone is tightly deformed, apparently because it slumped while it was still soft mud. Although the rock sequence here is poorly understood, it appears to record a basin that filled with muds while the water was relatively deep, then with sands as it became shallower.

Cross-bedding

Cross-bedding is common in Pennsylvania's sandstones. As the name implies, it is inclined to normal bedding, which typically originates as nearly horizontal layers. Cross-beds form in such places as sand dunes or deltas. For example, dunes may build wherever a plentiful supply of sand blows in a more or less constant direction. Each dune tends to have a gentle windward slope and a steep lee slope. The wind blows sand onto the dune, and up the gentle windward slope. Then the sand goes over the dune crest and cascades down the lee side of the dune, coming to rest in sloping layers, cross-beds. If dunes build upon dunes, a thick horizontal layer of dune sand accumulates, which contains within it countless inclined cross-beds.

Delta cross-beds form in virtually the same manner, but are deposited by water instead of wind. Where a sediment-laden stream

Cross-bedding in upper Devonian Catskill sandstone by U.S. 6 near Meshoppen.

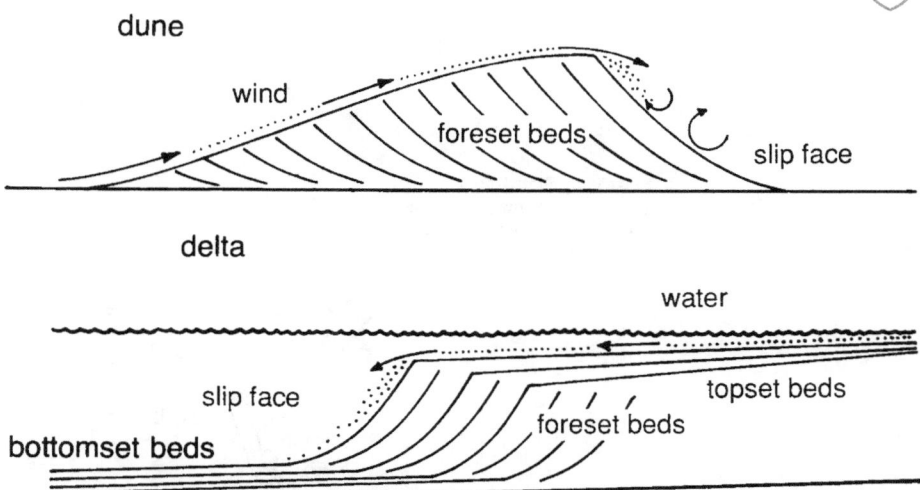

These schematic cross-sections show the two most common geologic environments that produce sedimentary cross-beds. The slip faces at the front of each are essentially landslide slopes.

flows into a lake, the stream suddenly slows, and drops its load. The sediment cascades down the relatively steep front slope of the delta to form steeply inclined cross-beds.

• Between the Pennsylvania 435 interchange and Crescent Lake, 26 miles, you see few roadcuts, all in upper Devonian Catskill formation. They expose flat or nearly flat reddish brown to greenish gray sandstone, siltstone, and shale.

Moosic Mountain approximately marks the boundary between the Valley and Ridge province and the Pocono Plateau. During the last glaciation, ice completely covered this section of the plateau and left abundant glacial features, including numerous kettle lakes — more than in any other part of Pennsylvania — as well as glacial erosional and depositional features of all kinds. None of the lakes is visible from I-380, but many bogs are — for example, near miles 16, 18, 19, and 22.

Near the I-80 intersection at Crescent Lake, an excellent view east profiles the steep east buttress of Camelback Mountain and, farther away, the deep cut of Delaware Water Gap in Kittatinny Mountain. The view spans the narrowest part of the Valley and Ridge province in Pennsylvania.

Camelback Mountain is really an erosional projection of the Allegheny Front. It is a ridge of resistant quartz-pebble conglomerates of the Catskill formation. Big Pocono Ski Area is on its north side; Big Pocono State Park at its summit is easily reached by an entry road that is open year round.

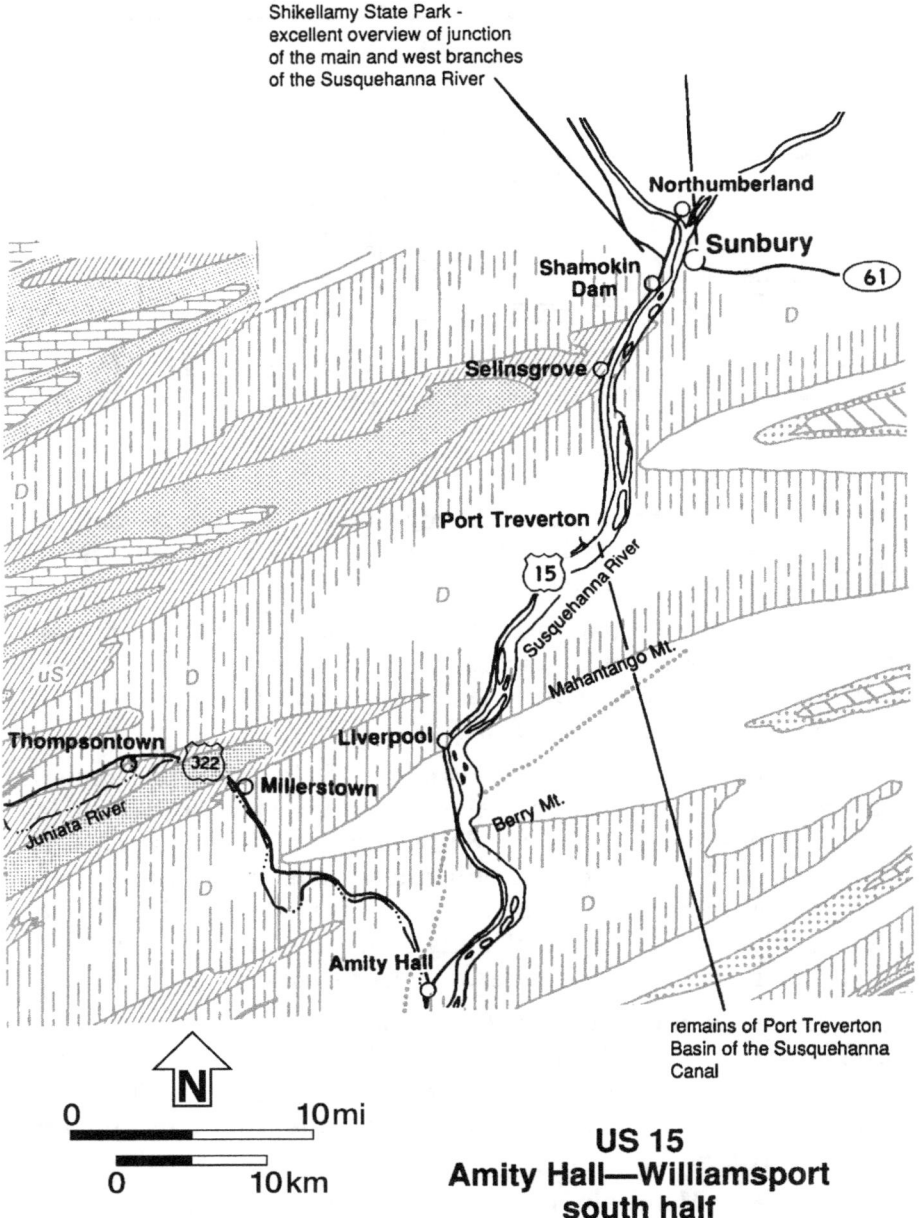

US 15
Amity Hall—Williamsport
south half

US 15
Amity Hall—Williamsport
67 mi./108 km.

The section of US 15 between Harrisburg and Williamsport follows the Susquehanna River in its remarkable traverse through the central part of Pennsylvania's Valley and Ridge province. Flowing generally southward, the river winds and weaves from one hard ridge to another across the structural grain of the bedrock. The principal supporting rocks of the ridges are the lower Silurian Tuscarora quartzite and lower Mississippian Pocono sandstone. These rocks form the backbones of the ridges that zig and zag on the limbs of alternating anticlinal and synclinal folds. Nearly all of the synclines form outcrop loops that point west, while all of the anticlines that separate them have their noses pointing east. The fold axes plunge east, as though the folds were raised at their western ends. Farther north, the west and main branches of the river avoid the hard ridges. From Liverpool south, the Susquehanna River cuts through them, creating some of the state's most striking water gaps.

Amity Hall lies at the northern tip of Duncan Island that separates the Juniata River from the now-abandoned Susquehanna Canal. The canal continues alongside US 15 for a mile and a half north of the village.

Between Amity Hall and Montgomery Ferry, eight miles, you cross a lowland eroded on upper Devonian Catskill and Trimmers Rock and middle Devonian Hamilton shales that offer little resistance to erosion. This section of highway is blessed with splendid views of the island-studded Susquehanna River, with even-crested Peters Mountain to the south, and Berry Mountain to the north. Watch for a spectacular large cut about four and a half miles north of Amity Hall that shows a textbook example of an anticline-syncline pair, a miniature version of the regional deformation. The rocks are brick red shales with thin to massive sandstone interbeds and minor green shales of the Irish Valley member of the Catskill formation.

In the four miles between Montgomery Ferry and Liverpool, you pass through the Berry Mountain and Mahantango water gaps. Berry Mountain continues west of the river for seven miles, makes a sharp turn, and returns to the river as Buffalo Mountain, having wrapped around the nose of a syncline. East of the river, Buffalo Mountain becomes Mahantango Mountain.

View down the Susquehanna at Port Treverton by U.S. 15.

Liverpool lies just inside the moraine that marks the southernmost advance of the earliest glaciation recorded in Pennsylvania. Liverpool and Dundore are 13 miles apart. The road follows the river across lowlands developed on arched Catskill, Trimmers Rock, and Hamilton shales. Six miles north of Liverpool, at the Mahantango Creek crossing, are well-preserved remains of a lock and aqueduct built in 1828-31 as part of the Pennsylvania Canal. Low ridges visible across the river from the villages of Mahantango and McKees Half Falls are ridges of middle Devonian Old Port sandstones that enclose the core of the anticline; the rocks of the core are weaker Silurian and Devonian shales and carbonates that eroded into a valley.

At Port Treverton, stone piers near the western bank of the river are remains of the Port Treverton Basin of the Susquehanna Canal. In the canal years of 1855-70, coal from mines 15 miles east of here crossed the river on the Treverton, Mahanoy, and Susquehanna Railroad bridge and was transshipped to canal boats at this point. The view south is unusually picturesque, encompassing many tree-studded islands and, in the distance, the first water gaps downstream.

At Dundore, two miles north of Port Treverton, you see red and greenish-gray sandstone and red shale of the Catskill formation. Just across the river is an 1100-foot high ridge capped by lower Mississippian Pocono sandstones at the nose of the west-pointing Centralia syncline, part of the Western Middle Anthracite Field.

In the ten miles between Dundore and Shamokin Dam, you cross lowland over weak shales in the core of another anticline. The main and west branches of the Susquehanna River join just two miles north of Shamokin Dam, at Northumberland, where the main branch completes a 40-mile reach along the southern flank of Mountour Ridge.

US 15 and 11 separate north of Shamokin Dam, and between there and Winfield, five miles, US 15 leaves the river to cross over a low ridge capped by Catskill sandstones. US 11 continues along the river, skirting a 400-foot high scarp at the end of the ridge, and then crosses the west branch of the Susquehanna River to Northumberland. Catskill formation exposures along the base of the scarp are excellent. Three-tenths of a mile south of the bridge, a road leads to Shikellamy State Park and Blue Hill picnic area on top of the ridge, where you see an excellent view of the river junction.

Two branches of the Susquehanna Canal, built in 1828-34, joined at a basin next to Northumberland. Coal boats came down the eastern branch from Nanticoke, near Wilkes-Barre; lumber boats came down the western branch from Williamsport. Both locked down to the river here and then re-entered the canal on the western bank to continue their southward journey to Baltimore, Philadelphia and Chesapeake Bay.

Along the four miles between Winfield and Lewisburg, you are in the shadow of Shamokin Mountain, which rises nearly 700 feet above the river. It is a ridge of Tuscarora quartzite tightly folded into the trough of a syncline. The same ridge continues across the river as Montour Ridge, which rises to about 1000 feet. This is an excellent example of a topographic high that corresponds to a structural low — the axis of a syncline.

Lewisburg, the site of Bucknell University, is a pleasant, small town noted for its Victorian and late Federal architecture. In the mid-1800s, it was a popular way-station for canal boatmen and land travelers alike.

The Montandon sand dunes along the eastern bank of the river opposite Lewisburg date from the last ice age. The ice halted several miles north of here, but meltwater flushed much of its sediment downstream to form outwash deposits. As the glacier retreated and the land rose, the river cut down into this sediment pile, leaving the well-defined terraces now visible west of the river. The lowered river level also drained the groundwater reservoir and left the terrace sediments high and dry. Wind, blowing across them in an easterly direction did the rest, piling the sand into dunes. The main field north, south, and east of the village of Montandon, is mostly farmland, and the dunes are not very conspicuous from the ground. They show up very well from the air.

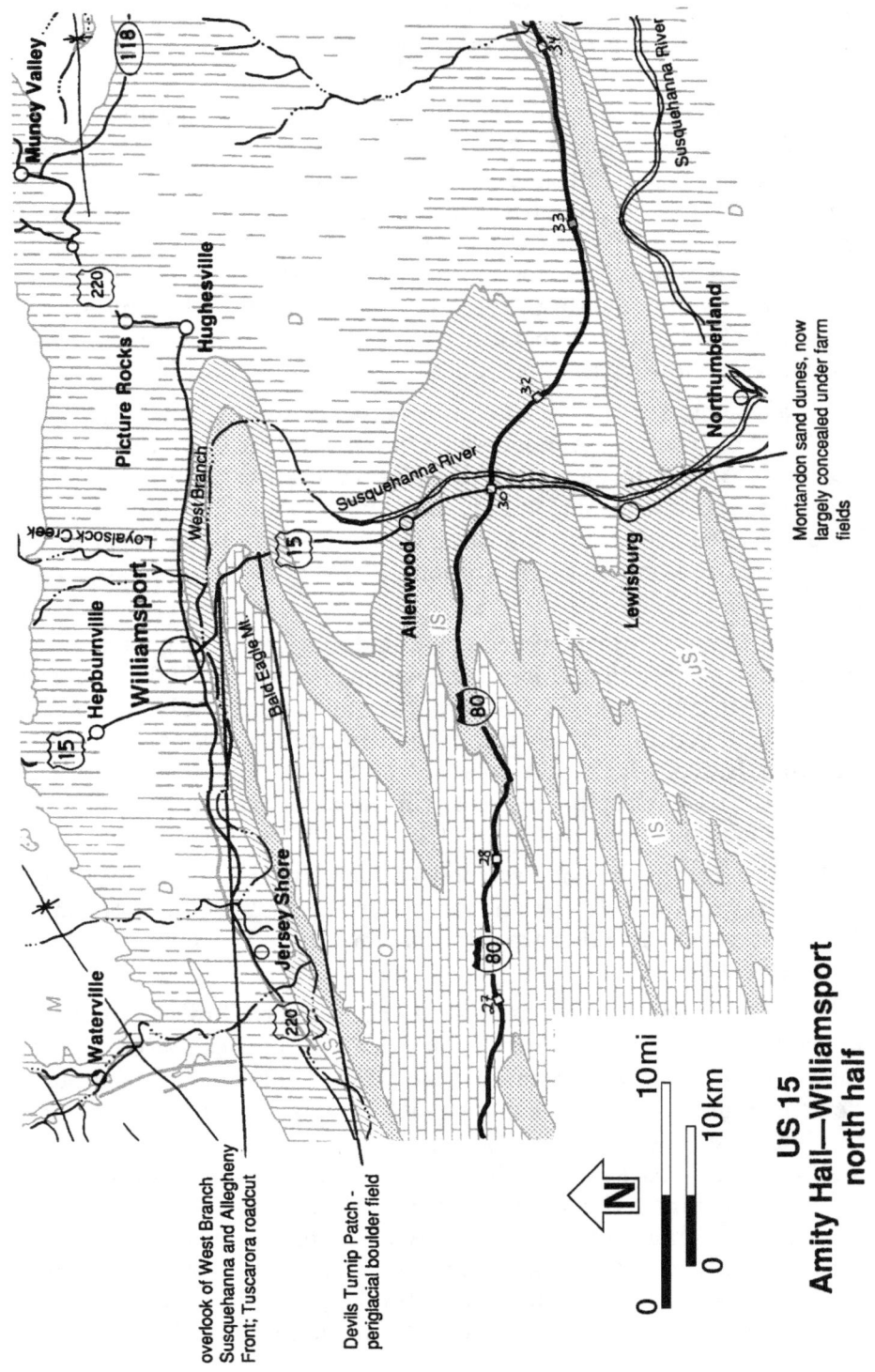

US 15
Amity Hall—Williamsport
north half

Between Lewisburg and Allenwood, 11 miles, you cross more lowland over Silurian shales with a few roadcuts in Bloomsburg and Mifflintown redbeds. To the west are five ridges of Tuscarora quartzite arranged like the toes of a giant right foot pointing east. The end of the big toe, called South White Deer Ridge, is beside the highway just south of Allenwood. A large cut there exposes gray shale of the lower Silurian Clinton group that overlie the Tuscarora quartzite.

Between Allenwood and the junction with Pennsylvania 54 near Montgomery, six miles, you cross another lowland over Silurian and Devonian shales. Bald Eagle Mountain directly to the north is the nose of the largest and most northwesterly anticlinal fold of the Valley and Ridge province, the Nittany Arch. Here it is capped by Tuscarora quartzite, but farther west, where it opens, its flanking ridges are double-crested; the second crest is a ridge of upper Ordovician Bald Eagle formation. The river makes a big loop around the nose of the fold, avoiding the hard rock and swinging six miles to the east. The highway between Pennsylvania 54 and Williamsport, 7.5 miles, goes north through a 700-foot deep wind gap in the ridge.

The pioneers referred to a boulder field by the road near the top of the pass as the Devil's Turnip Patch because some of the rocks are as purple as turnips. Most are reddish to white sandstones and conglomerates derived from the Tuscarora quartzite and transported to their present position during the last ice age. Several fields like this exist in the glaciated parts of Pennsylvania; all probably formed the same way.

Devil's Turnip Patch alongside U.S. 15 on Bald Eagle Mountain, five miles southeast of Williamsport. Boulders are lower Silurian Tuscarora sandstone and conglomerate from the caprock of the ridge.

Hard Tuscarora quartzite beds dipping north on the north slope of Bald Eagle Mountain by U.S. 15 overlook, 3.8 miles southeast of Williamsport.

A small roadside park north of the Turnip Patch has an interesting standing slab of Tuscarora quartzite with glacial polish and striations on one side, positive evidence of glacial transport. Boulders encased in moving glacial ice scrape against each other and the bedrock. This grinds the rocks into powder, called glacial flour, that serves as a polishing compound.

A scenic overlook about a mile north of the roadside park and four miles from Williamsport, offers an outstanding panoramic view of the west branch of the Susquehanna valley, Williamsport, and the Allegheny Front rising to the north beyond the town. On the opposite side of the road an excellent cut exposes well-bedded, Tuscarora quartzite dipping about 30 degrees north, the caprock of Bald Eagle Mountain.

US 209
Matamoras—Delaware Water Gap
35 mi./56 km.

This short route follows a peculiar segment of the Delaware River. At the northern end, the river flows southeast from the Allegheny Plateau, cutting across the northeast-trending structural grain of the bedrock, across the axes of the folds. At Matamoras, it makes an abrupt 90 degree turn to the southwest; then continues along the Port Jervis trough for approximately 25 miles to Wallpack Bend near Bushkill. This segment of valley is in the middle Devonian Mahantango formation, brownish shales and sandstones quite well exposed in numerous roadcuts. The river is hard against the northwest valley scarp, the Allegheny Front. To the southeast, the valley wall slopes more gently parallel to the upper Silurian Bloomsburg beds that overlie the Shawangunk formation. This resistant conglomerate supports the crest of Kittatinny Mountain; its continuation in New

Dingman Falls on middle Devonian Mahantango formation.

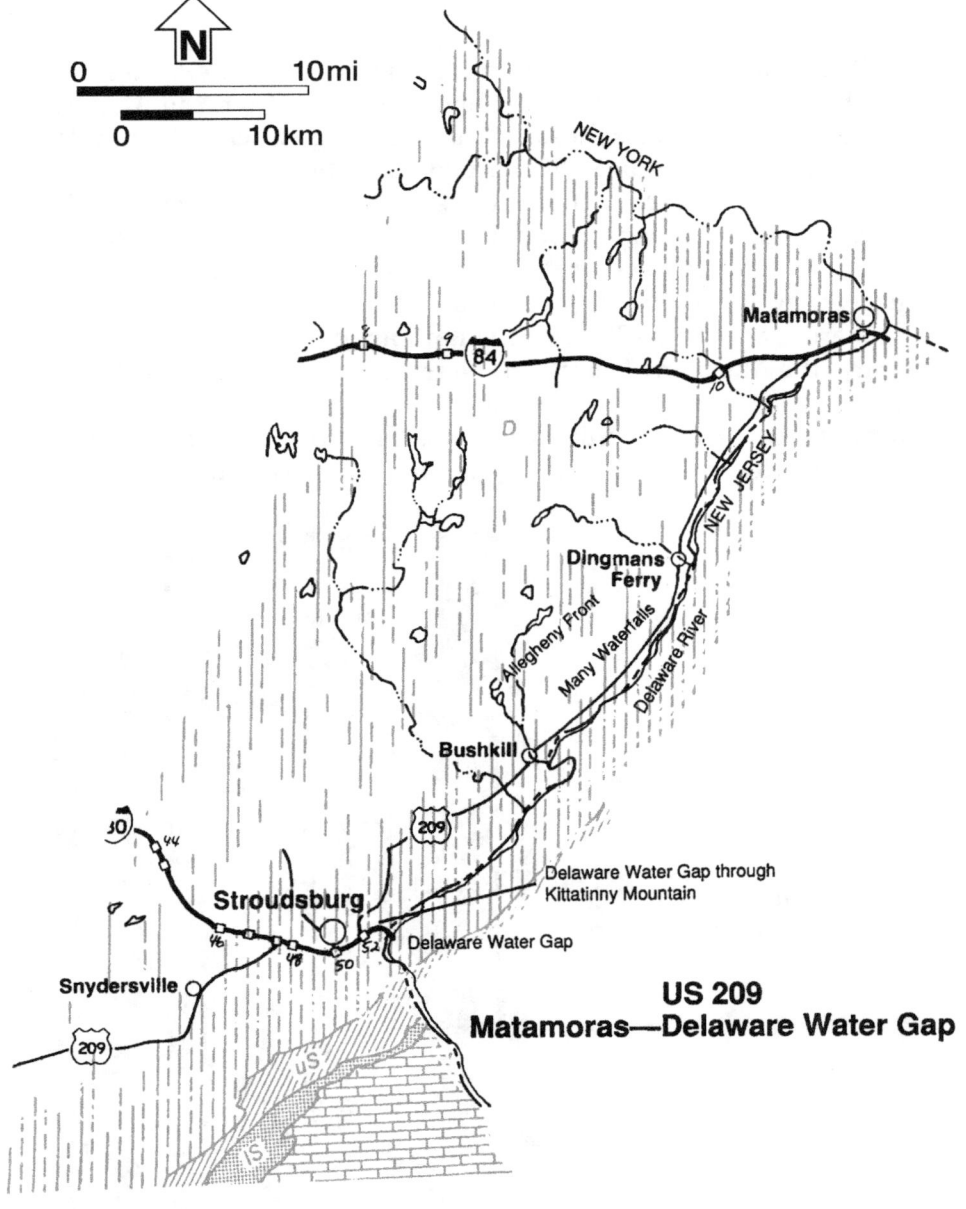

US 209
Matamoras—Delaware Water Gap

Silverthread Falls cut into flat Mahantango shales and sandstones along vertical joints.

York is referred to as the Shawangunk Mountains. Kittatinny ridge lies between three and seven miles distant, so you see it only in a few places.

Rivers eroding tilted sedimentary layers commonly bank against the scarp side of the valley. All of the sedimentary strata of the valley floor and sides dip northwest, though at variable angles. As the river cut down, it migrated down the dip of the more resistant beds, maintaining the steepness of the scarp face through continual undercutting at its base. Many beautiful waterfalls descend the Allegheny Front, mostly across resistant ledges of Mahantango sandstone beds. They include Pinchot or Sawkill falls near Milford; Factory, Fullmer, and Deep Leap falls three miles west of Dingmans Ferry; Dingmans and Silverthread falls near Dingmans Ferry; and Bushkill and Winona falls, near Bushkill.

At Wallpack Bend, the Delaware River makes a double hairpin turn to cross Wallpack Ridge, cutting through resistant Devonian strata. In so doing, it empties into the valley to the southeast, then

Bushkill Falls spill over resistant sandstone in the middle Devonian Mahantango formation on part of the Allegheny Front near U.S. 209.

continues southwest along it to the Delaware Water Gap. This lower reach of the river is narrowly confined between the Allegheny Front and Kittatinny Mountain, but US 209 does not follow it. Instead the road continues southwest along the Echo Lake lowland, also known as the upper Delaware valley. It is probably an abandoned river valley that stands about 150 feet above the modern Delaware River. The floor of the upper Delaware valley is abundantly veneered with glacial drift and full of kettle ponds.

At Delaware Water Gap, the river makes another willful right-angle turn to the southeast and cuts right through the extremely hard Shawangunk quartzite of Kittatinny Mountain.

To get a feel for the Delaware Water Gap and its history, you should first pass through it along Pennsylvania 611, where at one point you can look across the river at the great cliff of Shawangunk quartzite dipping about 50 degrees to the northwest. Then continue four miles farther to the village of Portland, where you can look back to the gap from the bridge over the river. From there, the Kittatinny crest is astonishingly even and the gap looks like a knife slash through it.

Erosion of steeply-dipping, tough Shawangunk sandstone and conglomerate beds produced the Kittatinny Mountain hogback, here magnificently cross-sectioned in the Delaware Water Gap. Note the extensive talus on the lower slopes that conceals the underlying upper Ordovician Martinsburg shaley beds.

According to some geologists, the even ridge crests may express a dynamic equilibrium between the resistance of the bedrock and the processes of erosion. The rocks that hold up the ridges all offer about the same resistance; thus they wear down to about the same level. Steep slopes develop on the hard rocks, gentle slopes on soft rocks. The difference in elevation between hard and soft rocks is established early in the degradation of the regional landscape and is maintained as erosion continues. That idea is an alternative to the theory that the even ridge crests are remnants of an old erosion surface.

The Delaware Water Gap cuts through a zone where the rock was weakened by a sharp fold. The flexed rock is largely eroded, but you can see evidence of its former existence. The Shawangunk beds match perfectly at river level; they crop out directly across from each other and dip at an angle of about 40 degrees to the northwest on both sides without any offset that might suggest a fault. The same beds at the top of Mt. Tammany, on the New Jersey side, dip 50 degrees to the northwest; those of Mt. Minsi, on the Pennsylvania side, dip only about 25 degrees northwest. Further, Mt. Tammany is offset 700 feet northwest of the trend of Mt. Minsi. Clearly the beds there have been shoved farther northwest and tilted up more steeply than those on Mt. Minsi. The intervening sharp flexure or fault weakened the brittle rocks, making them easily vulnerable to stream erosion.

Other water gaps in this region also follow zones of bedrock weakness. This suggests, but does not prove, that stream piracy caused them. Tributary streams draining the cliff on the southeastern side of Kittatinny Mountain would have an advantage over those that drain the hard rock of the gentler backslope, more energy because the slope is steeper. They cut into the cliff faster by undercutting the soft Martinsburg shales beneath the harder Shawangunk quartzite, causing its periodic collapse.

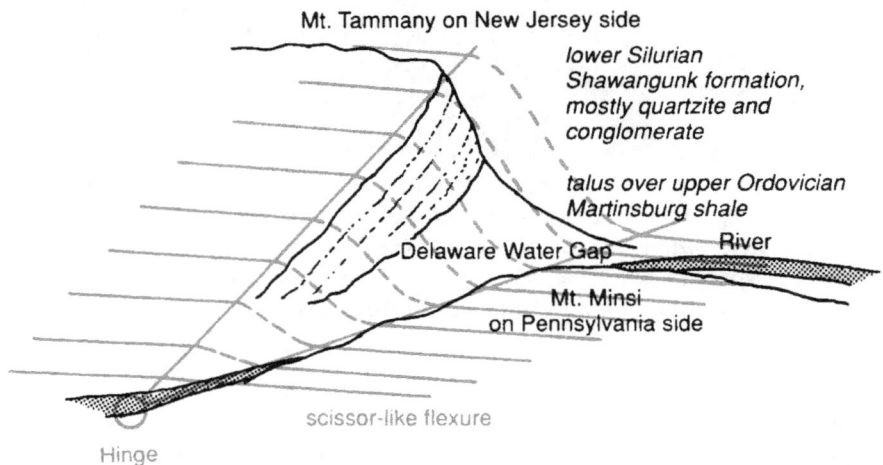

Sketch view looking northeastward across the Delaware Water Gap, illustrating a scissor-like flexure that may have crushed and weakened the brittle rocks, allowing the Delaware River to cut through the ridge. —After Epstein, J.B., 1966, Structural Control of Wind gaps and Water gaps in The Stroudsburg Area, Pennsylvania and New Jersey

Limestone quarry at Snydersville by U.S. 209. Crushed rock is used principally for asphalt aggregate.

US 209
Stroudsburg—Jim Thorpe
34 mi./55 km.

The route between Stroudsburg and Lehighton generally parallels the northeast-trend of the upper Devonian strata on the southern limb of the Lehighton anticline, which is also the northern limb of the Parryville syncline. Black shales in roadcuts along the four miles between Stroudsburg and Snydersville, belong to the middle Devonian Marcellus formation. They are deep-water sediments full of organic matter that were deposited in the Acadian foredeep basin just before the main collision between North America and Europe.

The large sand and gravel quarry on the north side of the road at Snydersville is in one of several glacial kames that dot the valley. They are shaped like mounds. The stone quarry in the side of the ridge almost directly across the highway from it produces crushed limestone, primarily for asphalt aggregate.

A few more stone and sand and gravel quarries lie near the five miles of highway between Snydersville and Brodheadsville. Cuts in this stretch also expose Marcellus black shales. Brodheadsville lies at

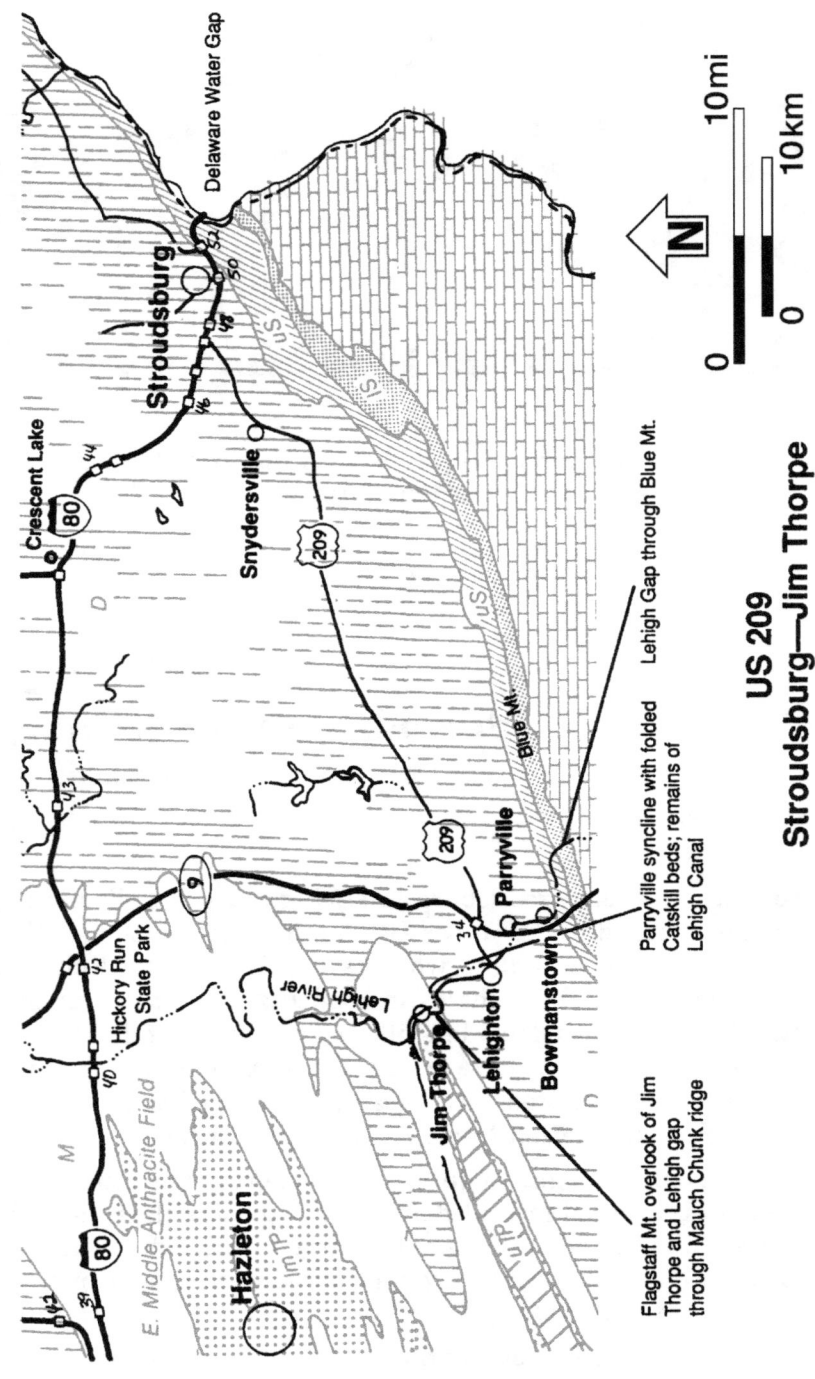

the bottom of a horseshoe-shaped segment of end moraine that crosses the valley between the Allegheny Front and Kittatinny Mountain. The horseshoe opens to the northeast, with its ends at Camelback Mountain and the Delaware Water Gap. The moraine outlines the margin of an ice lobe that filled the valley northeast of it during the last ice age. Radiocarbon dates show that organic material in glacial lake clays at the outer margin of the moraine just southwest of Brodheadsville is about 12,760 years old.

The four and one half miles of valley between Brodheadsville and Kresgeville is broad and flat because it is filled with glacial outwash and lake sediments. Note the large sand hill at Kresgeville.

Between Kresgeville and the Pennsylvania 9 intersection, ten miles, you cross diagonally over the low ridges of upper Devonian Trimmers Rock shales and siltstones on the northern limb of the Parryville syncline. At one high point along the winding road, look north to Lehigh Gap through Mauch Chunk ridge at the eastern tip of the Southern Anthracite syncline. Long, narrow Beltzville Lake, formed by damming of Pohopoco Creek, is just to the north over most of the way; the dam is at Beltzville, not far from Pennsylvania 9.

In the two miles between the Pennsylvania 9 intersection and Lehighton, you see dark middle Devonian Mahantango shales on the southern limb of the Lehighton anticline. Near the western end of the bridge, where the road turns north, are high cuts in Mahantango and Marcellus shales in the core of the fold.

Sketch of roadcut between Parryville and Bowmanstown along new route Pennsylvania 29, on the east side of the Lehigh River with vertical exaggeration. —Adapted from Frakes and others (1963, p. 25)

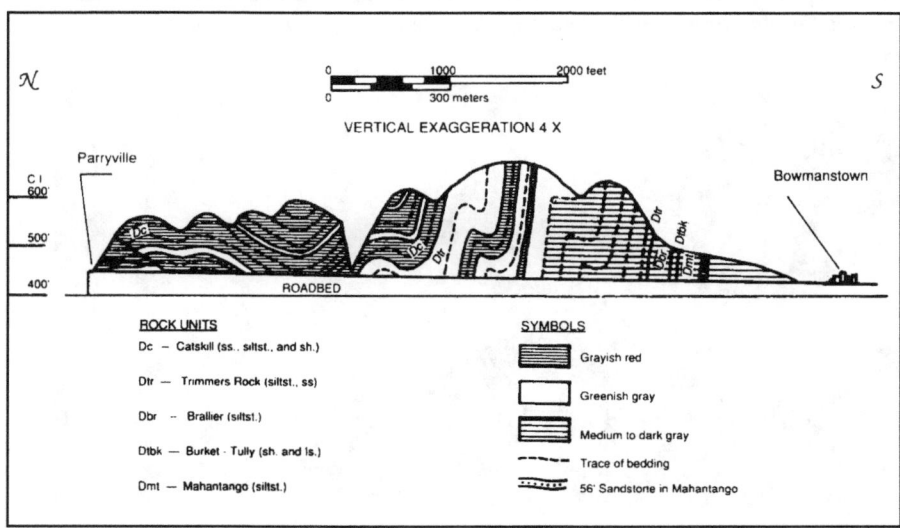

Folded upper Devonian Catskill beds in the Parryville syncline on Pennsylvania 248, showing strong axial plane cleavage dipping steeply right.

Parryville Syncline Cross-section

Watch for an unusual and colorful mile-long cut on Pennsylvania 248 between Parryville and Bowmanstown on the eastern side of the Lehigh River, with its northern end near the Pennsylvania 9 bridge. Two long roadcuts with a short gap between them provide a cross-section of the Parryville syncline. Beds at the northern end lie almost flat and at the southern end, dip steeply north making the syncline asymmetrical. In between are numerous smaller anticlines and synclines that add complexity to the larger fold.

Close-up of axial plane cleavage in Catskill beds of the Parryville syncline cut on Pennsylvania 248. Bedding dips gently left, and cleavage dips steeply right.

In places, as at the southern end of the nearest Parryville cut, a closely-spaced, pervasive rock cleavage cuts obliquely across the sedimentary layers. Each cleavage plane is a surface along which the rock on one side shifted slightly relative to that of the other side. Similar rock cleavage forms the extremely flat surfaces of slate, a low-grade metamorphic rock. Its natural function in rock deformation is really the same as the flexing of the beds into folds — to accommodate the lateral, sidewise compression of the layers. In some cases, the cleavage planes are so closely spaced and movement along them is so minute, that the bedding deformation caused by it appears smooth, as does the distortion of any fossils the rocks may contain. You can see such a case at the southern end of the nearest Bowmanstown cut, where brachiopod shells in Mahantango shales are distorted by cross-cutting cleavage.

• The four miles of road between the Lehighton bridge and Jim Thorpe, follow the river through the winding Lehigh gap in Mauch Chunk Ridge, past progressively younger Catskill beds in almost continuous outcrop on the northern limb of the Lehighton anticline. On the long, straight decline to Jim Thorpe, north of the ridge-end, the road parallels nearly vertical Catskill beds.

Abandoned Lehigh Canal lock near Parryville.

Jim Thorpe and Lehigh River looking north from Flagstaff Mountain. Bear Mountain is at lower right, and Mt. Pisgah is the sunlit ridge behind the west bank section of Jim Thorpe. The Lehigh Canal came through the floodplain near the railroad tracks at lower right.

Flagstaff Mountain Overlook

You can get to this overlook, 800 feet above the Lehigh River on Mauch Chunk Ridge in a ten-minute drive from Jim Thorpe. The ridge is held up by Catskill beds which, at the overlook, are quartz pebble conglomerates. The village of Jim Thorpe, at the northern gate of the gap, was formerly called Mauch Chunk, but was renamed in the 1950s for the famous 1912 Olympic athlete. On the left horizon just beyond the village is Mt. Pisgah, a ridge of Pennsylvanian Pottsville beds at the eastern tip of the Southern Anthracite syncline. Slightly to the southwest, the mountain opens to two even-topped ridges, with Mt. Pisgah on the southern limb and Nesquehoning Mountain on the northern limb of the fold. At the core of the structure are Pennsylvanian Llewellyn beds, so pock-marked with mines that they look as if giant termites have been at work. Mt. Pisgah is the site of the famous Switchback Railroad, America's oldest railroad, laid out in 1818 to carry coal from the mines to the Lehigh River for shipment.

Outcrops visible along the present Lehigh Valley railroad at the sharp bend in the river are mainly Catskill formation. Watch for the remains of the old Lehigh Canal alongside the railroad.

US 220
Williamsport—Port Matilda
72 mi./116 km.

Williamsport, on the west branch of the Susquehanna River, has been a trade and travel hub for more than a century. Now a manufacturing center, the city was mainly known for lumbering until the 1890s, when the forests were depleted. From 1846-89, a six-mile long boom built on piers in the Susquehanna River was used to store logs during Spring drives down the river. An 1889 flood badly damaged the boom and led to its eventual abandonment. Today artifical levees on either side of the river provide flood protection. During the canal era, logs were barged downriver from here to Baltimore and Philadelphia. Victorian mansions once owned by the lumber barons line West 4th Street.

In the 24 miles between Williamsport and Lock Haven, you follow the Susquehanna Valley. The river descends the Allegheny Front to Lock Haven, then turns northeast and continues past Williamsport to Muncy, where it turns south to cross the structural grain of the Valley and Ridge province to Harrisburg. Within the valley it meanders widely in a fairly broad flood plain between Bald Eagle Mountain and the Allegheny Front. Soft rocks, mostly Silurian and Devonian shales, underlie the valley.

This attractive section of highway is open, with especially good views of the Allegheny Front and Bald Eagle ridge. The lower slopes of the Front rise in a series of subtle steps developed over the more resistant sandstone interbeds of the upper Devonian Lock Haven and Catskill formations that dip into the slope. These culminate in the relatively steep scarp held up by lower Mississippian Burgoon sandstone with the Huntley Mountain formation at its base. Younger Mississippian and Pennsylvanian beds crop out beyond the rim. In general, the dip of the strata along the Allegheny Front steepens toward Bald Eagle Mountain, thus grading from the gentle Allegheny Plateau structures to the more intense Valley and Ridge folding. The upturned beds of this part of the Allegheny Front are the eroded stumps of a great arch called the Nittany anticline.

If you could reconstruct them, the Mississippian beds now exposed at the plateau rim would continue curving upwards to the crest of the Nittany anticline a few miles southeast of you and more than 10,000 feet overhead. From there, they would curve back down on the

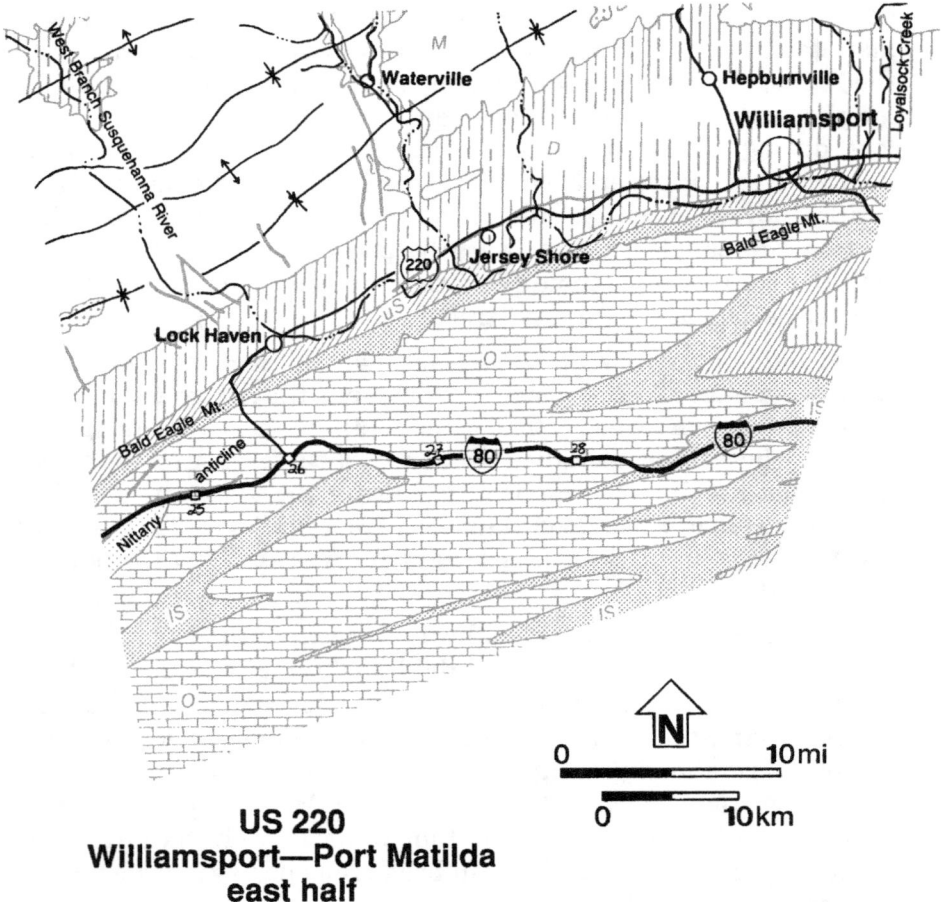

**US 220
Williamsport—Port Matilda
east half**

opposite limb of the fold. All of that enormous mass of rock is now gone, presumably lost to erosion.

The Nittany anticline is more than 100 miles long; it stretches from Muncy, east of Williamsport, to Hollidaysburg, with a maximum width of about nine miles. Beds are arched at the Allegheny Front along its entire course across Pennsylvania. The arch formed as a rock package flexed as it rode over a sole thrust fault. In this general region, the sole thrust slices through weak Cambrian beds about 20,000 feet beneath the Valley and Ridge province. Near the Allegheny Front, the thrust breaks free of the Cambrian beds, rises steeply to less than 10,000 feet, then levels off in weak Silurian shales beneath the plateau. The riding strata are folded into a syncline at the ramp base, and an anticline at its top, with another, plateau-style broad syncline beyond.

Near Lock Haven, where the highway is on the south side of the river, watch for an abandoned flagstone quarry in the middle Devonian Old Port formation. Layers of rock in the quarry illustrate the steepness of the bedding dip on this side of Bald Eagle ridge.

Between Lock Haven and the I-80 junction, eight miles, you pass through Bald Eagle Mountain along Fishing Creek Gap. In the gap look for pale talus derived from the Tuscarora quartzite caprock. This is so commonplace near the crests of ridges of Tuscarora quartzite that you can, with considerable confidence, identify the formation from afar.

Hypothetical reconstruction of the eroded Nittany anticline places Mississippian rocks of the Allegheny Front thousands of feet above Nittany Valley. Note that the anticline is where the sole thrust ramps up and then levels off beneath the Allegheny Plateau. —Adapted from Berg and others (1980, Plate 2)

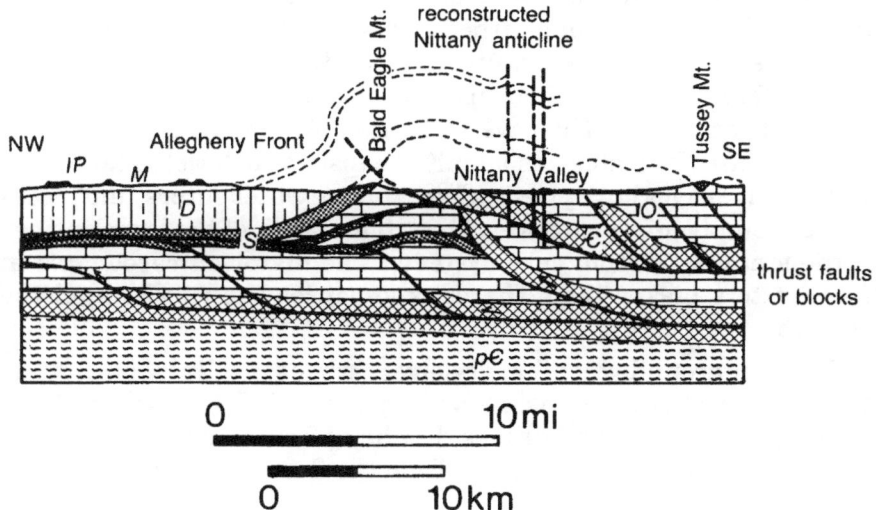

Steeply-dipping middle Devonian Old Port formation flagstone on the northern slope of Bald Eagle Mountain near Lock Haven.

Nittany Valley is underlain by Ordovician limestones. This is a region of many caves, sinkholes, and springs where streams sink into the bedrock and underground water resurfaces. Big Spring, on private land at Salona, produces water from flat-lying beds of the Bellefonte dolostone at a rate in excess of 5000 gallons per minute.

Caves

Caves are common in many parts of Pennsylvania where the dominant bedrock is carbonate, meaning limestone, dolostone, or their metamorphic equivalent, marble. Slightly acid groundwater passing through such rocks along bedding planes or fractures dissolves the carbonate. Rain absorbs carbon dioxide from the atmosphere to produce carbonic acid, like carbonated water, but much weaker. For large caves to form, the slightly acid groundwater must pass through the rock for hundreds or thousands of years, continually carrying the dissolved rock away. The cave will cease to grow when circulation stops because the water has a limited capacity to dissolve carbonate rocks. Active caves have places where surface water enters, and places where groundwater resurfaces. Most, but not all, of the large springs in the state issue from cave systems. The dissolved carbonates in these waters makes them "hard," and causes the ring around the bathtub.

The same groundwater that dissolves carbonate will also precipitate it under the right conditions, depositing new rock. It forms the weirdly beautiful, stalactites and drapes that hang from many cave ceilings, stalagmites that rise like stumps from the floors, columns composed

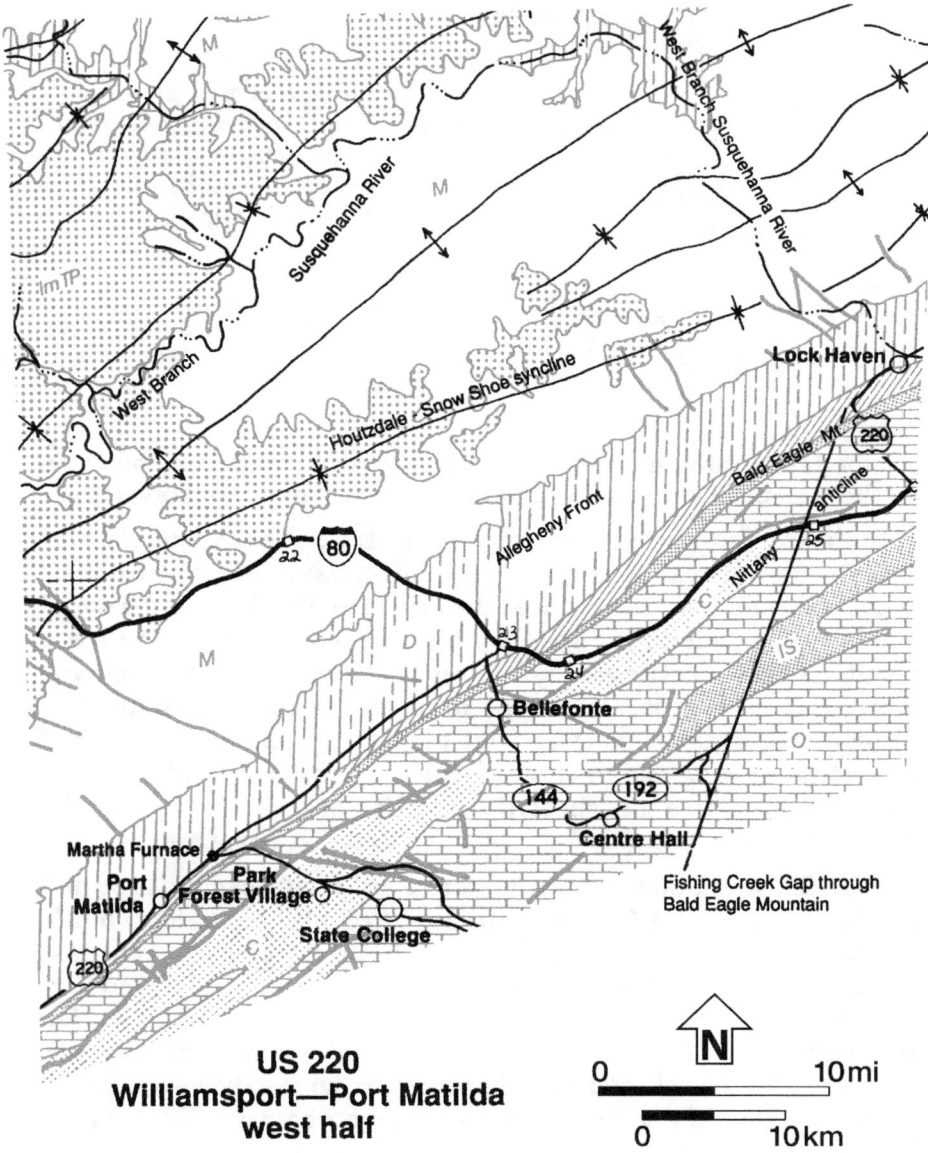

US 220
Williamsport—Port Matilda
west half

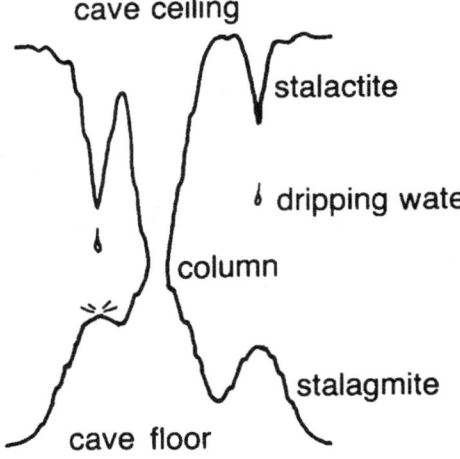

Formation of dripstone in caves. Each drop releases a small amount of calcium carbonate that builds up in time to the forms shown.

of joined stalactites and stalagmites, and rounded masses of flowstone on the walls. All such deposits of calcium carbonate are travertine.

Stalactites grow where water drips from the ceiling. A drop of water hanging from the cave ceiling loses part of its dissolved carbon dioxide to the atmosphere. That reduces the acidity in the drop, and a small amount of carbonate precipitates. Evaporation produces the same result. As the stalactite grows, the drops descend to its point and hang there. And so it goes. Below, the drops hit the floor and splatter, accomplishing the same thing, but producing stumpier deposits of travertine. Flowstone forms where water issues from wall fractures and spreads out in a thin sheet, thus offering a larger surface area for evaporation or carbon dioxide exchange to the atmosphere.

• US 220 and I-80 run together for 19 miles between Exits 26 and 23. This section is described under I-80: Brookville—Bellefonte and I-80: Bellefonte—Bloomsburg. Between Exits 24 and 23, you pass through Bald Eagle ridge along Curtin Gap. Between Exit 23 and Port Matilda, you again follow Bald Eagle Creek in the narrow transition zone between the Allegheny Front and the Valley and Ridge province. The road and the railroad, like the river, are where they arc because the rocks make it the easiest way to go. This route between Williamsport and the Maryland border demonstrates, perhaps better than any other, the close relationship between bedrock and human activity, even in the modern age.

Patches of pale Tuscarora sandstone talus appear again near the crest of Bald Eagle ridge. Near Port Matilda you see brown shales of the middle Devonian Hamilton group that lie above the Tuscarora quartzite.

US 220
Port Matilda—Maryland Border
86 mi./139 km.

The 15 miles between Port Matilda and Tyrone follow Bald Eagle Creek valley, where a few roadcuts expose brown shales of the middle Devonian Hamilton group. The lower slopes of the Allegheny Front rise immediately to the west, in a series of indistinct steps developed on the more resistant sandstone interbeds of the upper Devonian Lock Haven and Catskill formations. The plateau rim is marked by a steep scarp, consistently about four miles from the road, held up by lower Mississippian Burgoon sandstone. The Rockwell formation separates the Catskill and Burgoon formations. All of the beds dip gently northwest into the slope. Numerous creeks that flow southeast crease the scarp along the lines of vertical fractures.

Cross-fold Joints

A large area near the eastern edge of the plateau, several miles from the highway, is chopped by a swarm of the northwest-trending fractures like those that cut the Allegheny Front near Tyrone. They

The divergent northwestward transport of Valley and Ridge rocks created a zone of northeast-southwest extension, leading to the development of the "pocket" of cross-fold joints. —Adapted from Faill (1981, p. 9)

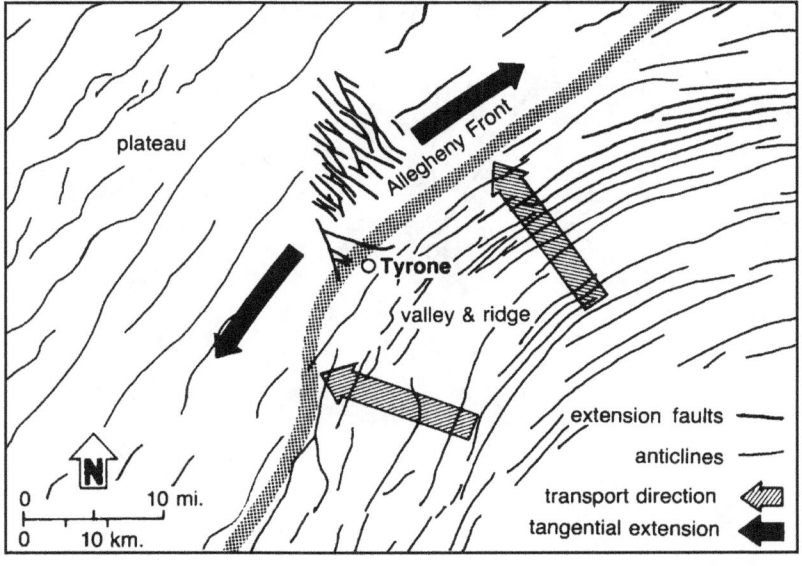

US 220
Port Matilda—Maryland Border
north half

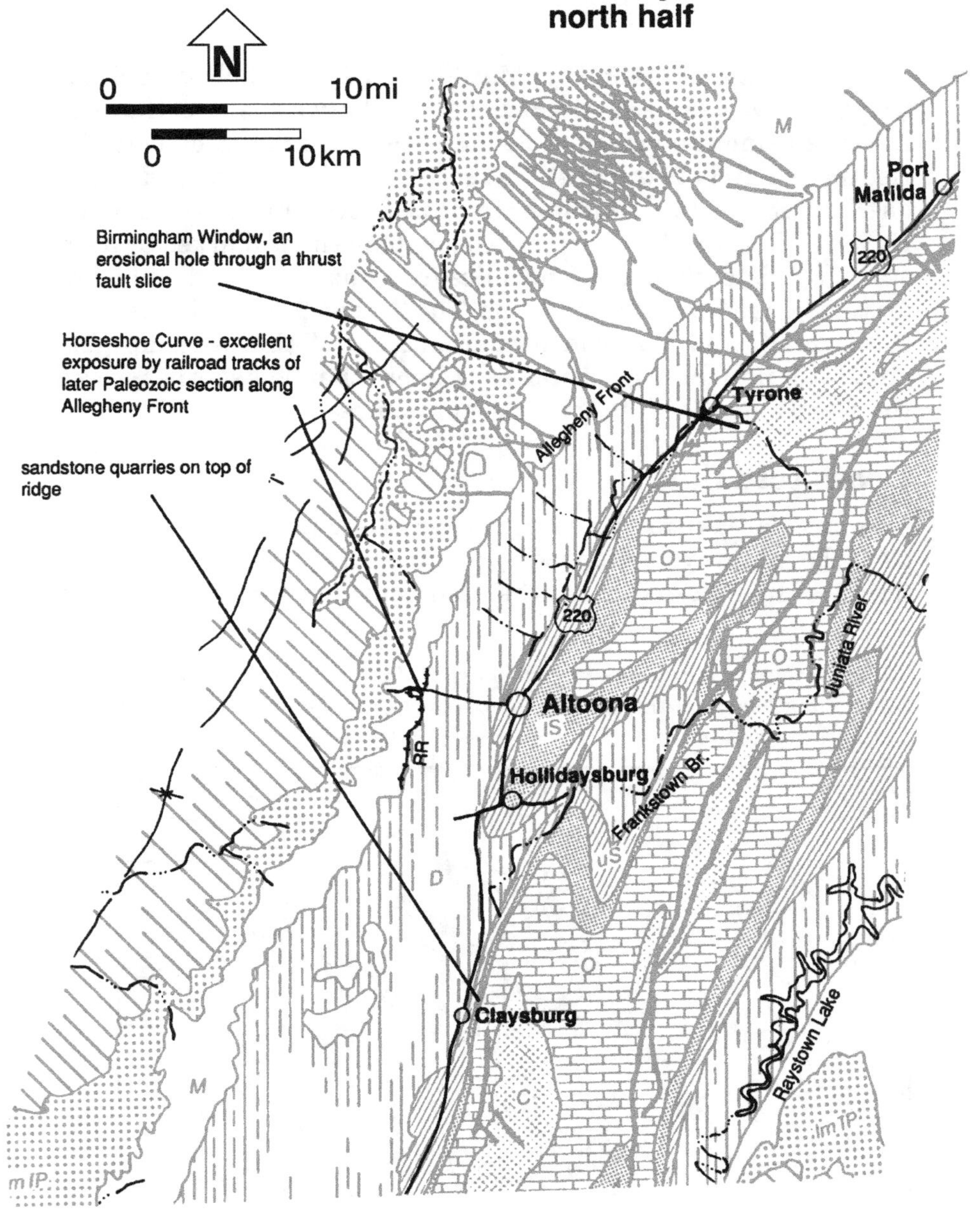
N
0
10mi
0
10km
Birmingham Window, an erosional hole through a thrust fault slice
Horseshoe Curve - excellent exposure by railroad tracks of later Paleozoic section along Allegheny Front
sandstone quarries on top of ridge
M
Port Matilda
220
D
Tyrone
Allegheny Front
O
220
Juniata River
Altoona
lS
RR
Hollidaysburg
Frankstown Br.
uS
D
O
Claysburg
Raystown Lake
M
C
lmTP
mIP

are cross-fold joints that trend across the Allegheny Plateau and Valley and Ridge folds. Such fractures occur in much of the Allegheny Plateau in Pennsylvania, New York, Ohio, and elsewhere, most conspicuously in shale, sandstone, and coal. They appear to form as the Earth's crust stretches perpendicular to the compression that produced the Alleghenian folding and thrusting. Northwest compression created northeast extension that pulled the rock apart along a set of northwest-trending fractures. In general, compression in one direction is mechanically equivalent to extension in the perpendicular direction.

• Southeast of the highway you see the remarkably even crest of Bald Eagle Mountain in the core of the great Nittany anticline. The Little Juniata River cuts through it in Tyrone Gap.

The Birmingham window, an interesting geologic feature on the other side of the gap, reveals something of the subsurface nature of the Nittany anticline. Erosion has cut through Cambrian rocks, exposing Ordovician rocks beneath them in the core of the fold — young rocks beneath old rocks, the reverse of the normal arrangement. The boundary between the window and encircling Cambrian strata is a single thrust fault. Thus, it appears that the anticline is within a thrust fault slice, and the window eroded through it exposes the rocks over which the slice moved.

The Birmingham window reveals structures typical of the whole Valley and Ridge province. The entire mass of folded and faulted rocks in this region rides on gently inclined sole thrusts, which moved it generally westward during the Alleghenian orogeny. The Cambrian and older strata beneath the sole thrusts are undeformed. Above them, the strata are cut by steeply inclined reverse faults that branch from the sole thrust and ascend toward the surface. Some break the surface where you can see them, pushing older rocks over younger. In all cases, the rocks arch into anticlines over these branching faults.

South of Tyrone Gap, Bald Eagle ridge becomes Brush Mountain. The ridge, still capped with Tuscarora quartzite, continues to Altoona, then makes a sharp bend, forming the south-pointing nose of the Nittany anticline. It then goes 14 miles northeast, makes another sharp bend and goes 20 miles back in a south-southwest direction as Lock Mountain to the area of Roaring Spring. It makes another bend there and goes north as Loop Mountain for five miles to Hollidaysburg, where it makes a final bend and continues south as Short Mountain. Thus, the continuous ridge almost completely encloses the Frankstown Valley. The structural configuration is a syncline, the center of which is floored with younger strata that lie above the ridge-capping Tuscarora quartzite. The younger strata crop out in a bullseye pattern, ranging from the lower Silurian Clinton group immediately above the Tuscarora quartzite to late Devonian Brallier and Harrell formations in the center of the valley. The distinctive topographic expression of the syncline, and of all the other folds in the Valley and Ridge province, is the result of erosion. Fold limbs stand high where they are composed of resistant rock.

Between Tyrone and Hollidaysburg, 20 miles, the road continues alongside Brush Mountain, crossing the Little Juniata River three times. Part of this highway is on the lower slopes of the ridge, providing good views of the Altoona Valley and the Allegheny Front. This section of Front is deeply creased by numerous small creeks that descend almost straight downslope to river. Several creeks have been dammed to supply Altoona and Hollidaysburg with water.

Horseshoe Curve

The Horseshoe Curve is a famous segment of railroad on the slopes of the Allegheny Front west of Altoona. It is well-known to geologists because rock exposures there provide a splendid display of later Paleozoic strata. Discontinuous outcrops along nine miles of line, between the eastern portal of the Gallitzin tunnel and the outskirts of Altoona, span a stratigraphic section about 7000 feet thick. It includes, from west to east, Pennsylvanian Allegheny and Pottsville groups; Mississippian Mauch Chunk, Loyalhanna, and Pocono formations; Devonian-Mississippian Rockwell formation; and upper Devonian Catskill and Lock Haven formations. The entire section spans a time interval of about 70 million years.

The base of the section lies where the railroad turns northeast and continues along the trend of the Lock Haven beds through Altoona. The whole section dips gently into the scarp and beneath the coal fields of a broadly synclinal region about 15 miles wide parallel to the Allegheny Front. Farther west, the Mississippian beds seen at Horseshoe Curve resurface on Laurel Hill anticline, on the other side of the syncline, and Catskill beds of the fold core crop out in a gorge cut through it by the Conemaugh River.

From west to east along the Horseshoe Curve section, the strata rise to the surface in gradually steepening dips that signal the transition from the mildly folded and faulted beds of the eastern Allegheny Plateau to the strongly folded and faulted beds of the Valley and Ridge province.

All of the upper Devonian rocks of the Horseshoe Curve section are part of the Catskill Delta complex. This great, southeastward-

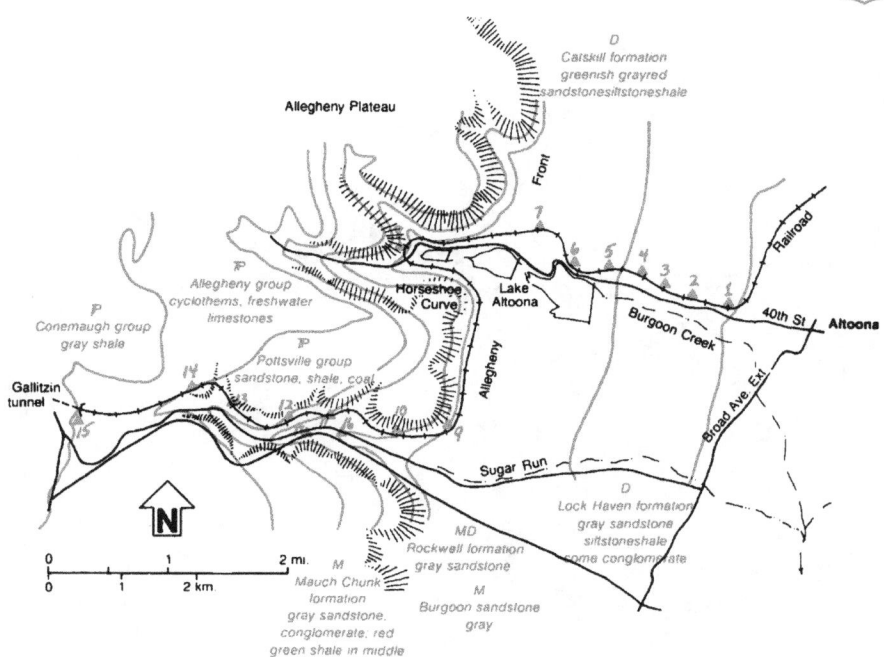

Geologic sketch map of the Allegheny Front west of Altoona, showing principal outcrops along and near the famous Horseshoe Curve of the Pennsylvania Central Railroad. —Adapted from Inners (1987)

thickening, wedge of continental sediments was laid down as the Acadian Mountains eroded during and after collision of North America with Europe.

The Rockwell formation contains the time boundary between the Devonian and Mississippian periods, about 365 million years ago. Its gray sandstones, some siltstones, and shales were deposited in a near shore and coastal plain environment after the Catskill Delta had ceased to expand northwestward. It records the transition from the dominantly terrestrial conditions of the Catskill Delta to the shallow marine conditions of the sandstones and conglomerates of the Burgoon formation, the strongly cross-bedded limy sandstone of the Loyalhanna formation and the red and green siltstones and shales of the lower part of the Mauch Chunk formation. The upper part of the Mauch Chunk formation was deposited as a new delta began to form against the rising Alleghenian Mountains while North America and Africa were closing during Pennsylvanian time.

• Hollidaysburg lies at the southern nose of the Nittany anticline, the end of the Frankstown Valley. Just east of the US 22 junction is a large cut in limestones of the upper Silurian Keyser and Tonoloway

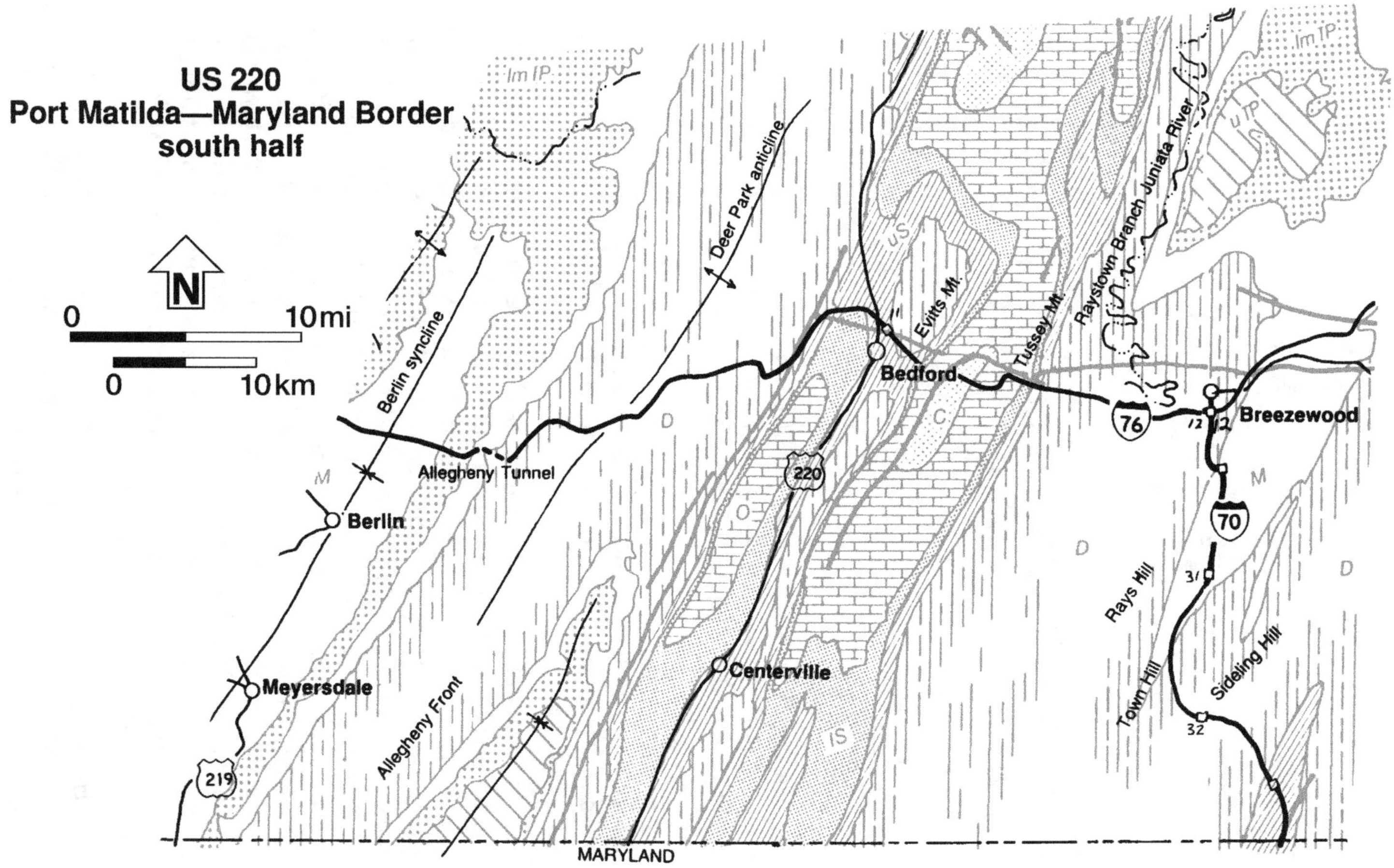
US 220
Port Matilda—Maryland Border
south half
N
0
10mi
0
10km
lmIP
uIP
Berlin syncline
Deer Park anticline
Allegheny Tunnel
Allegheny Front
Berlin
Meyersdale
219
M
D
O
US
C
IS
Evitts Mt.
Tussey Mt.
Raystown Branch Juniata River
Bedford
220
Centerville
76
70
Breezewood
12
31
32
Rays Hill
Town Hill
Sideling Hill
MARYLAND

formations that wrap around the nose of the anticline. They are shallow marine limestones deposited on the continental shelf that developed in the tectonically quiet interval between the Taconian and Acadian orogenies.

Between Hollidaysburg and Claysburg, nine miles, you follow the Frankstown branch of the Juniata River alongside Short and Dunning mountains. These are parts of the same ridge separated by McKee Gap at East Freedom. Numerous, mostly inactive, quarries follow the crest of Dunning Mountain along the four miles between East Freedom and Claysburg. They are visible from the highway as white scars and piles of waste rock. These quarries produced Tuscarora quartzite used to make silica brick. Conveyors transported the stone downslope to a mill and kilns at Claysburg. Active quarries at the base of the ridge work high-calcium limestones used in making lime from the middle Devonian Oriskany formation. A road to the ridge-top quarries leads upslope from Sproul, two miles south of Claysburg.

You cross the lower slope of Dunning Mountain in the nine miles between Sproul and St. Clairsville, with splendid views over the valley to the Allegheny Front, about ten miles away. Much of this section of highway crosses upper Silurian Keyser and Tonoloway formations, but they are poorly exposed.

Between St. Clairsville and Bedford, nine miles, you pass into Bedford Valley between the southern end of Dunning Mountain and the northern end of Wills Mountain. Dunning and Wills mountains are aligned and appear to be part of the same anticlinal fold. The ridges end here because a saddle in the fold axis causes the Tuscarora caprock to dip and come back up again. A low pass has developed in the soft rocks that overlie the Tuscarora quartzite. Hamilton shales floor the center of Bedford Valley.

The road for 25 miles between Bedford and the Maryland border follows a rather narrow synclinal valley crowded between Wills and Evitts mountains, both high and imposing barriers supported by Tuscarora quartzite. Wills Mountain reaches the highest elevation of the two at Hyndman Peak, 1600 feet above the valley at Centerville. The highway goes up and down, affording many fine views of rolling farmland and neat villages. The youngest rocks in the synclinal core are dark Hamilton shales, but most of the roadway crosses Silurian shales and limestones on the flank of Wills Mountain.

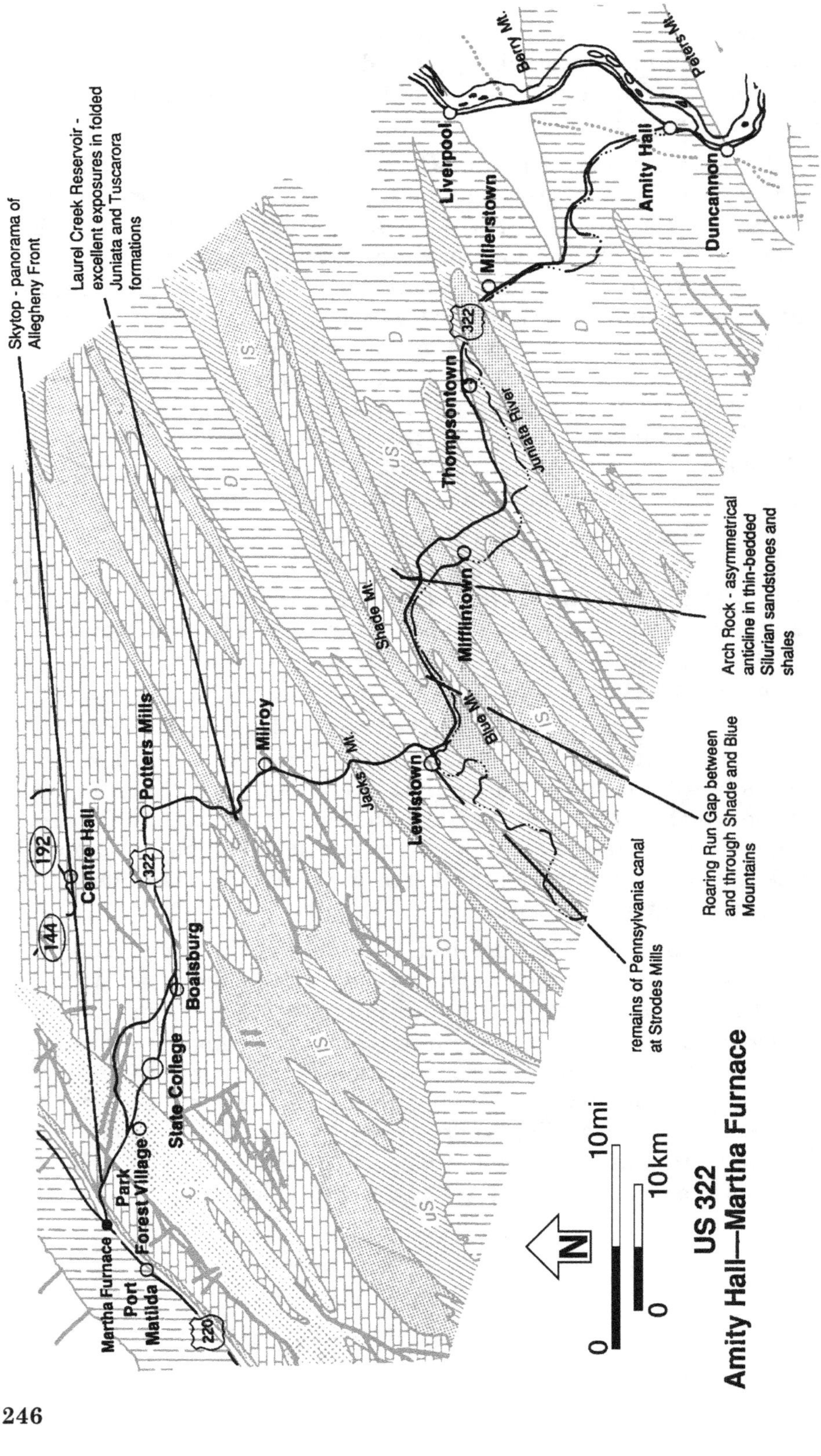
Skytop - panorama of Allegheny Front
Laurel Creek Reservoir - excellent exposures in folded Juniata and Tuscarora formations
Liverpool
Berry Mt.
Peters Mt.
Millerstown
Amity Hall
Duncannon
Thompsontown
322
Juniata River
Mifflintown
Shade Mt.
Arch Rock - asymmetrical anticline in thin-bedded Silurian sandstones and shales
Blue Mt.
Potters Mills
Milroy
Jacks Mt.
Lewistown
Roaring Run Gap between and through Shade and Blue Mountains
remains of Pennsylvania canal at Strodes Mills
192
Centre Hall
144
322
Boalsburg
State College
Park Forest Village
Port Matilda
Martha Furnace
220
0 10mi
0 10km
N
US 322
Amity Hall—Martha Furnace

US 322
Amity Hall—Martha Furnace
76 mi./123 km.

This route spans the widest and structurally most complex part of Pennsylvania's Valley and Ridge province. Numerous roadcuts, some very large, offer spectacular displays of folded and faulted rocks that range in age from Cambrian to upper Devonian. The rocks are part of a thick package of sedimentary strata overlying a sole thrust that transported them northwest during Alleghenian mountain building.

The sole thrust follows weak Cambrian beds as it climbs gently and uniformly from nearly 35,000 feet below Harrisburg to about 20,000 feet below State College near the Allegheny Front. Rocks below the thrust are undeformed. Rocks above it are faulted and folded into wave after wave of anticlines that are above faults that branch from the sole thrust and ascend steeply. The rocks you now see were deep below the surface when they were deformed. Thousands of feet of overburden have since eroded off them.

Amity Hall is at the junction of the Juniata and Susquehanna rivers, just north of the Susquehanna gap through Peters Mountain. During the last century two branches of the Pennsylvania Canal system terminated at Duncan Island just south of town. One came down the Juniata River, the other followed the Susquehanna River.

Between Amity Hall and Millerstown, you follow the Juniata River for 11 miles as it winds through folded Catskill redbeds and olive-gray siltstones of the Trimmers Rock formation — both of late Devonian age, and brownish shales of the middle Devonian Hamilton group. One large cut three miles north of Amity Hall displays sandstone and shale arched into an anticline between the two western prongs of the great Southern Anthracite syncline.

Seven miles north of Amity Hall, an excellent overlook provides a sweeping view of a large river meander. Across the river, a broad floodplain lies inside the curve and well-defined terraces rise above it. The Juniata Canal ran through the floodplain near the river, and the railroad now traces a similar path a little farther from the river. In the near distance, Hickory Ridge rises nearly 500 feet above the river. It is the outcrop of the resistant Montebello sandstone, part of the Mahantango formation.

A large cut in upper Devonian Catskill redbeds south of Millerstown is on the axis of the northern fork of the Southern Anthracite syncline.

The Pocono sandstone-supported ridge splits to the east, where the Susquehanna River cuts through both limbs. Millerstown lies in the Pfoutz Valley, just south of the northeast-pointing nose of Tuscarora Mountain. The route of the old Juniata Canal crossed the river south of Millerstown and continued upstream along this side of the river. The canals followed the floodplains as much as possible because they provided the easiest and most economical routes.

Between Millerstown and Thompsontown, five miles, you follow the river through a gap in Tuscarora Mountain. For nine miles upstream from the gap, the river hugs the northern flank of this prominent ridge of Tuscarora quartzite. This is the Little Pfoutz Valley, developed along a belt of weak shales of the upper Silurian Wills Creek formation and carbonates of the Keyser and Tonoloway formations. In the gap, watch for the huge cut on the northwestern limb of the anticline in brownish shales of the lower Silurian Clinton group that grade downward to sandstones. The upper shale beds dip about 30 degrees away from the fold axis, whereas the sandstones nearer the axis are almost flat.

Between Thompsontown and Mifflintown, nine miles, you cross part of the Tuscarora Valley over shales and limestones. Lost Creek Ridge east of Mifflintown is Montebello sandstone rimmed by Keyser and Tonoloway limestones. Many springs issue from the carbonate rocks of its flanks. A fine example of an a small anticlinal fold called Arch Rock is near Mifflintown.

Arch Rock, an asymmetrical anticline in thin-bedded lower Silurian Keefer sandstones. Located behind Arch Rock Grange on Arch Ridge Road, 2.5 miles north of Mifflintown, and 0.5 miles east of U.S. 322.

Between Mifflintown and Lewistown, the road follows the river closely for six miles through lovely Lewistown Narrows, also called Roaring Run Gap. The river follows Clinton shales in a synclinal crease between overlapping Tuscarora-capped anticlinal ridges, Shade Mountain on the west with its nose at Lewistown, and Blue Mountain on the east with its nose at Macedonia, north of Mifflintown. The gap is nicely profiled in the approach from either the upstream or downstream side. Pale talus of Tuscarora quartzite appears near the crests of the ridges.

Along the streams of the Lewistown region are the remains of many charcoal iron furnaces and forges built between 1790 and 1850. At that time, Juniata iron was the best in America. Its reign ended with the rise of the use of coal and coke in iron making. Just north of Lewistown at Burnham is Freedom Forge, built in 1795, the site of the third Bessemer steel plant in the nation.

Between Lewistown and Reedsville, six miles, you cross the Ferguson Valley and pass through a gap in Jacks Mountain along Kishacoquillas Creek. Just south of the gap are cuts in the colorful, variegated shales, siltstones, and sandstones of the upper Silurian Bloomsburg and Mifflintown formations. The gap is well-profiled in both the northern and southern approaches, and the Tuscarora quartzite is exposed in several places along the ridge crest. You see steeply-dipping beds of Tuscarora quartzite and of the underlying upper Ordovician Juniata and Bald Eagle formations in the gap.

In the five miles between Reedsville and Mt. Pleasant, you cross Kishacoquillas Valley, mainly over Keyser and Tonoloway limestones. The carbonate rocks were deposited in the shallow water of a stable shelf that existed after the Taconian orogeny and before the Acadian.

About four miles northeast of Reedsville is Mammoth Spring, the source of Honey Creek. This is the third largest spring in Pennsylvania with a flow of 14,000 gallons per minute.

The six miles of road between Mt. Pleasant and Potters Mills crosses a complex series of northeast-trending mountain ridges collectively referred to as the Seven Mountains. All are either Tuscarora quartzite or Bald Eagle sandstone. Most intervening valleys are in shales and siltstones of the upper Ordovician Reedsville or Juniata formations. En route, you pass Laurel Creek Reservoir in Coxes Valley.

This area of the Devil's Elbow contains an almost continuous exposure that includes the upper part of the Juniata formation, all of the Tuscarora formation, and most of the overlying lower Silurian Rose Hill shale, equivalent to the lower Clinton group. Most of the

Large-scale kink folds in upright lower Silurian Tuscarora quartzite beds on U.S. 322 by the Laurel Creek Reservoir.

beds are vertical or nearly so. Tuscarora quartzite crops out by the road near the dam on the end of Spruce Mountain as well as on the opposite side of the lake on Front Mountain. It contains numerous small folds known as kink bands. The colorful redbeds of the underlying Juniata formation crop out downhill from the Tuscarora quartzite and the overlying grayish Rose Hill shales are on the uphill side.

The kink bands in the Tuscarora quartzite appear to have formed before folding rotated the beds to a vertical position. That is puzzling. Actually, the large folds are final products of horizontal compression of the crust — most large folds, as well as thrust faults, originate this way. The kink bands are small expressions of the same thing, but they

Kink zones are outlined in this tracing of a photograph of part of the rock cut near the Laurel Creek Reservoir. Such sharp-angled folds at this scale are commonly referred to as chevron folds. Area shown is about 12 feet wide.

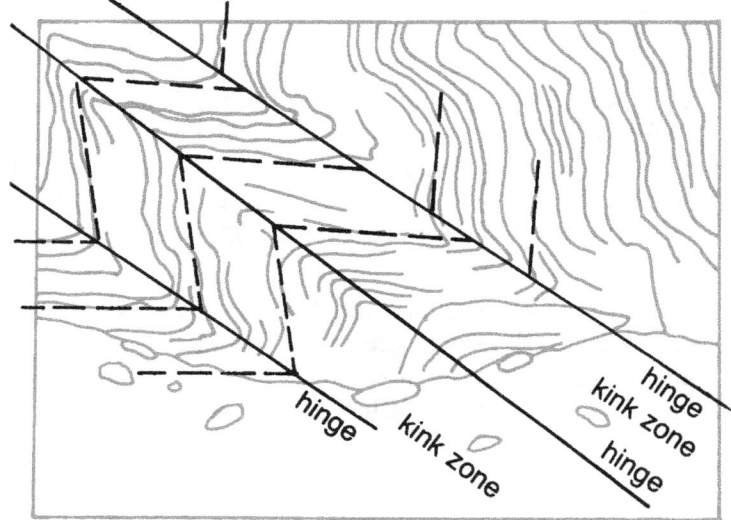

Upper Ordovician Bald Eagle sandstone beds near Potters Mills on U.S. 322.

had to form while the beds lay flat. They could not form while the beds were upright because the horizontal compression at that stage is perpendicular to the bedding and would tend to flatten small folds.

You cross Penns Valley in the nine miles between Potters Mills and Boalsburg, on bedrock of Ordovician age all the way, Reedsville shales in the central part and carbonates elsewhere. Look south to Tussey Mountain, a ridge of Bald Eagle sandstones.

Between Boalsburg and Park Forest Village, ten miles, you follow the State College Bypass in the shadow of Nittany Mountain. The mountain is Bald Eagle formation folded into a syncline with Tuscarora sandstone exposed in its core several miles to the northeast. Magnificent

Lower Silurian Tuscarora quartzite talus on Tussey Mountain.

251

Middle Ordovician Bellefonte dolostone on State College bypass near the Pennsylvania 26 intersection.

new cuts near the Pennsylvania 26 intersection in the middle of the bypass, display well-bedded Bellefonte limestone beds that dip about 30 degrees towards Nittany Mountain on the northwestern limb of the syncline. At the western end of the bypass, you cross mostly Cambrian carbonate rocks in the Nittany Valley. The valley to northeast and southwest is pockmarked with numerous small sinkholes and ponds, evidence of carbonate underpinnings.

Between Park Forest Village and Martha Furnace, at the US 220 junction, you complete the Nittany Valley traverse and cross Bald Eagle Mountain. The ridge is double-crested, with the southeastern crest held up by Bald Eagle sandstone, and the northwestern crest by Tuscarora quartzite. The shallow crease between is in weaker rocks of the Juniata formation. Skytop overlook on the Tuscarora crest offers a splendid panorama of Bald Eagle Valley and the Allegheny Front.

PA 61
Sunbury—Frackville
40 mi./65 km.

Sunbury and its sister city Northumberland lie at the junction of the west and east branches of the Susquehanna River, a dramatic and historically important spot. Enjoy the excellent views from overlooks in Shikellamy State Park, atop the high scarp just across the west branch from Northumberland. The scarp exposes interbedded siltstones, shales, and sandstones of the Sherman Creek member of the Catskill formation in the core of the Northumberland syncline. You can see the scarp from the Northumberland bridge over the west branch.

Fort Augusta

The fort, one of the largest and most important frontier strongholds of the upper Susquehanna, was on the river bank north of Sunbury. Built during 1756-57, and in use until 1785, the site now contains a museum and fort model. Nearly all of the forts used in the Revolutionary War in Pennsylvania were on rivers, then principal travel routes. Major junctures, like this one, were particularly strategic.

Pennsylvania Canal

Two branches of the canal system, built in 1828-34, joined in Northumberland at a canal basin. Boats with coal came down the eastern canal from Nanticoke near Wilkes-Barre, and lumber boats came down the western branch from Williamsport. Both locked down to the river here and then re-entered the canal on the western bank to continue their journey south to Chesapeake Bay, Baltimore, and Philadelphia.

• Between Sunbury and Paxinos, Pennsylvania 61 goes through 12 miles of rolling farmland a few miles north of Little Mountain ridge. Little Mountain is the northern limb of the Western Middle Anthracite syncline, with a nose that points west. It is held up by lower Mississippian Pocono sandstone.

In the five miles between Paxinos and Shamokin, you go north-south through two Shamokin Creek gaps. The northern gap is in Little Mountain, and the southern gap is in Big Mountain, a ridge of Pennsylvanian Pottsville sandstones and conglomerates.

(61)

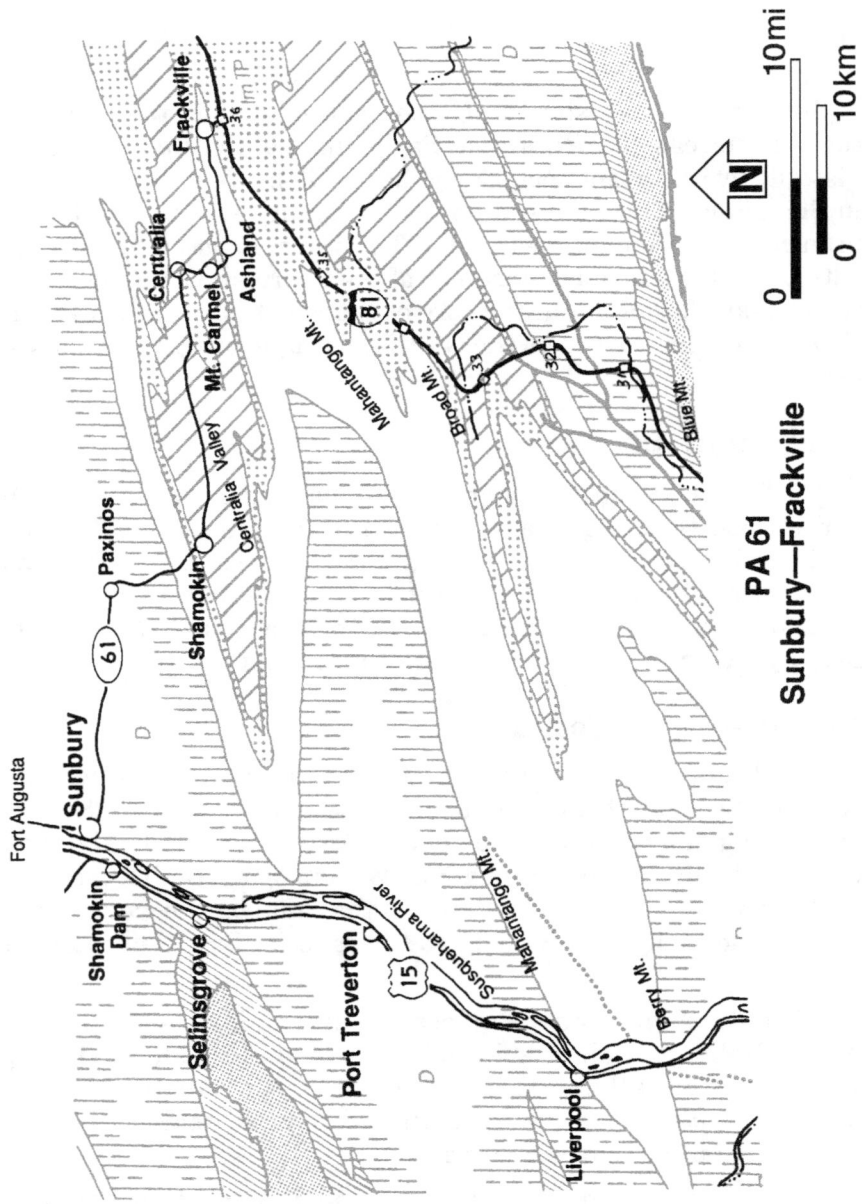

PA 61
Sunbury—Frackville

Centralia Valley

Shamokin, an Indian name meaning "place of the horns," was the original name of Sunbury. At Shamokin, you are in the famous Centralia Valley in the heart of the Western Middle Anthracite Field. This is hard coal country and it has come on hard times. The ruins of coal-mining are all around — huge piles of barren rock waste from old strip mines, derelict buildings with broken windows, and rusting equipment. Beneath the surface is even greater devastation in this region of underground mining.

Hard coal mining in Pennsylvania is in a state of suspended animation, and it is uncertain whether it will ever revive. The industry has been in general decline since 1917, its peak year, when the anthracite region as a whole produced about one-fourth of all coal in North America. Plenty of accessible coal remains — an estimated 20 billion tons or more. Only about seven billion tons have been removed in more than 100 years of mining.

The hard coal seams of eastern Pennsylvania were originally horizontal beds of bituminous coal continuous with those of the Allegheny Plateau, but the heat and pressure associated with Alleghenian deformation converted them to anthracite. The seams have survived erosion only in deep synclinal folds like that of the lower cross-section. —Adapted from Edmunds and Koppe (1968, p. 10)

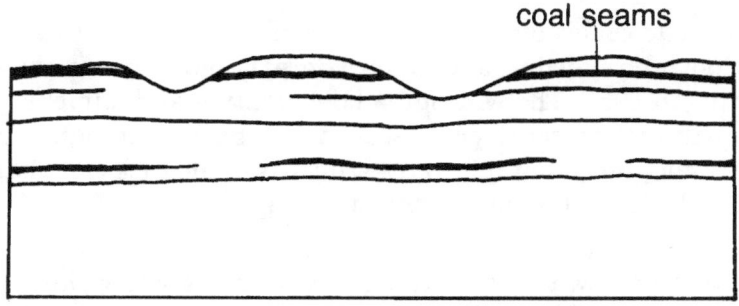

soft coal region of the Allegheny Plateau

hard coal region of east Pennsylvania

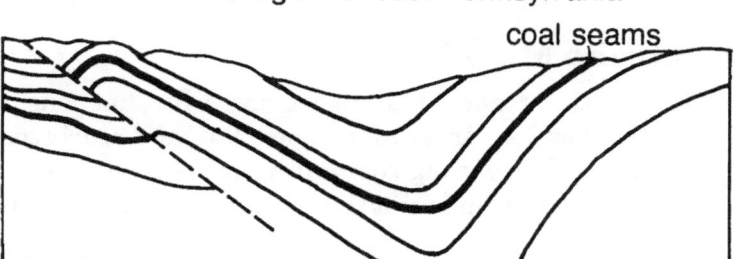

The ruins of hard coal mining are all around the village of Shamokin in the Centralia Valley.

One problem is the expense of mining this coal. All of it here, as well as in the other fields of the anthracite region, occurs in generally thin seams in synclinal folds of the lower Pennsylvanian Pottsville group and upper Pennsylvanian Llewellyn formation in synclinal folds. Following and excavating the coal in steep, narrow inclines is difficult, costly, and extremely hazardous. Oil and gas have become the much preferred and cheaper fuels. Meanwhile, soft coal, which in Pennsylvania only occurs in the Allegheny Plateau region, is far superior for coking. Its use in the steel industry sustains the market.

Anthracite coal seams come to the surface on the upturned edges of the Centralia syncline, and this is where the old strip mines worked, beginning in 1957. The widespread devastation and immense waste piles testify to the intensity of past activity. The trip through Centralia Valley is not pretty, but it is interesting to see the rocks laid bare and consider the human triumph and suffering involved.

• Between Shamokin and Kulpmont, five miles, you see a large roadcut in the Llewellyn beds that floor Centralia Valley, and contain most of the coal. They are shales, sandstones, and conglomerates, with some thin coal seams in nearly vertical beds.

Between Kulpmont and Centralia, seven miles, you go east-west along the valley with abandoned mines and waste rock all around.

The Centralia Mine Fire

Centralia is the scene of a disastrous fire that has burned underground in abandoned coal mines since 1962, and nearly converted this once-thriving coal-mining community into a ghost town. Many houses have been razed, and most that remain are uninhabitable and

Coal seam in near-vertical Pennsylvanian Llewellyn beds on Pennsylvania 61 east of Shamokin.

boarded up. Centralia in 1980 had a population of about 1000, by 1988 only 23 families remained. The others moved away to escape the poisonous gases rising from the fire.

Fires are a danger wherever coal is mined underground. Once started, they can burn for years, as this one has, drawing air through multiple man-made openings, or through the ground itself. A large part of the problem lies in the way coal occurs in generally thin, widespread seams interstratified with other sedimentary rocks. Most is extracted by the room-and-pillar method, which leaves about 50% of the coal in place as columns to support the roof. The intense mining activity of the past under Centralia Valley created a honeycombed subterranean world covering tens of square miles, with abundant coal left to go on burning practically forever.

The Centralia mine fire began in an old open-pit mine on the southeastern side of the village next to the Odd Fellows cemetery. The pit was excavated in 1935 and had been used as a landfill since early 1962. In a clean-up operation preparatory to Memorial Day celebration that year, the Centralia Council decided to burn the trash in the pit, and accidentally ignited the exposed coal. The fire spread to the underground network of mines and has continued to burn an ever-widening area despite numerous efforts to put it out. The fire spreads mainly by burning coal in front of it, and it can go uphill or downhill.

It may also leap from one place to another by hydrogen explosions. Temperatures above 1000 degrees Fahrenheit have been measured in bore holes.

The principal hazard to humans lies in the release of deadly poisonous carbon monoxide and suffocating carbon dioxide. The gases work their way to the surface through fractures or other natural openings, as well as mine entrances. Both are odorless and nearly impossible to detect without special instruments. The extreme temperatures in the mines also boil off the groundwater, creating steam that often pinpoints gas vents. Several such steam holes are visible alongside Pennsylvania 61 in Centralia. The fires eventually destroy coal pillars and weaken the overlying ground, promoting collapse that may affect structures on the surface and provide new passageways for gas escape. The longer the fire burns, the more serious the hazard becomes.

Why hasn't the Centralia mine fire been extinguished? Perhaps the best answer is too little, too late. If adequate action had been taken when the fire started, it would have been a relatively simple matter. Inaction allowed the fire to spread down the dip of the coal seam, making it increasingly difficult even to delineate, let alone control. Now it is so deep and so widespread that it is virtually impossible to put out. Finally, realizing the hopelessness of the situation, the U.S. Congress, in 1983, appropriated 42 million dollars for the relocation of Centralia and nearby Byrnesville.

• At Ashland, two miles south of Centralia, near the Pennsylvania 54 junction, you see a large cut in Pottsville and Llewellyn sandstones of the southern limb of the syncline, here called Mahanoy Mountain. The same ridge is called Big Mountain on the north side of the fold. Ashland has an Anthracite Museum. A little south of town are massive detached blocks of the gray Pottsville conglomerate that caps the ridge throughout. Farther south are roadcuts of the distinctive redbeds of the underlying upper Mississippian Mauch Chunk formation. You go east-west between Ashland and Frackville, eight miles, over poorly exposed Mauch Chunk beds.

III
Southeastern Pennsylvania

This region encompasses about one-sixth of the total land area of Pennsylvania. Geologically very diverse, it includes the Great Valley that borders the Valley and Ridge province; South Mountain, the Precambrian-cored northern tip of the Blue Ridge province; Reading Prong, the Precambrian-cored tip of New Jersey's Ramapo Mountains; the Triassic Gettysburg and Newark rift basins; and the Piedmont province. These provinces are adequately covered within the route descriptions and also in the introductory chapter, the Pennsylvania Landscape.

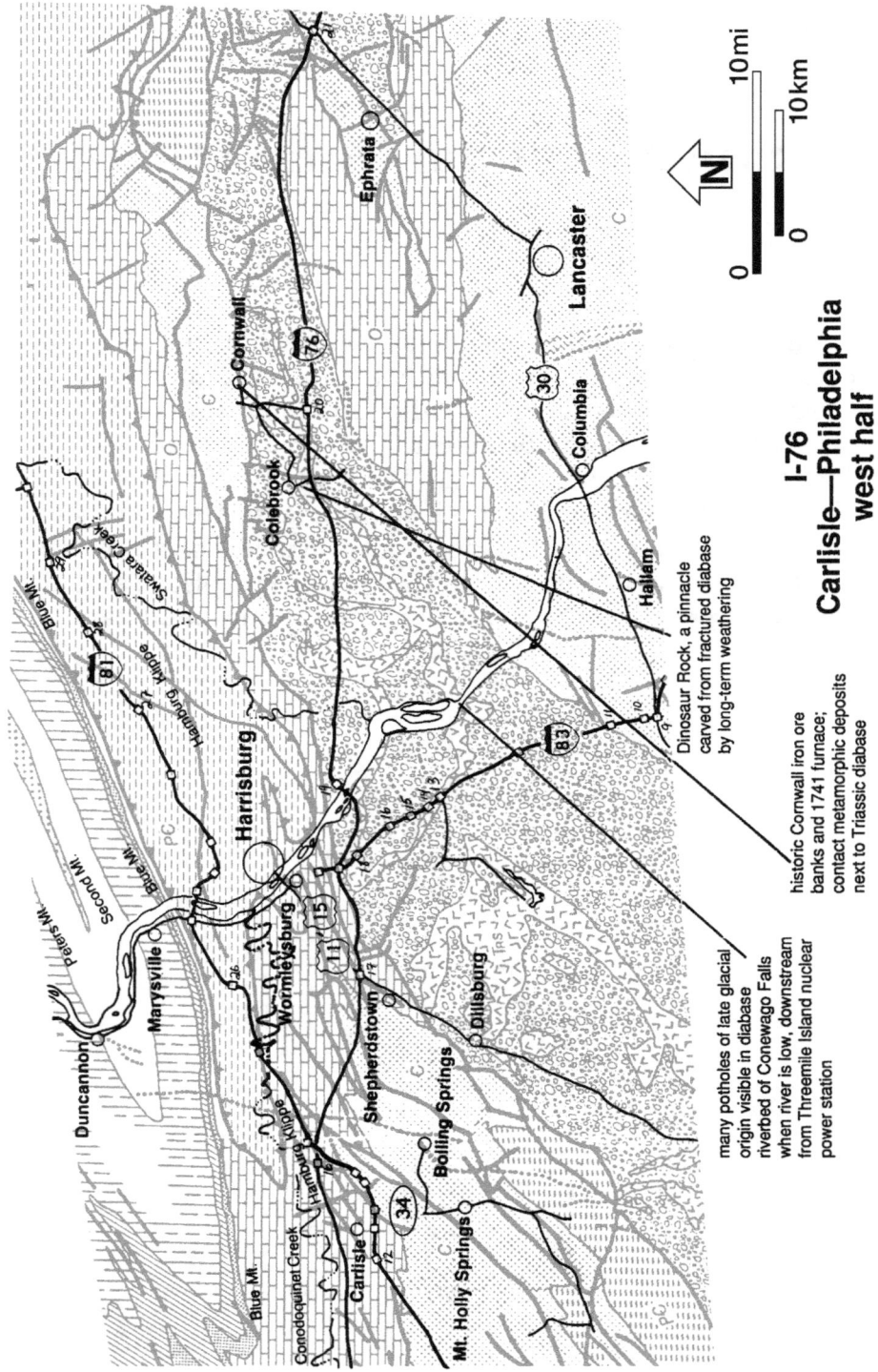

I-76
Carlisle—Philadelphia
west half

260

I-76
Carlisle—Philadelphia
110 mi./177 km.

The Pennsylvania Turnpike

Between the Carlisle exit and mile 242, you cross diagonally over the eastern half of Cumberland Valley with the flat top of Blue Mountain ridge rising a few miles north. Practically all bedrock near the road is Cambrian and Ordovician limestone and dolostone with some shaley interbeds at the northeast-pointing nose of South Mountain. The rocks are tightly folded and much broken along faults, although these structures are not very evident in the few low roadcuts along the highway. Many outcrops exist in pastures alongside the highway as scattered and irregular pinnacles of light-gray carbonate rocks that jut through the soil. This is limestone country with numerous sinkholes, caves, and streams that sink into underground passageways and reappear elsewhere as springs. The pinnacled outcrops are simply projections of the uneven limestone surface through the soil cover.

The highway cuts through Stony Ridge diabase dike at mile 229.5. The dike appears as a vertical band of dark gray rock flanked by the light gray limestone it intrudes.

Triassic and Jurassic Dikes

The Stony Ridge diabase dike has been traced continuously for 25 miles along a north to south direction. It is remarkably thin, with a length to width ratio of at least 5000 to 1. Furthermore, it is just one of several aligned, but separate diabase dikes that continue south into Maryland and north to beyond Millersburg in the Valley and Ridge province. Their combined length from the Maryland border is more than 70 miles. Near the turnpike, the Stony Ridge dike stands up like a wall because it is more resistant to erosion than the carbonate rocks it intrudes.

Numerous diabase dikes cross the Triassic Gettysburg Basin of Pennsylvania, many extending into the Piedmont and the Precambrian-cored ridges of South Mountain and Reading Prong. The Stony Ridge dike is the only one known to cut across the Great Valley and continue deep into the Valley and Ridge province.

All of the diabase dikes injected fractures that opened during the breakup of the supercontinent Pangaea. Apparently, basalt magma formed in the upper asthenosphere as the overlying lithosphere thinned, then rose along the faults. Normally, the superheated mantle rocks remain solid because of high confining pressure; when pressure is relieved, they melt, and the melt rises to intrude the crust. The same crustal stretching caused the Gettysburg and similar east coast basins to subside and fill with sediments.

Our growing knowledge of plate tectonics enables us to see the broader significance of small outcrops like Stony Ridge. The intricate deformation of Cumberland Valley with its tight folding and extensive faulting results from compression of the crust and continental collision during assemblage of Pangaea whereas the basin subsidence, volcanism, and dike intrusion result from crustal stretching as the supercontinent split into the modern continents.

Deformation here is more complex than in the Valley and Ridge province in part because it results from the combined effects of Taconian and Alleghenian mountain building, possibly also the Acadian event. Valley and Ridge folding and faulting are only Alleghenian because that region was farther away from the rising mountains.

• Exit 18 lies almost on the fault that dropped the northwestern edge of the Gettysburg Basin. The basin is best described as a half-graben because only this side of it is faulted; the other side is not faulted. It hinged down as Triassic and Jurassic sediments accumulated on the erosional Piedmont surface.

Grabens

A graben is an elongate trench formed when a fault-bounded crustal block drops in response to crustal stretching. A half-graben has a fault on one side, and the other bends down as though it were hinged. Most of the Triassic basins of eastern North America, including the Gettysburg Basin, are half-grabens that opened as Pangaea split to open the Atlantic Ocean basin. All filled as they subsided with sediments washed in from the flanking highlands.

Grabens and half-grabens form in response to crustal stretching, as when a continent begins to pull apart. The mid-Atlantic rift is made up of graben structures.

Threemile Island nuclear power plant in March 1988 with only one reactor working.

Most of the sedimentary rocks we now see in the Triassic basins are red shales, siltstones, sandstones, and conglomerates. The two mineral pigments that contribute their warm color to the rocks are brownish limonite, a hydrous iron oxide, and red hematite, anhydrous iron oxide. Very small percentages of these minerals can give deep color. The Triassic redbeds may have formed as stream-laid, alluvial fan deposits in a semi-arid climate.

• You cross the Susquehanna River between Exits 18 and 19. This is a particularly beautiful section of the river with many wooded islands visible upstream and downstream from the bridge. Infamous Threemile Island, with its nuclear power facility lies just six miles downstream.

Conewago Falls Potholes

These beautiful potholes become visible in the Susquehanna River channel only during low water. They are in diabase bedrock at Conewago Falls, a short distance downstream from the Threemile Island Nuclear Power facility and the York Haven Dam, near the village of Falmouth. Reach them from Exit 19 of the turnpike. The falls cascade over a ledge of diabase, which is much harder than the Triassic redbeds it intrudes.

The river sculptured the rocks at Conewago Falls during glacial recession, when it carried much more water and sediment than it does now. Sand and gravel carve potholes when they grind against bedrock

Large potholes in Triassic diabase at Conewago Falls of the Susquehanna River south of Harrisburg. —photo by W.D. Sevon

in swirling eddies beneath waterfalls or in rapids. Most potholes form where fractures or intersections of fractures caused eddies.

The site also contains small, intersecting channels, well-developed ripples and cross-trending grooves, all cut in solid rock. The origin of at least the grooves is clear. They formed where sediment-laden water spilled over a ledge in a small fall, carving a linear plunge basin. Floodwaters have since plucked away the ledges, leaving the grooves.

Random channels at Conewago Falls. —photo by W.D. Sevon

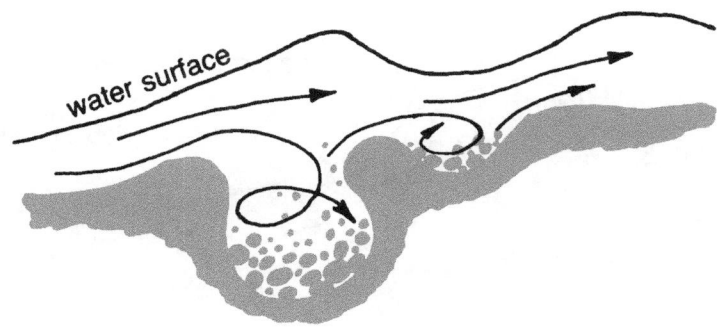

Cross-section of potholes containing grindstones that swirl in eddies and wear away the bedrock.

Origin of the rock ripples is less certain. One suggested mechanism is depicted in the drawings below. In A, sediment-laden water passes over a slight ridge on the diabase, creating enough turbulence immediately downstream for sand to begin cutting a groove oriented perpendicular to flow. In B, the downstream edge of the new groove similarly generates turbulence, and another groove is carved out. The result is a series of more or less parallel grooves separated by sharp crests that look very like oscillation ripple marks in sand along the seashore. Here, of course, they are carved in solid rock. The ripple marks apparently are unstable and deteriorate in time to a chaos of small channels. The channels formed because uneven ripple crestlines concentrated the flow downstream.

• Between Exits 19 and 20 are numerous cuts in Triassic red sandstone, shale, and mudstone and several cuts in dark gray to black diabase. The different colors alone make the distinction between these sedimentary and igneous rock types obvious.

Watch for the cut at mile 245.5 west of the river. It exposes a dark diabase sill sandwiched between redbeds, all dipping about 40 degrees north. A sill goes between beds, and a dike cuts across beds. The really

Suggested origin of ripple marks at Conewago Falls. Note that the ripples get larger downstream, apparently because turbulence increases in that direction as the number and size of the troughs increase.

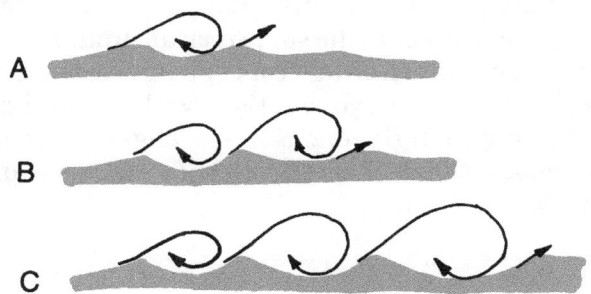

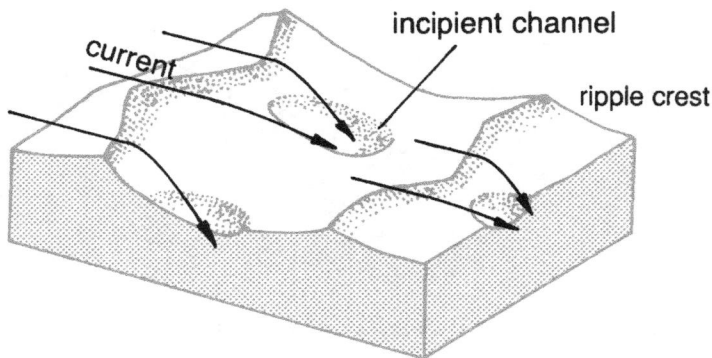

Suggested origin of channeling at Conewago Falls. Irregularities in the ripple crests concentrate flow at points ahead of them, thus excavating the bedrock; in time these new hollows extend and become elongate, wavy channels.

large diabase intrusions in the Gettysburg Basin are thick sills, all probably fed by dikes.

Cornwall Iron Ore Banks and Furnace

This site in Cornwall village is just a few minutes drive from Exit 20 of the turnpike. The ore banks, historically some of the most important in this country, produced more than 40 million tons of iron. The furnace was built in 1741 and operated until 1883. In the early days, the iron was made into stoves, kitchenware, and farm tools. During the Revolutionary War, the furnace produced cannon and ammunition for the Continental Army. Structures include the original furnace stack, the blast machinery with an early 18th century engine, blowing tubs, the Great Wheel measuring 76 feet around, wagon and blacksmith shops, and the ironmaster's mansion.

The ore at Cornwall is a contact metamorphic deposit developed between a thick Triassic diabase sill and the Cambrian limestone it intruded. The diabase is on the north side of the Gettysburg Basin where sills, sheets, and dikes of the same rock abound. The magnetite ore, subordinate amounts of iron and copper sulfides, and some cobalt formed in the limestone through reactions with mineral-rich fluids and gases emanating from the magma. The slightly larger western body measures about 5000 feet long by 230-650 feet wide and dips 25 degrees south, thinning with depth.

Dinosaur rock is an odd-shaped pinnacle of spheroidally weathered Triassic diabase boulders. Compare the character of weathering with that of Ticklish Rock, shown in U.S. 220: South Waverly—Williamsport.

Dinosaur Rock

Dinosaur Rock stands in the woods just north of I-76 between Exits 19 and 20, southeast of Colebrook. It is an erosional remnant of a very large diabase sill that intruded between sedimentary beds. Erosion has now removed the cover rocks and cut the surface down to the diabase, which stands high because it is more resistant than the other rocks. Throughout the Gettysburg Basin, diabase crops out in remarkably abrupt ridges and hills on the otherwise rather even surface.

Diabase is a granular rock mainly composed of the minerals plagioclase feldspar and pyroxene. It commonly fractures along numerous criss-crossing joints, which are one of the main causes of odd erosional features like Dinosaur Rock. Note that all of the components of the dinosaur, as well as boulders and outcrops around it, are well-rounded. This is a product of spheroidal weathering along the joints.

Trap rock is an old term applied to diabase that probably originated from the German word, Treppe, meaning step. German quarrymen used this term because the rock separated into large, rectangular blocks and could be worked in the face of the quarry in a series of steps. Diabase has also been called ironstone, an allusion to its hardness and dark color.

- Between Exits 20 and 21, you are just north of the southern margin of the narrow strip of Triassic lowland that connects the Gettysburg and Newark basins.

Cushion Peak

Cushion Peak, a well-known landmark near Fritztown, is an 800-foot hill a few minutes drive north of Exit 21 of the turnpike. A road to the summit provides a good view of this section of the Great Valley.

Cushion Peak is carved from tough Hardyston quartzite formed from sediments deposited about 570 million years ago at the edge of a Cambrian sea. A quartz pebble conglomerate near the base of the formation crops out near the top of the hill. The Hardyston quartzite and the Precambrian granitic gneiss upon which it rests, are within the uppermost of a stack of thrust fault slices that moved up and north, possibly during Taconian mountain building. No one knows exactly how far they moved because the steep east-west-trending border fault of the Newark Basin chops them off just one mile south of Cushion Peak.

Cushion Peak is at the eastern end of a klippe with a core of Precambrian gneisses that was thrust northward, possibly during Taconian mountain building, then cut off on its south side by down-faulting of the Triassic basin. The core gneisses are the most westward exposures of the Reading Prong, from which they are separated by erosion. —Map and cross-sections adapted from MacLachlan and others (1975, Plates 1 and 2)

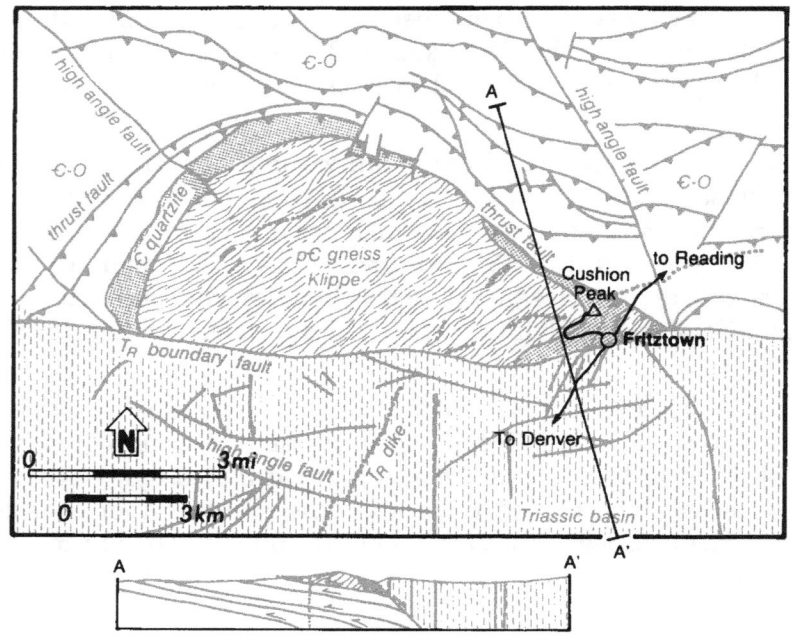

268

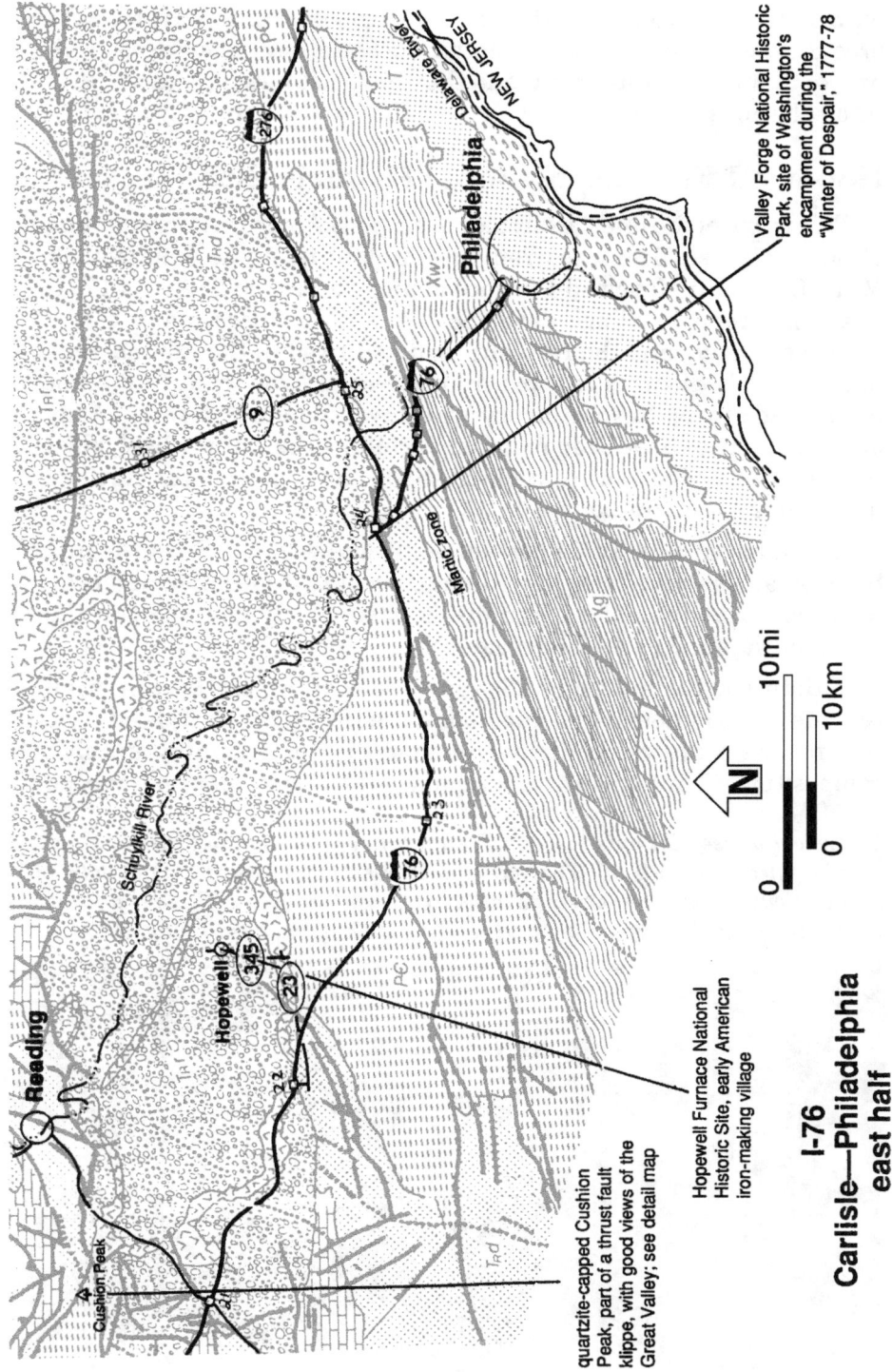

I-76
Carlisle—Philadelphia
east half

- Between Exits 21 and 22, you cross a section of Newark Basin with much diabase. Note that the strong jointing in several cuts breaks the diabase into angular blocks. These are man-made exposures, unlike that of Dinosaur Rock, and they have not had time to weather into spheriodal forms.

Hopewell Furnace National Historic Site

This is one of the finest examples of an early American iron-making village. It is just 15 minutes from Exit 22 of the turnpike. Englishman Mark Bird built Hopewell Furnace alongside French Creek in 1771. A small company village developed around it where many of the employees lived, and the resident manager lived in the ironmaster's mansion. The furnace cast pig iron, stoves, hollow-ware, and other items, and during Revolutionary War days, cannon and shot. Many of the structures have been restored, including the water wheel, blast machinery, bridge house, cooling shed, barn, store, ironmaster's mansion, and tenant houses.

Iron-making was an important early industry in Pennsylvania because all the necessary ingredients, iron ore, limestone, and coal, were here close at hand. Iron ore is particularly abundant in the Precambrian and early Paleozoic rocks.

- Between Exits 22 and 23, you cross the largest Precambrian terrane of the Pennsylvania Piedmont, the Honey Brook Upland. The few roadcuts are all in light-colored gneiss of generally granitic composition. Throughout the region, numerous metamorphosed

Triassic diabase dike cuts through Precambrian gneiss of the Piedmont province near mile 308 of I-76. Note one strong set of joints approximately perpendicular and another set parallel to the contact walls. These result from shrinkage during cooling and solidification of the basalt magma.

diabase dikes, probably of Precambrian age, cut through the gneisses. One of these, about 15 feet thick, is near mile 308 where its dark color and tabular form make it stand out against the surrounding pale gneiss. Note the strong fractures, columnar jointing at right angles to the dike contacts. Mild metamorphism, during Paleozoic mountain building converted the diabase partly to a green mineral called chlorite, which gives it greenish color.

Columnar Jointing

Columnar joints are a common, and much-studied feature of dikes, sills, and lava flows. They form as the melt crystallizes and contracts. A basaltic lava, the volcanic equivalent of diabase, for example, will cease to flow when its temperature drops below about 1650 degrees F. This first occurs at the top where heat is rapidly lost to the air. The lava crystallizes and crusts over while the center cools more slowly and may continue to flow. This contraction creates tensional stresses parallel to the surface. Columnar joints form as the lava shrinks and propagate as cooling progresses down from the top of the flow. Similarly, the bottom of the flow is also chilled against the underlying surface, columnar joints open at the contact, and extend themselves upward from there. Because the tension is directed parallel to contact surfaces, whether air or solid, the cracks tend to orient perpendicular to them. In dikes and sills, most of which are intruded at shallow depths, columnar joints extend inward from, and more or less perpendicular to, contact walls, whatever their orientation.

- You see almost no bedrock between Exits 23 and 25. This section of highway lies near a profound linear geologic boundary known as the Martic zone. Recent research suggests that this may be an ancient shear zone along a plate boundary like the great San Andreas fault of California, on which movement was horizontal and the northwestern side moved northeast relative to the southeastern side. Such movement, if it did occur, would most likely be Alleghenian, and may have followed an old Taconian thrust fault zone. The geologic explanation for the Martic zone is still uncertain. Rocks along it are metamorphically recrystallized and have granular textures produced by shearing. The rocks of the northwestern side are granitic gneiss. Those on the southeastern side are very different—schists, quartzites, and phyllites formed through recrystallization of sedimentary rocks.

Valley Forge National Historic Park

You can reach this 2250-acre park between the Schuylkill River and the turnpike in a few minutes from Exit 24. It commemorates Washington's encampment in the "Winter of Despair" of 1777-78, during the Revolutionary War. Washington led his troops here after his defeats at Brandywine and Germantown left the British in control of Philadelphia. He needed a winter camp close enough to Philadelphia to contain the British in the city. Valley Forge was militarily defensible and logistically practical.

Valley Forge was then a tiny, iron-making settlement with numerous charcoal furnaces and a forge near the junction of Valley Creek with the Schuylkill River. The forge, with a wheel turned by the creek, was about a mile upstream. In its last two miles to the river, Valley Creek flows almost straight north through a small gorge between two hills, Mt. Misery on the west and Mt. Joy on the east. With its northern side protected by the river and western side by the creek, an attacking force would be compelled to approach from the southeast, where it would face fire from the readily fortified crest of Mt. Joy.

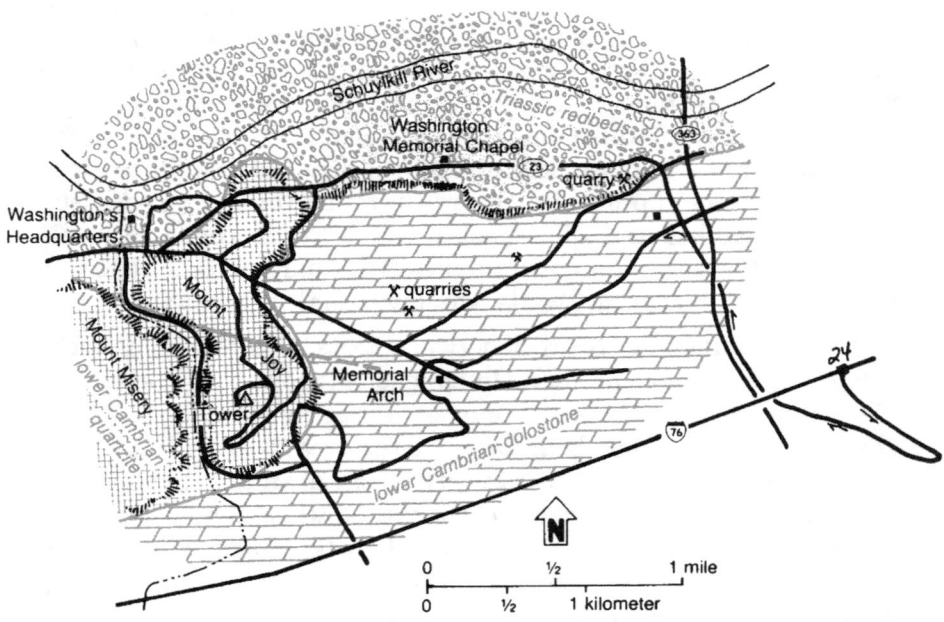

Geologic sketch map of part of Valley Forge National Historic Park.
—Adapted from Swanson (1974)

The park straddles the boundary between the Piedmont province and the Newark Basin. The oldest bedrock south of the boundary is Cambrian Antietam and Harpers quartzites, which form Mts. Misery and Joy. The rest of the area south of the boundary is lowland eroded on Cambrian Ledger dolostone that overlies the quartzite. These units dip east and the quartzite hills rise north of the line where their contact comes to the surface, reflecting the difference in erosional resistance of the two rock types.

The boundary with the Newark Basin follows a low scarp bordering a terrace eroded on Triassic redbeds. All three of Washington's headquarters buildings are made of reddish sandstones from this unit. A small quarry in the scarp in the northeastern section of the park exposes the contact between the dolostone and the redbeds. The contact is an unconformity representing a gap in the geologic record of about 400 million years! We have no rock record here of Earth history between about 600 and 200 million years ago, the ages of the rocks below and above the unconformity. The irregularity of the unconformity is a ragged topography, developed by solution of the dolostone before deposition of the redbeds.

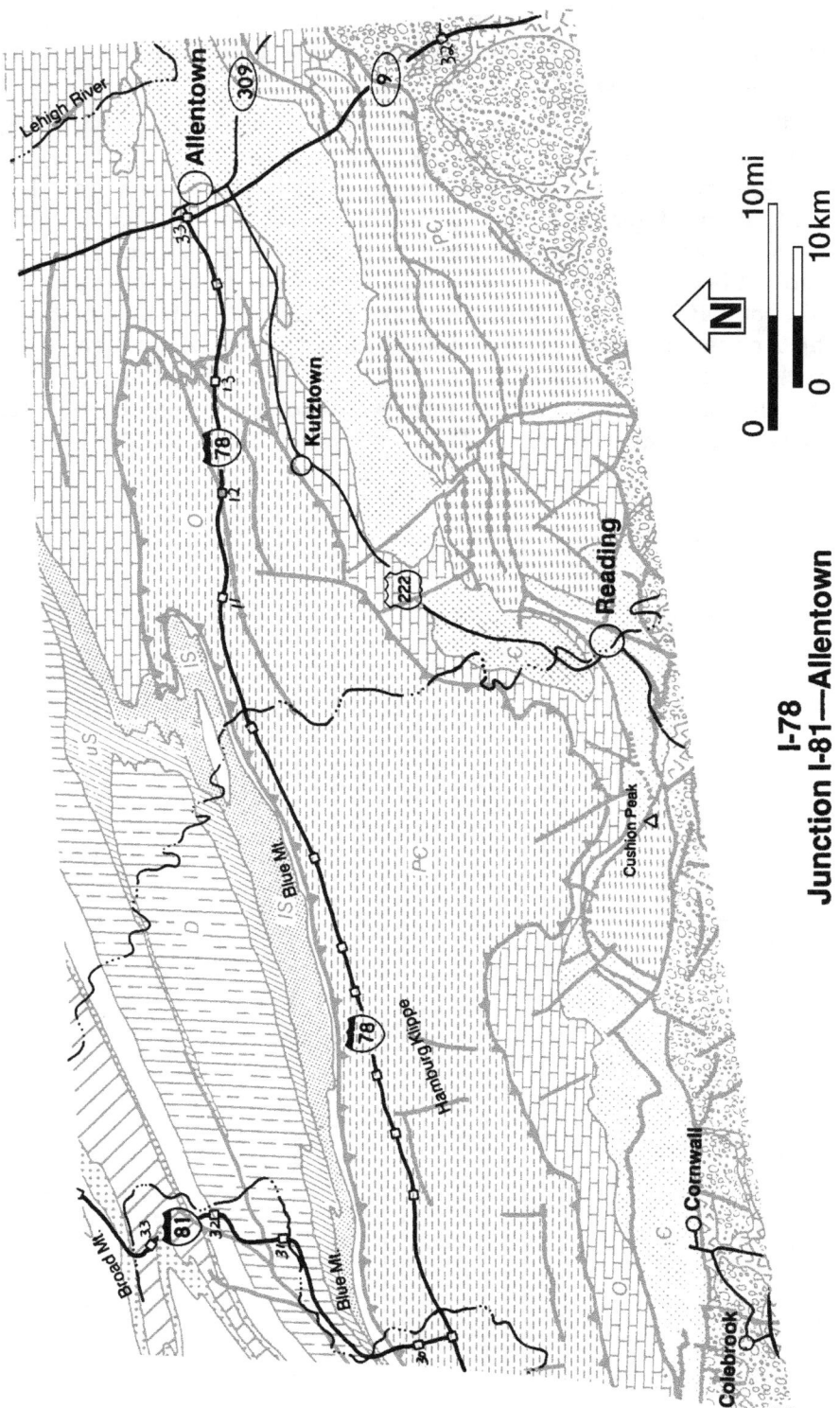

I-78
Junction I-81—Allentown

274

I-78
Junction I-81—Allentown
54 mi./87 km.

This entire route lies within the Great Valley. All but the easternmost eight miles of it crosses the Hamburg klippe, an enormous complex of thrust fault slices that moved many miles from its place of origin during the Taconian orogeny and then was isolated by erosion. It is eight to twelve miles wide north-south by 80 miles long east-west. Rocks of the Hamburg sequence are mostly highly deformed, mildly metamorphosed, dark shales and graywackes.

Graywacke

Graywacke is muddy sandstone. Many graywackes apparently form from turbidity currents; which is why many people call them turbidites.

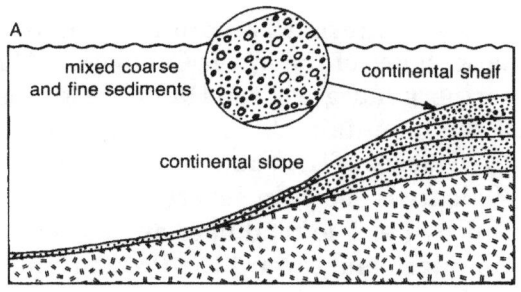

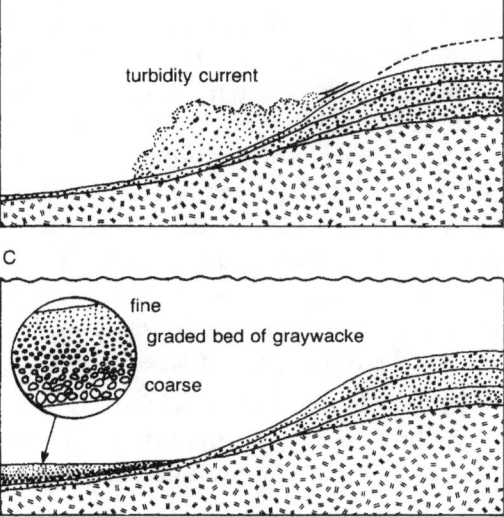

Schematic sequence illustrating how the graded bedding that is so common in graywacke is formed. In A, sediments pile up under water at the outer edge of the continental shelf. In B, storm, earthquake or other disturbance upsets the sediments and they plunge downslope in a turbidity current. In C, the turbid sediments have settled out, coarse first, increasingly fine upwards, forming a graded bed.

The relatively steep slopes at the outer edge of the continental shelf or alongside oceanic trenches are common sources of turbidity currents. Sediments accumulate there, and when the pile becomes too steepened or is disturbed by an earthquake, it collapses in a cloud of muddy water that sinks because it is denser than the surrounding clear water. When the current slows, the sediments settle. The coarser, heavier particles drop out first, followed by progressively finer material. Each current, therefore, produces a single graded bed in which the grains become finer upward. Most graywackes appear gray and dirty because they contain dark-colored rock or mineral fragments.

• Between the I-81 junction and Hamburg, 30 miles, you follow the northern margin of the Great Valley, never more than three miles away from Blue Mountain ridge. Tuscarora quartzite caps the ridge in the west, and its coarser equivalent, the Shawangunk formation, in the east. Note the remarkably even crestline at 800-1000 feet above the valley. The Appalachian Trail follows that crest. The highway crosses the Schuylkill River at Hamburg; look north for a good view of the water gap where the river cuts through Blue Mountain.

Lower Silurian Tuscarora quartzite dips into the ridge and covers dark shales of the upper Ordovician Martinsburg shale. The faulted northern margin of the Hamburg klippe lies concealed near the base of the mountain. Along the highway you see olive-gray shales and graywackes of the Hamburg sequence. The layers dip between 45 and 70 degrees south in several exposures between Hamburg and New Smithville. The road here angles across the valley, away from Blue Mountain and toward the Precambrian Reading Prong.

New Smithville lies just inside the eastern end of the Hamburg klippe. Between there and the Pennsylvania 9 junction at Allentown, eight miles, you cross a broad lowland over carbonate bedrock of the lower Ordovician Beekmantown group and Jacksonburg formation. These are part of the immense shelf sequence of late Precambrian to middle Ordovician carbonates and sandstones that accumulated on the eastern edge of a smaller North America during the long period of tectonic quiet that accompanied opening of the Proto-Atlantic Ocean. Erosion of the continental margin and development of the shelf began when the Precambrian Grenville continent first rifted some 650 million years ago. Sediments piled up on the shelf for 200 million years. Meanwhile, deep water sediments accumulated along the seaward edge of the continent, the continental rise and slope.

In middle to late Ordovician time, about 445 million years ago, the ocean began to close, initiating the Taconian orogeny. The crust broke east of the continent, and the western slab plunged beneath the eastern. Heating of the sinking slab at depth caused melting, and magma rose to create and fuel a volcanic island arc, while a deep oceanic trench developed on its landward side and filled with sediments eroded from the islands. Continued subduction gradually consumed the oceanic crust between the islands and continent; they converged and eventually collided, compressing the sediments between them. The sediments were crumpled, sheared, cleaved, heated, and metamorphosed. Most importantly, they were cut by numerous thrust faults, and the slices stacked up landward on the continental shelf.

The Hamburg klippe is just one of many east coast klippe formed this way and later isolated by erosion.

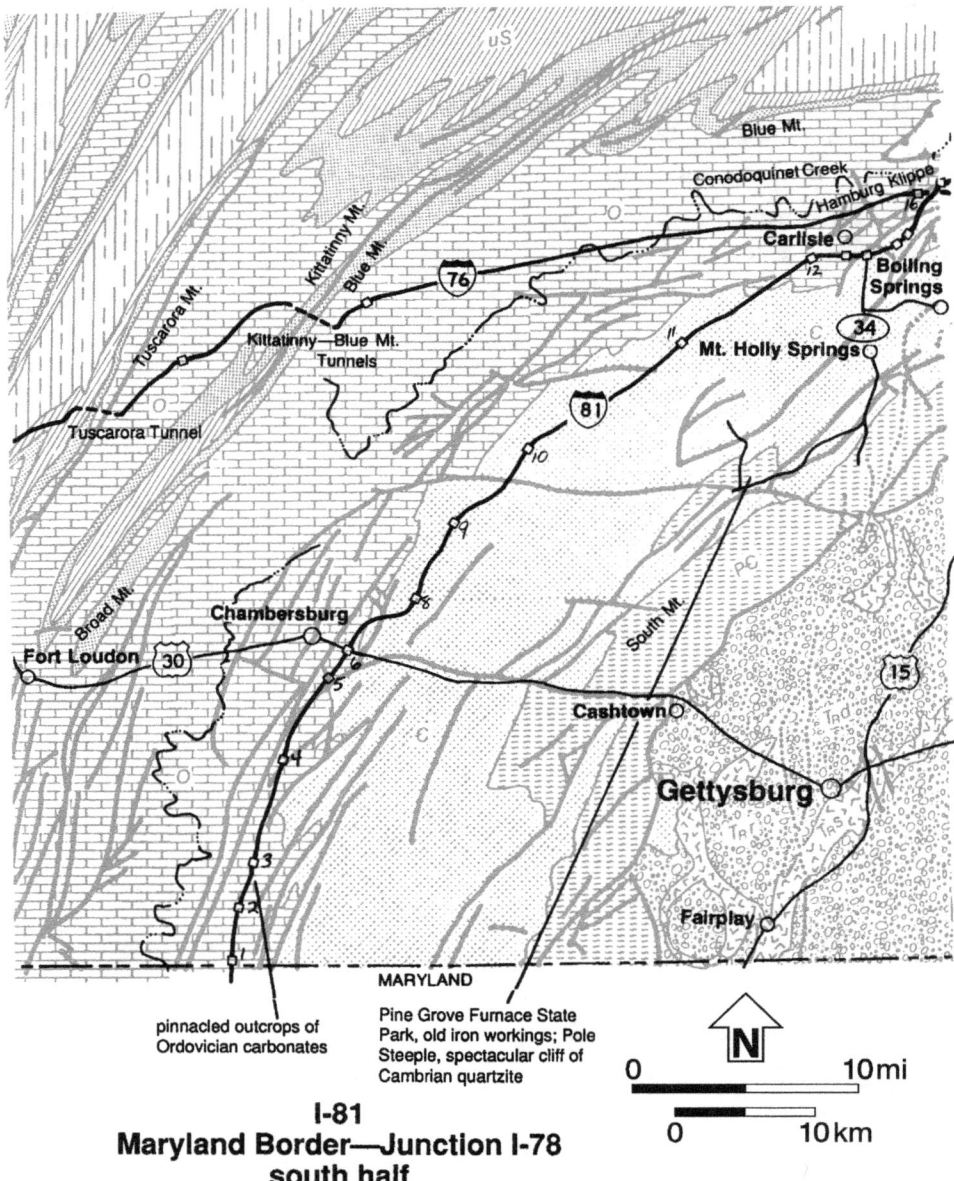

I-81
Maryland Border—Junction I-78
south half

Pinnacled outcrop of lower Ordovician Rockdale Run limestone by I-81 near Maryland border. These are not loose boulders, but projections though the soil of an uneven bedrock surface eroded by groundwater solution.

I-81
Maryland Border—Junction I-78
88 mi./142 km.

Cumberland Valley Section

Rock formations traversed by this section of I-81 are schematically illustrated in the Geologic-Time Travel Map on page 197.

Between the Maryland border and mile 17 you see peculiar outcrops of light gray limestone of the lower Ordovician Rockdale Run formation in the fields bordering the highway. Pinnacles of the limestone rise as scattered small knobs of rock projecting through the soil, their bleached appearance contrasting strikingly with the green of the turf. Most such exposures are on gentle swells or low hills; the same bedrock is buried under soil in the low areas. Pinnacle weathering occurs only in soluble carbonate rocks, and the resulting landscape is a kind of karst topography, named for the Karst region of Yugoslavia, but also developed on the Florida peninsula and many other places. Karst landscapes, like this one along the east side of the Cumberland Valley, typically have pinnacled outcrops, sinkholes, caves, disappearing streams, large springs, and underground drainage.

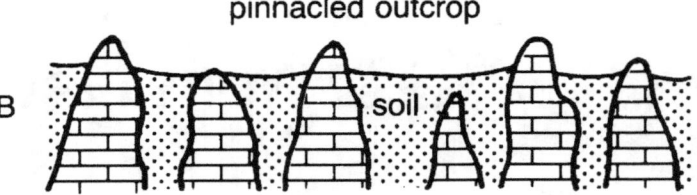

These schematic cross-sections show how pinnacled outcrops form. In A, numerous solution pockets develop along vertical joints in an incipient karst. In B, further solution leaves only slender pinnacles that project through soil.

The apparent simplicity of the geology in the Cumberland Valley segment of I-81 is deceptive. Hidden beneath the soil are numerous imbricate faults that branch from a deep-seated, flat-lying sole thrust and rise steeply toward the surface. Faulting, complex folding, and general fracturing all facilitated karst development. Infiltrating water, slightly acidic from small amounts of carbon dioxide absorbed from the atmosphere, slowly dissolved the rock.

East of the highway at mile 17, not far from Chambersburg, are sinkholes in upper Cambrian Shadygrove and Zullinger limestones. This side of the valley is lower and flatter than the western side because it contains so much soluble carbonate rock. Many old homes in the region are built of these attractive limestones. The hillier, western valley is underlain by more resistant shale, siltstone, and sandstone of the Ordovician Martinsburg formation.

South Mountain is visible a few miles south of the road between mile 17 and the Carlisle Exit 14. The ridge has stood above sea level and eroded to its present condition during the past 230 million years. It is the northern tip of the Blue Ridge that extends to Georgia and has a maximum width of 50 miles. The level-topped ridge visible on the other side of the valley near Carlisle is Blue Mountain, at the eastern

"Sandbank" quarry in lower Cambrian Antietam quartzite on Mt. Holly. Deep weathering of the quartzite facilitates crushing to make fine masonry sand.

edge of the Valley and Ridge province. Only a few pinnacled limestone outcrops expose rock along this 30-mile stretch of highway with almost no roadcuts.

Sandbanks of South Mountain

At Mt. Holly Springs, six miles south of Carlisle on Pennsylvania 34. watch for conspicuous large scars on the mountainside west of the village. These are deeply weathered, crumbly lower Cambrian Antietam quartzite that is being quarried and crushed for use principally as high quality masonry sand. Crushing quartzite to make sand illustrates the enormous importance of this mineral resource. The low intrinsic value of sand makes long transport too expensive, so it must be produced close to where it is needed.

This quartzite cliff, called Pole Steeple, is actually on the downthrown side of a fault. More rapid erosion of the upthrown block apparently reversed the relief produced by fault movement.

Pole Steeple

This striking cliff in the Montalto quartzite member of the Precambrian Harpers formation overlooks Mountain Creek valley. It stands 526 feet above and south of Laurel Forge Pond and may be reached by a short trail leading from near the dam at the east end of the pond. The Appalachian Trail passes east-west, a third of a mile south of the cliff. The rock is light gray, hard quartzite with well-defined beds that dip gently southeast into the northwest-facing cliff.

The beds contain a superabundance of the trace fossil *Skolithos*, worm burrows. In cross-section they appear as straight sand-filled tubes normal to bedding, and on bedding surfaces, as round or oval rings less than a quarter inch in diameter. *Skolithos* suggests that the original Montalto sands were deposited in an offshore bar environment — about 600 million years ago!

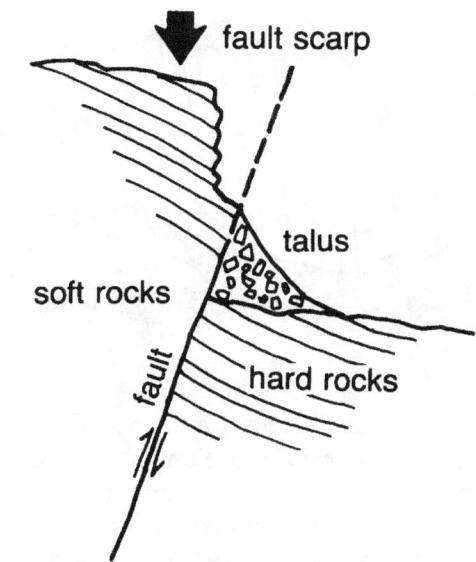

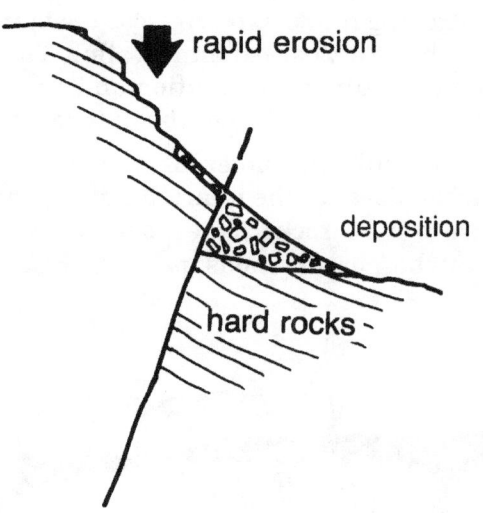

Hypothetical sequence of erosion at Pole Steeple that produced a reversal of the relief resulting from up-down fault movement.

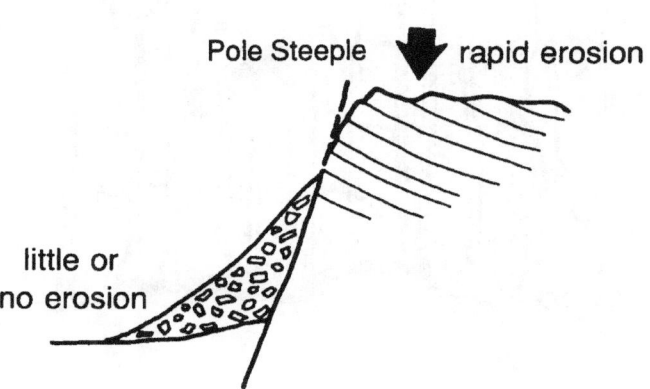

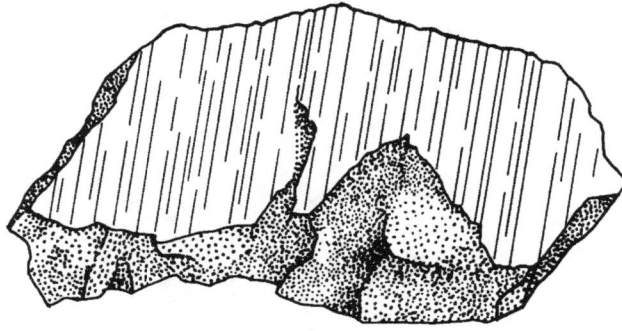

Hand specimen of quartzite containing a polished, striated surface, called slickensides, created by fault movement.

Precambrian metamorphosed volcanic rock and dolostone exposed around Laurel Forge Pond are separated from the quartzite by a fault on which the north, or pond, side moved upward. The present relief, therefore, is the reverse of that created by fault movement. This is common in faulted terrains where a resistant caprock, such as this quartzite, is more vigorously eroded on the upthrown fault block, destroying it and exposing less resistant rocks underneath that may then be easily carved into a valley. The dropped block, meanwhile, is in a less vulnerable position and even receives a protective cover of sediments derived from the upthrown side.

You will see numerous scratched and polished surfaces, called slickensides on the quartzite along the hike to Pole Steeple. These form where rocks slide past and grind against each other. The scratches, or striations, parallel the direction of fault movement.

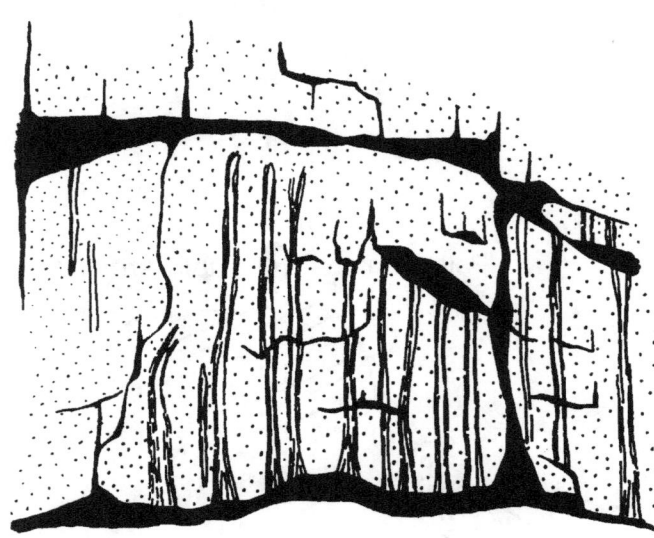

Skolithos worm burrows in the quartzite beds of Pole Steeple form brownish stripes normal to bedding. Area shown is 12" wide.

Pine Grove Furnace State Park

Numerous iron ore deposits on South Mountain were mined extensively in the 18th and 19th centuries. Pine Grove Furnace, southwest of Mt. Holly Springs, is a relic of a mining and smelting operation that lasted from 1773 until 1893, and played an important role in both the Revolutionary and Civil wars. The ore came from various open pits and shallow underground workings on the northwestern flank of South Mountain or in Mountain Creek valley. Most came from one large, 90-foot-deep open pit that now contains Fuller Lake. Because no coal exists in the region, charcoal made from local timber was used instead. The process required approximately 24 cords of wood to produce the charcoal to smelt enough ore to yield just one ton of pig iron. At the height of the industry, men cut and burned timber day and night to make charcoal, and a smoky haze covered the region constantly. In the end the lack of trees forced Pine Grove and other mining-smelting operations to close.

Limestone or dolostone for use as a flux, is plentiful in Cambrian and Ordovician strata of Cumberland Valley adjacent to South Mountain. In the blast furnace, the limestone combined with the nonmetallic impurities in the ore to form a molten slag that floated to the top of the furnace charge. Every hour or so, a worker raked the slag from the surface and discarded it. Slag is still scattered around the old furnace sites. It is glassy, varies greatly in color, and commonly contains holes formed from trapped gas bubbles.

The South Mountain ores, called brown ores because of their rusty appearance, are principally composed of hydrous iron oxides collectively referred to as limonite. They are essentially the same thing as rust. Most deposits are in or above the contact between the lower Cambrian Antietem quartzite (below) and the Tomstown dolostone (above) in sandy or clayey zones that are commonly also crushed. They are thought to originate from groundwater that percolated through the dolostone dissolving the existing iron compounds.

Boiling Springs

Four miles northeast of Mt. Holly Springs is the village of Boiling Springs, where a group of springs issue from folded middle Cambrian Elbrook limestones with a combined flow of about 16.5 million gallons per day.

Springs in cavernous limestones often discharge tremendous volumes of water because it flows through solution channels, caverns, that are like open pipelines. For the same reason, however, the water is often polluted and must be treated for drinking. Porous sand or

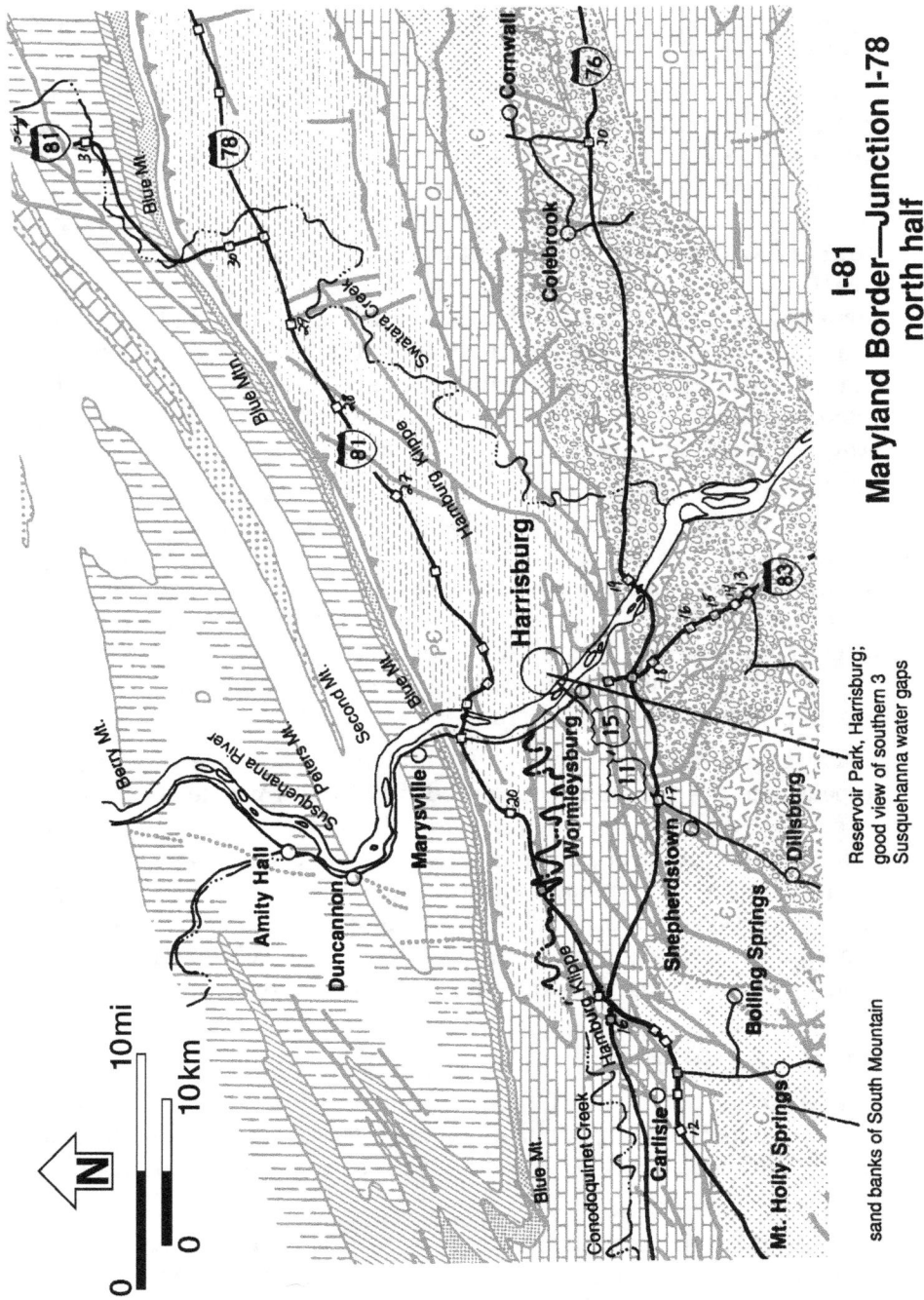

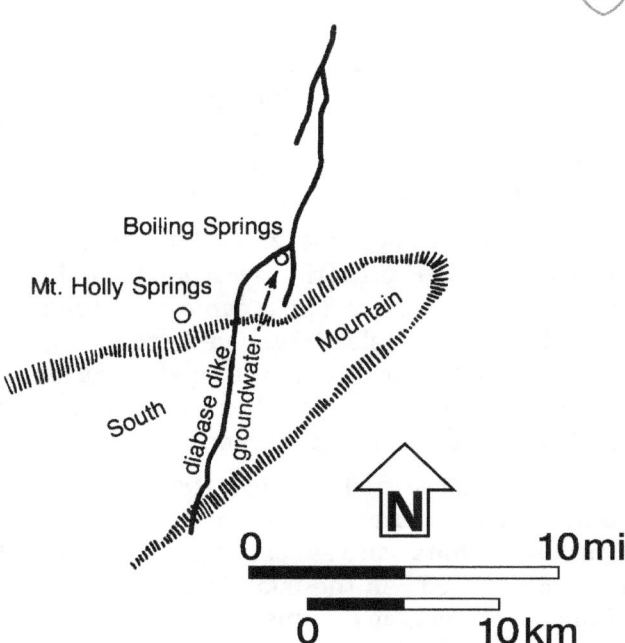

Northward progress of ground water flowing through Cambrian carbonates on the flank of South Mountain is blocked by a Y-junction in a diabase dike, causing it to surge to the surface at Boiling Springs.

sandstone aquifers are better filters so spring water issuing from them is more likely to be safe to drink.

The exceedingly heavy discharge stems from an unusual geological phenomenon. Folded carbonate rocks of the Elbrook formation contain a thin and impervious diabase dike that splits to form a Y junction. Water entering the Y under pressure is quickly forced to the surface. The same dike, trending north-south with branches, has been traced continuously for more than 25 miles.

• At mile 54, I-81 cuts through the Boiling Springs dike where it stands in moderate relief as Stony Ridge. The dike continues north across the Cumberland Valley and the first three ridges of the Valley and Ridge province.

Between miles 57 and 88, at the I-78 junction, the highway follows the Hamburg klippe, an enormous complex of thrust fault slices that traveled many miles from its place of origin and was then erosionally isolated. The klippe is 8 to 12 miles wide north-south and 80 miles long east-west. The rocks belong to the lower to middle Ordovician Hamburg sequence, mostly dark shales and graywackes with subordinate sandstones and limestones, all highly folded, faulted and sheared.

The dominantly carbonate rocks that floor the Cumberland Valley between the Maryland border and Carlisle are part of an immense shelf sequence of late Precambrian to middle Ordovician age. This time of tectonic quiet came as the Precambrian Grenville supercontinent rifted to open the Proto-Atlantic Ocean. The edge of the continent eroded and submerged to form a shelf on which sediments, mostly carbonates and sands, accumulated for about 200 million years. Meanwhile, dark muds, flysch, were deposited in the deeper water off the new continental margin.

In late Ordovician time, about 445 million years ago, plate motions shifted and the Proto-Atlantic Ocean basin began to close, initiating the Taconian orogeny. The oceanic crust broke east of the continent, and the western slab plunged beneath the eastern. A volcanic archipelago developed over the subducting slab, with a deep oceanic trench on its western side. Flysch sediments, derived from erosion of the islands, piled up in the trench.

The islands drifted closer to the continent as the oceanic trench continued to consume oceanic crust that separated them, and the dark flysch sediments were squeezed as if in a vice. They were folded and metamorphosed. As the islands collided with the continent at the climax of the orogeny, numerous thrust faults carried them in slices westward, and up onto the shelf. The Hamburg klippe is an erosional remnant of those thrust fault slices.

At mile 58.3, enjoy an excellent view of a large meander in Conodoquinet Creek just south of the highway. From here to the Susquehanna River, which it joins at Harrisburg, six miles straight east, the creek travels 18 miles in extremely broad meanders with remarkably straight sides that trend north-northwest, roughly parallel to the river and its famous water gaps in the ridges north of here. This strong alignment suggests that north to northwest trending joints or faults guided the stream courses as they eroded the bedrock. Between miles 62 and 65, the highway is hard against Blue Mountain ridge.

View of the Susquehanna water gaps north of Harrisburg.
—Drawing by Theresa Jancek (1988)

The Susquehanna Water Gaps

Of all the water gaps in Pennsylvania, those north of Harrisburg are among the most striking. The George N. Wade bridge carries I-81 across the river just south and in clear view of the southernmost gap through Blue Mountain. A better perspective is available from an overlook in Reservoir Park, in the city east of the river by US 22. From there, the flat-topped ridges appear stacked rank after rank in the distance, like opposing armies confronting each other across the river.

Even crestlines and water gaps are principal players in a long controversy over the origin of Pennsylvania's Valley and Ridge landscapes. In one interpretation, the even crestlines are seen as relics of an old erosion surface that cut across hard and soft bands of the folded bedrock. The land was then broadly and gently arched, causing the streams to erode more vigorously. The resistant ridges emerged in erosional relief as the streams preferentially eroded the softer formations. Water gaps formed where the original streams trended across a resistant formation and maintained those courses during the general degradation of the landscape.

Schematic landscape evolution leading to stream piracy and development of water gaps.

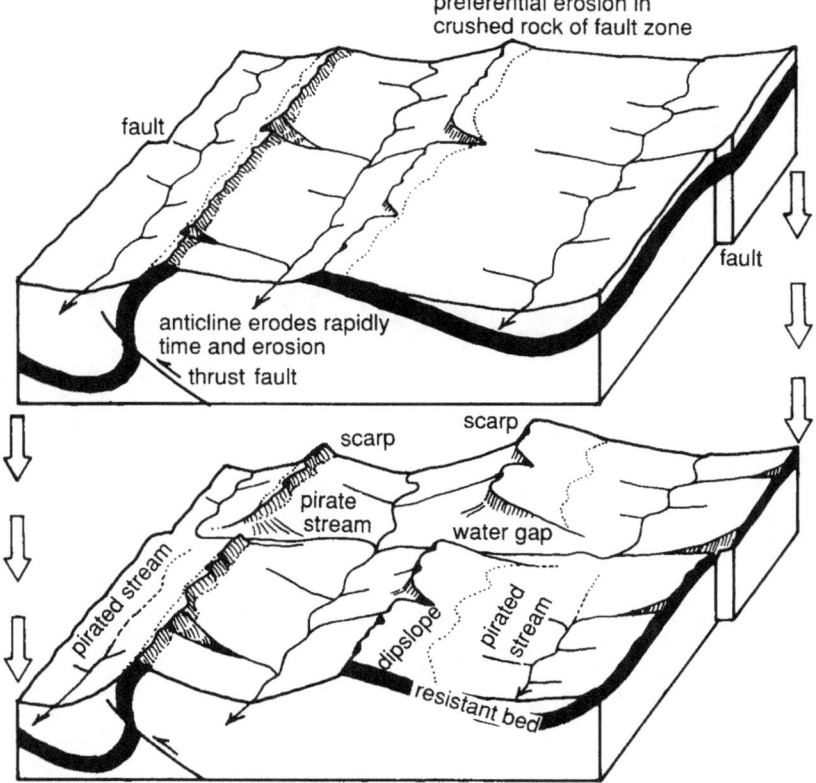

At least some of the water gaps appear to result from stream piracy. This can happen, for example, when a tributary stream cuts headward through a ridge, and intersects another stream on the other side. It pirates the upstream discharge of the second stream, increases its own discharge, and establishes itself as a major drainage route. Certain tributaries may have had an erosive advantage because they follow weak zones of crushed rock where faults cross-cut hard ridges. This may be the case in the Susquehanna gap through Blue Mountain where outcrops beside the river display extensive fracturing.

• The evenness of Blue Mountain ridge is quite apparent between miles 65 and 85, where I-81 runs parallel to it and two to three miles south of it. Pilktown and Manada gaps slice through the ridge near miles 73 and 75, Indiantown gap near Exit 29, three miles southwest of the I-78 junction. All the bedrock southwest of that junction belongs to the Hamburg sequence.

I-83
Maryland Border—Junction I-76
39 mi./63 km.

Between the Maryland border and Emigsville Exit 11, 24 miles, I-83 crosses a segment of Pennsylvania Piedmont province. The poorly exposed metamorphosed rocks consist mostly of flakey schists that weather to a rusty color, along with gneisses, greenschists, and quartzites derived from original sandstones. The rocks are complexly folded and locally faulted, but the erosion surface cuts across these structures in a landscape of only moderate relief. The rocks were deformed and metamorphosed in the core of the rising Taconian Mountains.

Triassic and Jurassic diabase dikes cut through the Piedmont in many places, trending mostly north or slightly east of north. Some continue for many miles. The highway crosses a dike near Loganville, Exit 3, that has a total length of 14 miles. It is one of three aligned dikes with a combined length of about 35 miles that extends into the Gettysburg basin. Many diabase dikes stand like ruined walls because the rock is more resistant than those they intrude, but the one near Exit 3 does not. All of the dikes were injected into cracks that opened during rifting of the supercontinent Pangaea, beginning nearly 200 million years ago. The same plate movement produced the Gettysburg basin and other half-grabens of eastern North America before the juvenile Atlantic Ocean basin opened wide enough to fill with water.

Exit 11 is on the southern boundary of the Gettysburg basin. This boundary is an erosional unconformity that represents a gap in the geologic record of about 400 million years. We have no record of what happened here during the period from nearly 600 to 200 million years ago, the approximate ages of the rocks below and above the unconformity, respectively. Apart from the diabase dikes and sills, the rocks of the Gettysburg basin are mostly red shales, siltstones, sandstones, and conglomerates. They were derived by erosion of the bordering highlands during basin subsidence. When the original sediments accumulated, this segment of continent was just one tiny piece of a vast interior region of Pangaea that may have had a semi-arid climate. The red color is rust. The iron was derived from weathering of iron-bearing minerals in the sediment source rocks. The two principal minerals that contribute to the rusty color are brownish limonite, a hydrous iron oxide, and reddish hematite,

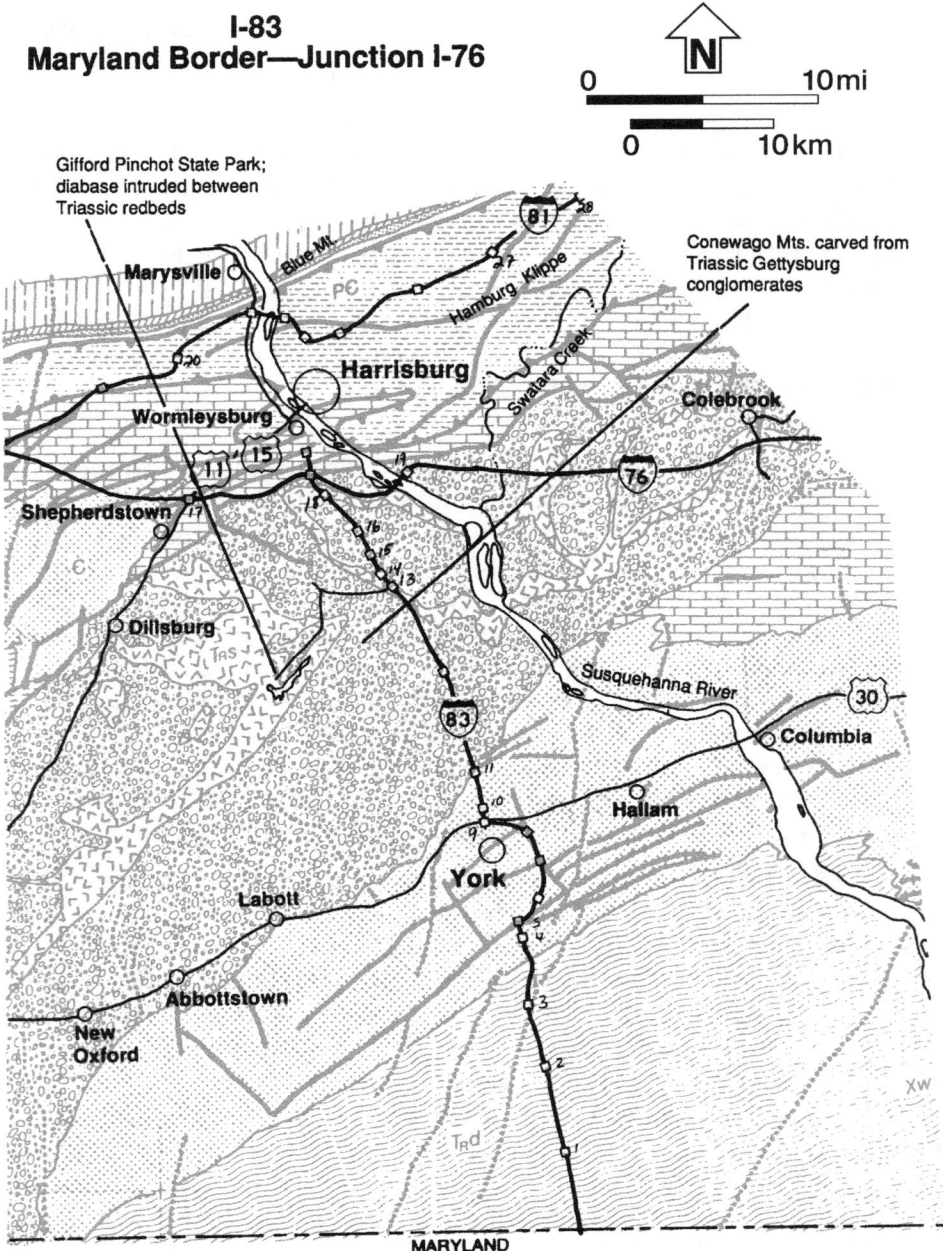

I-83
Maryland Border—Junction I-76

Triassic red shale near mile 30 just north of Exit 12 on I-83, largely broken down to a rubble slope.

anhydrous iron oxide. The association of diabase dikes and red sediments is a worldwide characteristic of continental rift basins. Geologists call it the rift facies.

Only the north side of the Gettysburg basin dropped along a fault. The sedimentary beds were originally horizontal, but they now dip toward the fault. The traverse across the Gettysburg basin between Emigsville and the I-76 interchange, 15 miles, is topographically more interesting and variable than that of the Piedmont section, and roadcuts are more numerous. Near Strinestown, Exit 12, you see the Conewago Mountains west of the highway rising more than 500 feet. The hills are held up by the Triassic Gettysburg conglomerate, which is more resistant than the surrounding redbeds. Other lower hills a little farther north and northwest, are held up by diabase, which stands high almost wherever it occurs, faithfully signaling its presence.

Gifford Pinchot State Park

This park is about seven miles southwest of Newberrytown, Exit 13, by the route shown on the map.

The diabase hills are part of a very large sill that formed as molten magma intruded between Triassic beds about 180 million years ago. The sill and beds dip about 35 degrees to the northwest. The intense heat of the magma baked the enclosing shale and sandstone in narrow zones adjacent to the upper and lower contacts, making those rocks more resistant than the diabase. Erosion left the contact zones standing high, and carved a basin in the less resistant diabase between them. Pinchot Lake, which now occupies the basin, is an artificial lake formed by damming of Beaver Creek.

The diabase is dark gray, massive, dense, and uniformly fine-grained. It consists mostly of plagioclase feldspar and pyroxene, both of which contribute to the dark color of the rock. Nearly every outcrop is broken along numerous criss-crossing shrinkage cracks, called joints, that form by contraction as the magma crystallizes and the rock cools. Probably the best example is at the section of park called Old Toboggan Run, where the rocks retain their angular, joint-bounded form; traces of joints are etched all over the faces of the boulders. In most other places, the original angularity is more or less subdued by spheroidal weathering.

The combination of abundant joints and spheroidal weathering have reduced many outcrops to piles of semi-rounded boulders, and created some odd erosional forms. Balanced Rock at Boulder Point, for example, perches on two small pedestals, all that remains of a deeply-weathered, once-continuous layer under it. Weathering of the boulder probably aided the balancing act by directing water to the underlayer, causing it to weather more rapidly and flushing out the rock waste.

• You can examine contact-metamorphosed rocks in places along Straight Hill, the ridge south of the lake where the redbeds were baked into a very hard dark gray rock. In this case, as in many instances of contact metamorphism, the original rock composition has been changed by reaction with hot, mineralized fluids that emanated from the intrusion.

You see more redbeds between Newberrytown and the I-76 interchange, 7 miles. At a rest area between Exits 13 and 14 are spheroidally weathered diabase boulders like those of Gifford Pinchot State Park.

US 15
Maryland Border—Amity Hall
61 mi./98 km.

Between the Maryland border and Gettysburg, you cross the central part of the Gettysburg Basin, twice passing through the knobby ridge of the Gettysburg sill. In the central part, you are east of the section of ridge that was the focus of the Battle of Gettysburg. The ridge is rocky and tree-covered, but surrounding it is low, gently rolling farmland with a lot of red soil derived from underlying Triassic redbeds. In the Gettysburg Campaign, the Union armies marched north over these same red plains all the way from Frederick, Maryland, avoiding the hilly country of the diabase. Earlier, they had marched the length of the similar Culpepper Basin of Virginia.

Between the US 30 junction at Gettysburg and Shepherdstown, 24 miles, you cross more of the plain. For some distance south of Shepherdstown, South Mountain hovers west of the highway. Erosion carved it from an elongate northeast-trending anticlinal fold with a core of Precambrian rocks. The western side of the anticline is mantled with sedimentary strata that become progressively younger westward, ranging from lowest Cambrian to late Ordovician. By contrast, the eastern limb of the fold in Pennsylvania is abruptly truncated by the steeply-dipping border fault of the Gettysburg Basin, which places metamorphosed Precambrian volcanic rocks against Triassic redbeds and diabase. The original volcanic rocks testify to Precambrian plate tectonic events similar to those that produced the Triassic basins, but much earlier. They involved rifting of the Grenville supercontinent and opening of the Proto-Atlantic Ocean, beginning about 650 million years ago. The continent had assembled in the plate tectonic convergence that produced the ancestral Adirondack Mountains about a billion years ago.

Rifting stretches the lithosphere, making it thinner. Extremely hot rocks in the underlying asthenosphere normally do not melt because the pressure exerted by the weight of the lithosphere is so great. They do melt when that pressure is relieved by rifting, and the magma ascends along the newly opened fractures to form dikes, sills, and other intrusions, as well as to fuel volcanic eruptions.

Thus, South Mountain occupies a geologic setting where two rift events happened side by side. In the 450 million years between them, one supercontinent broke up; an ocean basin opened to its maximum

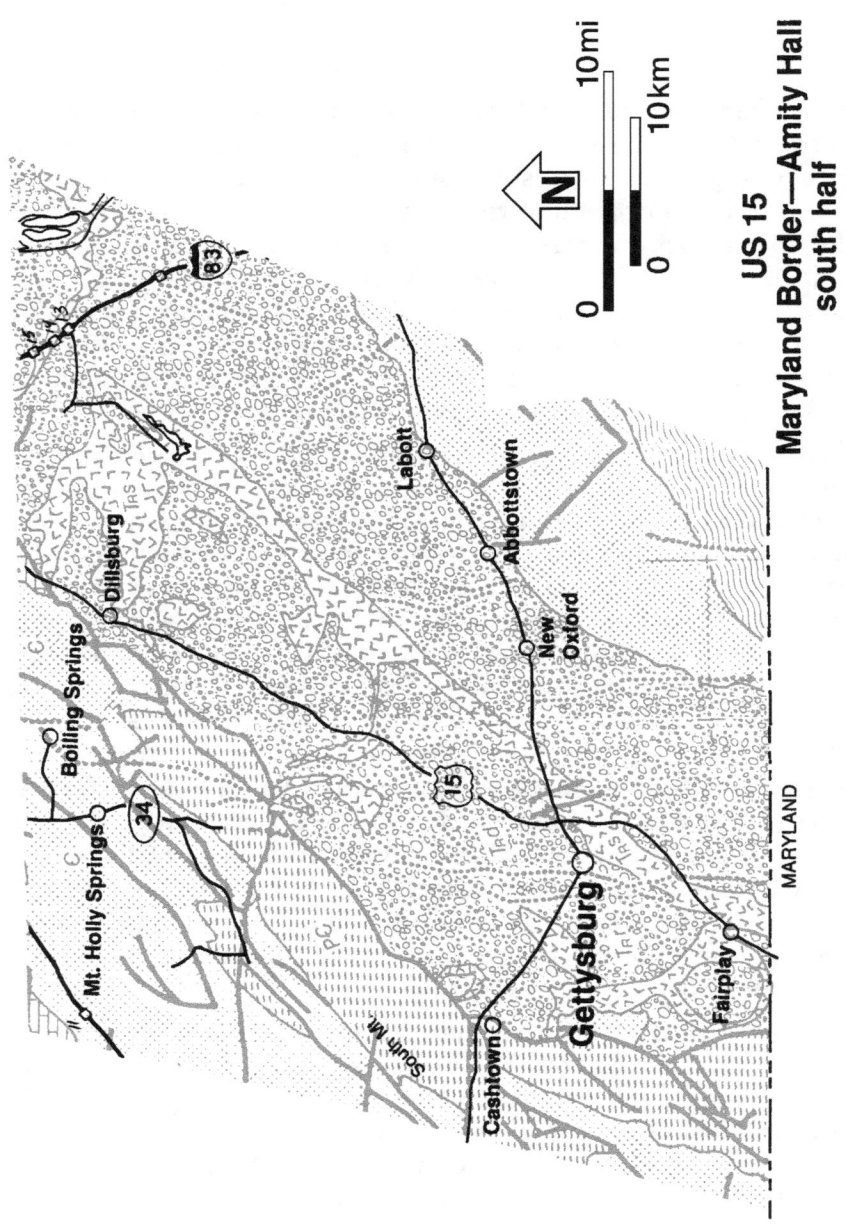

US 15
Maryland Border—Amity Hall
south half

width, then closed, causing three major mountain building events and assembling the supercontinent Pangaea; finally Pangaea rifted. Only the simplest forms of life existed as those events began, all of them marine. By Triassic time when Pangaea rifted, life had become incredibly diverse and had proliferated on land and in the sea; the age of dinosaurs had begun.

Shepherdstown lies just outside the Gettysburg Basin on weak Cambrian rocks at the edge of the Cumberland Valley and northeast of the tip of South Mountain. The lower Paleozoic strata of the valley on the northern and western flanks of South Mountain are complexly folded and faulted. This part of the valley lies on Cambrian and Ordovician limestones; late Ordovician Martinsburg shales and graywackes of the Hamburg klippe underlie the northern side.

The route takes you through the sprawling metropolitan area of greater Harrisburg west of the Susquehanna River. Between Wormleysburg and the I-81 junction, 4.5 miles, the route follows the west bank of the river through West Fairview, Enola with its railroad yards, and Summerdale, past excellent views of the island-studded river, and the water gap through Blue Mountain.

Conodoquinet Creek, which you cross between Wormleysburg and West Fairview traces a remarkably winding course on its way east to join the Susquehanna River. Its meanders have unusually straight sides that trend northwest, parallel to the Susquehanna water gaps

Sketch of part of the Rockville cut in the Susquehanna gap through Blue Mountain on US 22/322 north of Harrisburg. Extremely numerous small faults, represented by heavy lines, suggest that the river chose this place to cut through because the rock was weakened by crushing. —Adapted from Theisen (1983. p. 8-9)

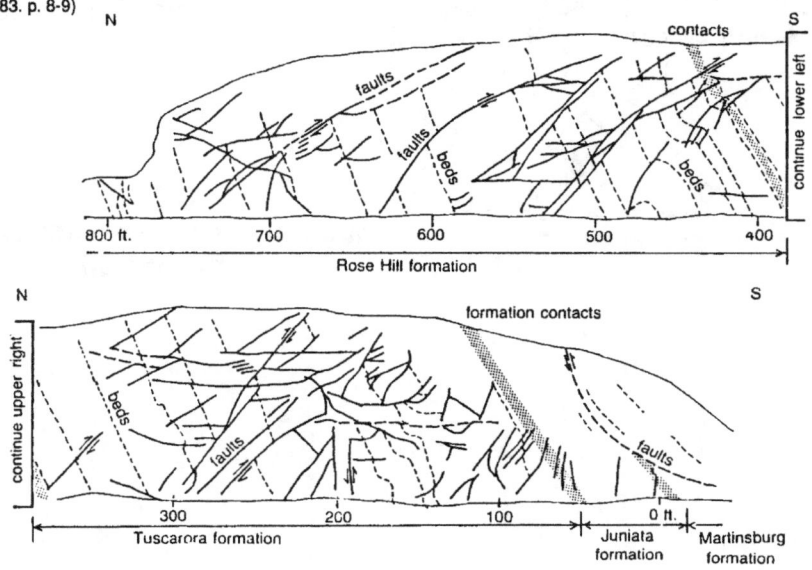

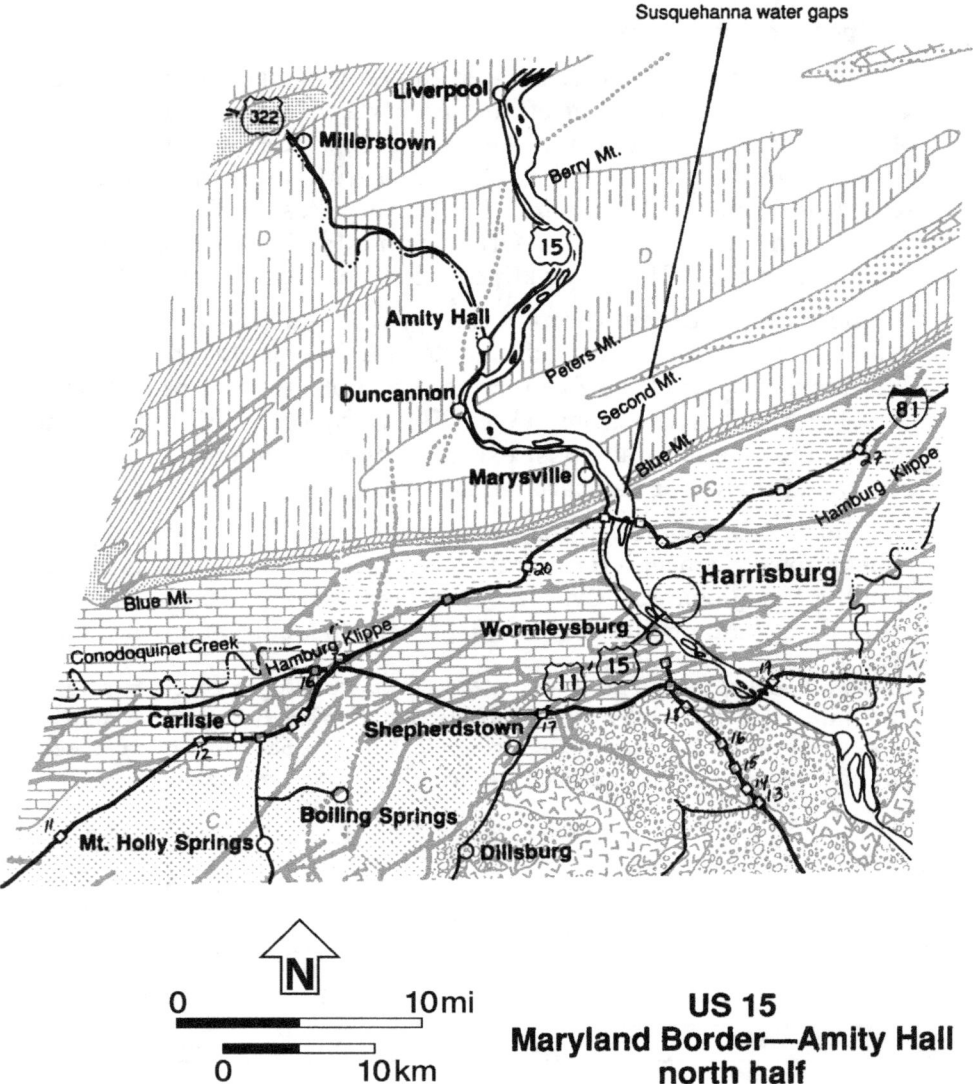

US 15
Maryland Border—Amity Hall
north half

through Blue, and Cove, or Second, mountains. The aligned meanders almost certainly follow fractures, as may the Susquehanna water gaps.

Between the I-81 junction and Amity Hall, you pass through all three of the Susquehanna water gaps. Roadcuts along US 15 and also along the River Relief Route on the eastern side of the river, shed some light on their origin. Major water gaps generally follow zones of weakness in the rocks. Rocks exposed in the Blue Mountain gap, often called the Susquehanna gap, include the Rose Hill, Tuscarora, Juniata and Martinsburg formations. On the eastern side of the river, the Rockville cut is 900 feet long, and exposes beds that dip steeply south at an average angle of about 60 degrees. They are chopped up by many small faults of variable orientation that may be related to a major cross-cutting fault along the course of the river. That weak zone of fractured rock may explain why the river chose this particular place to cut through.

The Susquehanna water gaps slice through the western tip of the Southern Anthracite syncline. Lower Silurian Tuscarora quartzite caps Blue Mountain. Lower Mississippian Pocono sandstone caps the next two ridges: Cove on the south and Peters Mountain on the north.

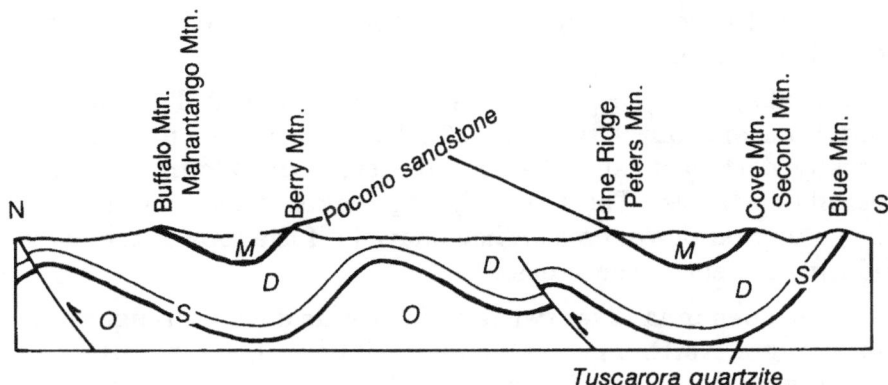

Schematic cross-section along the Susquehanna River through the two major synclines north of Harrisburg. Only Blue Mountain is held up by Tuscarora quartzite; the other four ridges are sustained by Pocono sandstone.

A few miles to the west, Cove and Peters mountains join at a remarkably sharp point where the synclinal axis rises to the surface. From the air, the dipslopes of these fold limbs are conspicuous, and the whole structure resembles a partly submerged canoe with its bow

Looking down to the northeast on the bow of the "canoe structure" through which the Susquehanna River has cut several gaps. Harrisburg lies beside the river just out of the right side of the picture. —Drawing by Theresa Jancek (1988)

breaking water where the ridges join. This topographic expression is typical of many large Valley and Ridge folds. The zig-zag outcrop patterns on the geologic maps express these structures; if synclines zig, anticlines zag.

Here, all synclines point west; anticlines point east. The patterns result from erosion of two sets of folds, the major one trending generally northeast, and the second, very mild folding, trending generally northwest. The second folding makes the axes of the first folds go up and down; erosion has shaved off the tops of the folds, producing the zig-zag outcrops.

The Susquehanna River twists and turns as it cuts through these several ridges. North of Peters Mountain, it flows directly south, then west for two miles alongside the ridge before turning 90 degrees to cut right through it. At Cove Mountain, the river turns east and follows the ridge flank for four and one half miles before turning sharply again to cut straight through Cove and Blue mountains. The long ridges exist because the rocks on their crests resist erosion.

The village of Duncannon lies on the riverbank just north of Peters Mountain, between the mouths of Sherman Creek and Little Juniata River on the south and the Juniata River on the north. Most of the village is high on the side of a hill that rises about 200 feet above the Susquehanna River, well above the reach of floods.

Between Duncannon and Amity Hall, two and one half miles, you cross the Juniata River and follow Duncan Island north. Two canals once terminated here. One followed the Juniata River. The other, the Susquehanna division of the Pennsylvania Canal, came 41 miles down the Susquehanna River from Northumberland, through 13 locks and seven aqueducts, past old Shamokin Dam, and across the West Branch Towpath Bridge. Traces of the canal, which was built in 1828-31 and operated until 1901, remain visible north of Amity Hall.

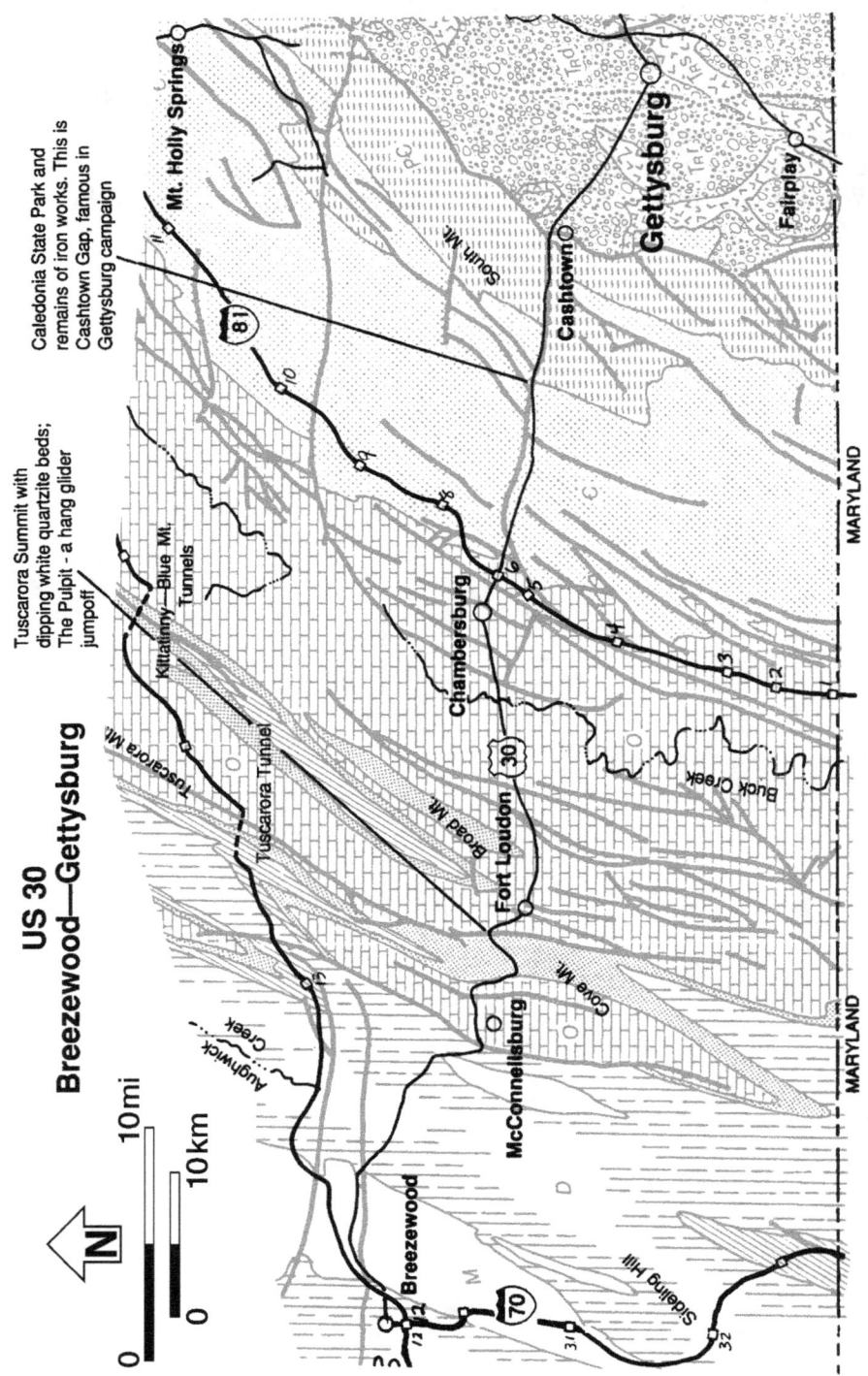

US 30
Breezewood—Gettysburg
66 mi./106 km.

The Lincoln Highway

The route between Breezewood and Saluvia, eight miles, begins and ends on reddish shales and sandstones of the upper Devonian Catskill formation as it crosses the Broad Top syncline. In the wide core section of the fold are buff sandstones of the lower Mississippian Pocono formation. Interstate 76, a short way to the north, crosses the identical section through spectacular large cuts. That segment of turnpike originally crossed a little farther north through the Rays Hill and Sideling Hill tunnels, now abandoned. US 30 goes over the tops of the hills on steep inclines, reaching a maximum elevation of 2195 feet on Sideling Hill. It has fewer roadcuts, but good views. The hills are ridges of resistant Pocono sandstones. Along the crest, the route follows a major east trending fault with an overall length of more than 30 miles that cuts across anticlines and synclines alike.

This is one of several large, vertical faults that cut across the structural grain of the Valley and Ridge province in south central Pennsylvania offsetting the ridges. They are tear faults that trend perpendicular to the trend of the folds and thrust faults, and moved horizontally. They formed where one part of the sole thrust ramped up at one place, and the adjacent part advanced farther before it also ramped up. The rock package carried to the farthest ramp was moved horizontally past the first ramp.

Between Saluvia and Breezy Point, three miles, you cross an anticlinal arch in Licking Creek Valley, over weak shales and siltstones of the middle Devonian Hamilton group and upper Devonian Brallier and Harrell formations. These dark, fine-grained rocks are deep water deposits laid down in the Acadian foredeep basin. Their deposition preceded the collision of North America and Europe in late Devonian time.

Between Breezy Point and the US 522 junction near McConnellsburg, five miles, the highway wends its way between the ridges, climbing slightly at one point to cross the bow of a topographic canoe eroded in the Meadow Grounds syncline. Immediately south of the bow, the canoe opens, with Scrub Ridge on the port side and Meadow Grounds Mountain on the starboard. The stern, also topographically

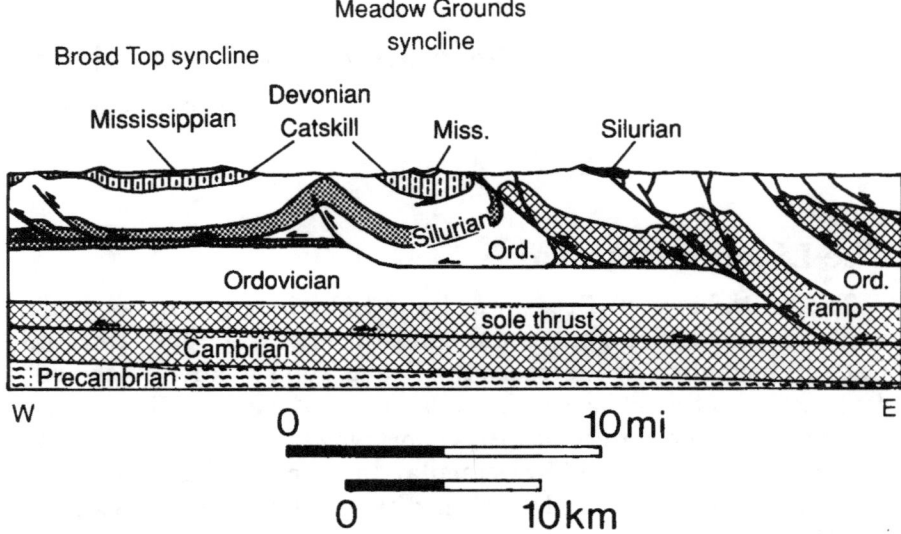

Geologic cross-section approximately east-west near US 30, through Broad Top and Meadow Grounds synclines, showing thrust fault control of folding. —Adapted from Berg and others (1980, Plate 2)

distinct, rises about seven miles to the south, where Roaring Run cuts through the starboard side and drains the interior. Streams created this peculiar form by excavating the relatively soft Mississippian Mauch Chunk redbeds in the core of the syncline, whose axis, the keel of the canoe, rises to the north and south. Meadow Grounds Lake now occupies the central basin. The harder lower Mississippian Pocono sandstone holds up the ridges. The structure is a basin because beds all around it dip inward.

The Meadow Grounds syncline lies between two steep thrust faults that branch upward from a flat-lying sole thrust in weak Ordovician shales some 15,000 feet below the surface. The western thrust is expressed on the surface only by arching of the beds above it. The eastern fault places Silurian and Ordovician rocks on the surface against and east of upper Devonian Catskill formation. You cross these faults at the base of Meadow Grounds Mountain, two miles south of the canoe bow, or one and a half miles west of the US 522 junction to McConnellsburg. US 30 traces a hairpin curve over the fault and across the south end of Little Scrub Ridge, an outcrop of Tuscarora quartzite.

Between McConnellsburg and Fort Loudon, eight miles, you cross a complex array of folds and faults and go over Tuscarora Summit. The McConnellsburg, or Big Cove Creek Valley is carved from an anticlinal arch cored with upper Cambrian Shadygrove limestone that crops out just south of the highway. This limestone contains stromatolites, evidence that the original sediments were deposited in a shallow, near-shore setting periodically exposed to the air. Middle Ordovician limestones and dolostones exposed near McConnellsburg rim the fold. The carbonates are part of a vast sequence of continental sediments deposited along the edge of a smaller North America in late Precambrian to middle Ordovician time, while the Grenville supercontinent rifted and the Proto-Atlantic Ocean basin opened. This was a long time of tectonic quiet on the ocean borders, when continental margins wore down and sank below sea level, creating the ideal shallow marine environment for carbonate and sandstone deposition.

On the western side of Tuscarora Mountain, you see olive-gray shales of the upper Ordovician Reedsville formation. Tuscarora Summit provides excellent exposures of the Tuscarora quartzite, particularly at a hang glider takeoff called The Pulpit, just north of the road. Look for a white, well-bedded sandstone that dips gently west. The Pulpit offers a sweeping overview of Cumberland Valley. The long, slanting road segment east of Tuscarora Summit follows Township Run between Tuscarora and Cove mountains. The route slants down along the dip slope of Tuscarora Mountain, where the quartzite crops out in several places. Near the base of the slant is a messy exposure of Tuscarora quartzite broken along a fault zone; more Reedsville dark shales appear nearby.

The Pulpit, a hang glider jumpoff on Tuscarora summit overlooking Cumberland Valley. Lower Silurian Tuscarora quartzite beds dip to left.

Fort Loudon is at the western edge of the Cumberland Valley. Along US 30 the valley is 19 miles wide, an attractive landscape of gently rolling farmland, fairly typical of the rest of the Great Valley in Pennsylvania, New Jersey, New York, Maryland, and Virginia.

Roadcuts are scarce in the 14 miles between Fort Loudon and Chambersburg. The upper Ordovician Martinsburg formation, which lies beneath most of this section of valley, is easy to recognize by its dark gray color, flakeyness, and interbeds of muddy sandstone. Recognize the Ordovician limestones by their pale gray color, and locally by their pinnacled outcrop pattern. The pinnacles appear as scattered patches of limestone projecting through the turf. They could easily be mistaken for separate boulders.

This southern section of the Cumberland Valley held several forts built about 1755 for protection from the Indians. They were Fort Loudon, one mile east of the village; Fort McCord, just west of St. Thomas; Fort McDowell at Markes; Fort Davis near Welsh Run; Fort Marshall, five miles south of Mercersburg; and Fort Waddel, west of St. Thomas.

In June 1863, General Robert E. Lee used Chambersburg as a concentration point in what began as a campaign to capture Harrisburg and ended in the Battle of Gettysburg. The choice was a highly strategic one, and the configuration of the land that made it so, reflects the bedrock.

Between Chambersburg and Fayettevile, five miles, you cross the eastern side of Cumberland Valley, underlain almost entirely by Ordovician and Cambrian limestones and dolostones. The carbonate rocks offer even less resistance to erosion than the Martinsburg shales and graywackes, so this side of the valley is very flat. Some farm fields are full of pinnacled outcrops formed by solution of the limestone bedrock. Caves and sinkholes also exist, their development undoubtedly facilitated by the presence of many faults that provided underground passageways for water.

Fayetteville is at the western end of Cashtown Gap, the only easy pass through South Mountain, and the one US 30 now follows. The gap was another reason for Lee's choice of Chambersburg as a concentration point; it gave him the freedom to move east, if circumstances demanded – as they did. Streams carved the gap along the crushed rock of the Cashtown fault, another east-west tear fault that lopped off the northern 30 miles of South Mountain and shoved it three miles west. The gap extends from Fayetteville to Cashtown, ten miles. Outcrops are rather nondescript, largely because faults have so severely broken the rocks. In the western six miles, the rocks are Cambrian quartzites, carbonates, and shales. In the eastern four miles you see metamorphosed Precambrian rhyolites, among the oldest rocks exposed on South Mountain.

The rhyolites have been interpreted as evidence of a period of volcanic activity initiated by rifting of the Grenville supercontinent that began about 650 million years ago. That ultimately created the Proto-Atlantic Ocean basin, and an arrangement of continents very different from that of today. The rhyolites were buried after they erupted, and metamorphosed as they recrystallized under the high temperatures and pressures at depth. They now look quite different from the original rocks, and consist of different minerals. In Cashtown Gap, the metamorphosed rhyolites appear in shattered outcrops of medium-gray, fine-grained rock that locally contains large, eye-shaped crystals of quartz. Beyond the highway, dark basalt interlayers appear in the rhyolite.

Caledonia State Park, which contains the ruins of the Caledonia iron ore furnace, is in the middle part of the gap. The furnace is a relic of a mining and smelting industry that flourished for more than 120 years on South Mountain. All the necessary ingredients were here: iron ore, limestone used as flux in smelting, forests to make charcoal, and water power to run the bellows. The industry died in 1893 when the trees were gone.

The Cashtown fault appears to end at Cashtown where the western border fault of the Gettysburg Basin truncates it, as well as the whole eastern flank of South Mountain. The border fault dips steeply to the east and drops the eastern side nearly 15,000 feet. Crustal stretching

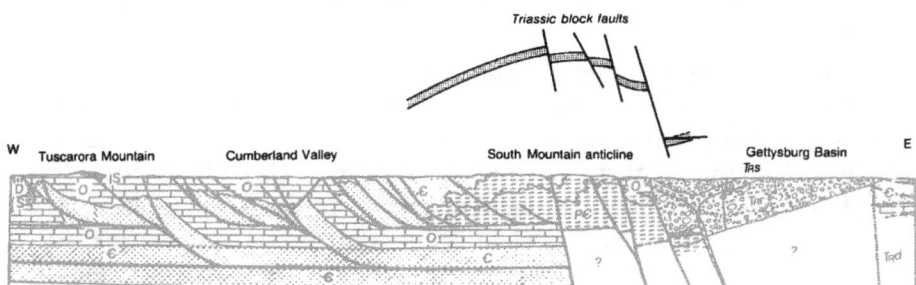

Cross-section near U.S. 30 with no vertical exaggeration shows the anticlinal form of South Mountain cut by Triassic border faults of the Gettysburg Basin, a half-graben that is hinged, but not faulted, in its east side. Also note the imbricated (branched) thrust faults in the region west of South Mountain. —Adapted from Socolow, A.A., 1980, Geologic Map of Pennsylvania, 1:250,000, Pennsylvania, Geologic Survey

during the breakup of the supercontinent Pangaea, caused this and related faulting, beginning about 200 million years ago. Sediments filled the basin as it subsided, and were intruded by magma that became diabase sills and dikes. The magma probably ascended from the upper part of the asthenosphere, where the rocks are hot enough that they would melt were they not under the high pressure exerted by the weight of the lithosphere above. They do melt when that pressure is relieved by stretching, thinning, and cracking of the lithosphere.

Along the seven miles between Cashtown and Gettysburg, you continue on Lee's 1863 route from Chambersburg that led to the Battle of Gettysburg. This is gently rolling, very attractive farmland, much as it was then, but more developed now. Red soils reveal the underlying Triassic redbeds. The hills are all carved from relatively resistant diabase, and though not very prominent, they greatly influenced the outcome of the Gettysburg battle.

Bedrock and the Gettysburg Campaign

The Gettysburg Campaign really began a month before the climactic Battle of Gettysburg in an area about 80 miles wide and 140 miles long, between Fredericksburg, Virginia and Harrisburg, Pennsylvania. Three major physiographic provinces parallel each other: the Great Valley, the Blue Ridge, and the Triassic lowlands of Virginia and Maryland whose extensions in Pennsylvania are the Cumberland Valley, South Mountain, and Gettysburg Basin. The topography of these provinces played a key role in the course of the campaign. Troops had to move along lines of least resistance, making use of the natural defenses.

The campaign commenced on June 3, 1863 near Fredericksburg when General Lee began a march against Harrisburg. Lee moved his troops west to the Blue Ridge, then through the wind gaps to the Great Valley, and finally north over easy terrain to Chambersburg. The ridge helped shield these movements from the Union armies, which were marching north through the Triassic lowlands on its other side. Lee chose to concentrate at Chambersburg where the landscape offered freedom of movement northeast and south along Great Valley and east through Cashtown Gap.

By June 28, Lee had troops at Chambersburg, Carlisle and Wrightsville in position to attack Harrisburg, when he learned that Union troops were concentrating at Frederick, Maryland, south of

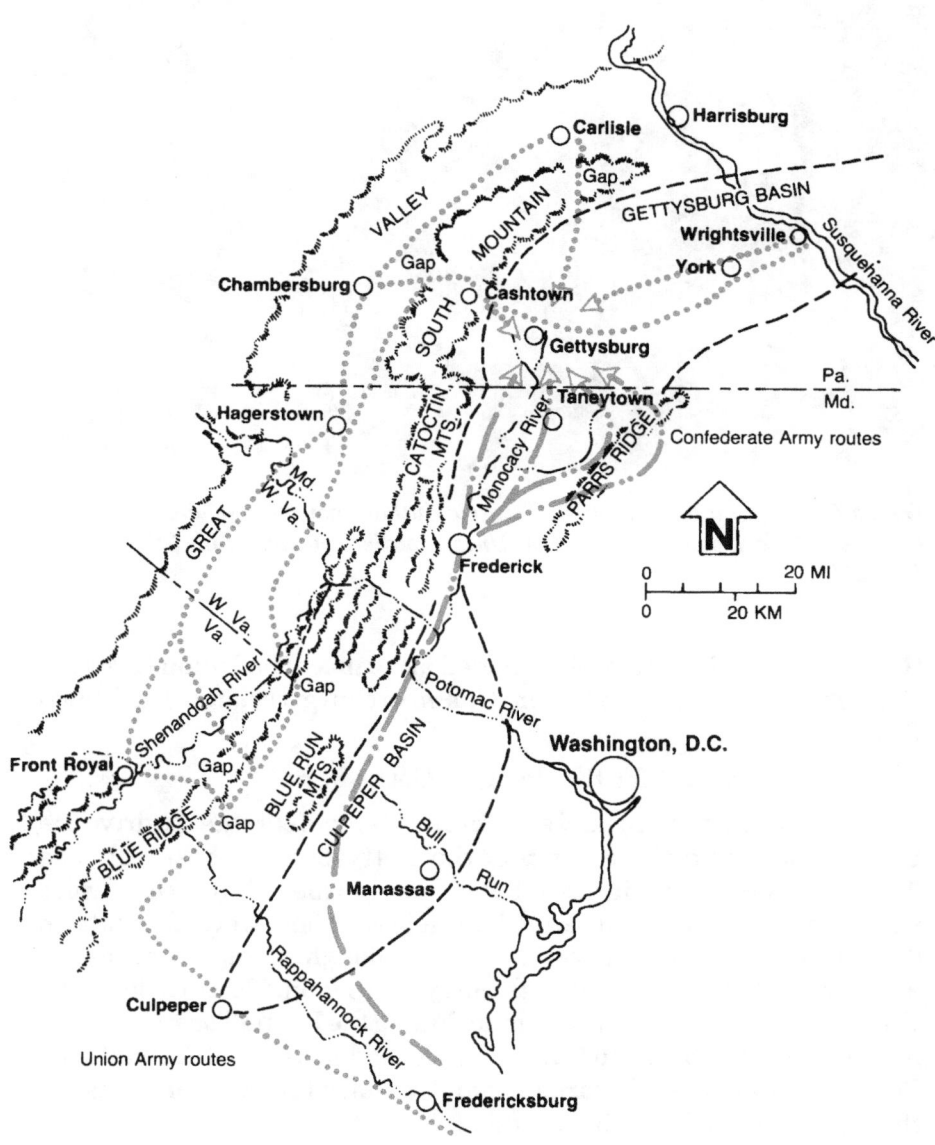

The marching routes of the Union and Confederate armies from Fredericksburg to Gettysburg between June 3-30, 1863. Both armies followed the valleys, and when they had to cross mountains, they followed natural gaps. —Adapted from Brown (1962, p. 3)

Devils Den, a jumble of rounded Triassic diabase boulders where Confederate sharpshooters fired on Union soldiers stationed on Little Round Top.

Gettysburg. He immediately directed all to meet the Union threat at Gettysburg and marched from Chambersburg through Cashtown Gap.

The stage was set for the Battle of Gettysburg.

The battle was essentially an effort by the Confederates to drive the Union Army from the outcrop of the Gettysburg diabase sill. The diabase everywhere in the battlefield formed hills that were simultaneously strategic and defenseless. They were difficult to defend because the soil was not deep enough to dig trenches, so boulders and outcrops offered the only protection. Nevertheless, the Union Army commanded the hills, including the Round Tops, Cemetery Ridge, Cemetery Hill, and Culps Hill, throughout the battle, while the Confederates mounted charges from Seminary Ridge, a gentle rise to the west, held up by a diabase dike.

US 30
Gettysburg—Lancaster
60 mi./97 km.

Between Gettysburg and the small village of Labott, 19 miles northeast, US 30 completes the traverse across Gettysburg Basin. The knobby hills near Gettysburg are all carved from the hard diabase of the Gettysburg sill. They are part of the same rocky ridge over which both sides fought so bitterly in the Battle of Gettysburg. Outcrops and roadcuts are scarce elsewhere, but the red soil indicates that Triassic redbeds are just below the surface. Some of the old stone homes in the region are built of red sandstone; others are made of diabase.

Labott lies on the boundary between the Gettysburg Basin and the Piedmont provinces. The underlying Cambrian dolostone is about 400 million years older than the Triassic beds that rest directly on top of it. We have no rock record here of Earth history between about 600 and 200 million years ago, the ages of the rocks below and above the unconformity, respectively. However, the deformation and metamorphism of the Cambrian rocks is a record of the Taconian, possibly Acadian, and probably Alleghenian orogenies.

In the ten miles between Labott and the I-83 junction at York, the highway is on Cambrian limestones or dolostones of the Piedmont. These are part of the vast expanse of shallow marine rocks, mostly carbonates and sandstones, formed on the North American continental shelf during opening of the Proto-Atlantic Ocean. Note the quarries on the north side near Thomasville and on the south, two miles east of Thomasville. Many old stone homes and barns visible from the highway reflect the local bedrock types. Most are limestone or dolostone since these are the most abundant rocks.

York played an important role in Pennsylvania's history. The town, laid out in 1741, was the first west of the Susquehanna River, the seat of the Continental Congress from 1777 to 1778, and the birthplace of the Articles of Confederation. In the Gettysburg Campaign, Confederate troops came through York on their way to Wrightsville in preparation for Lee's planned attack on Harrisburg.

Between York and Columbia, 12 miles, you cross more highly deformed Cambrian limestones and dolostones. You descend to the Susquehanna, then cross the river between Wrightsville and Columbia.

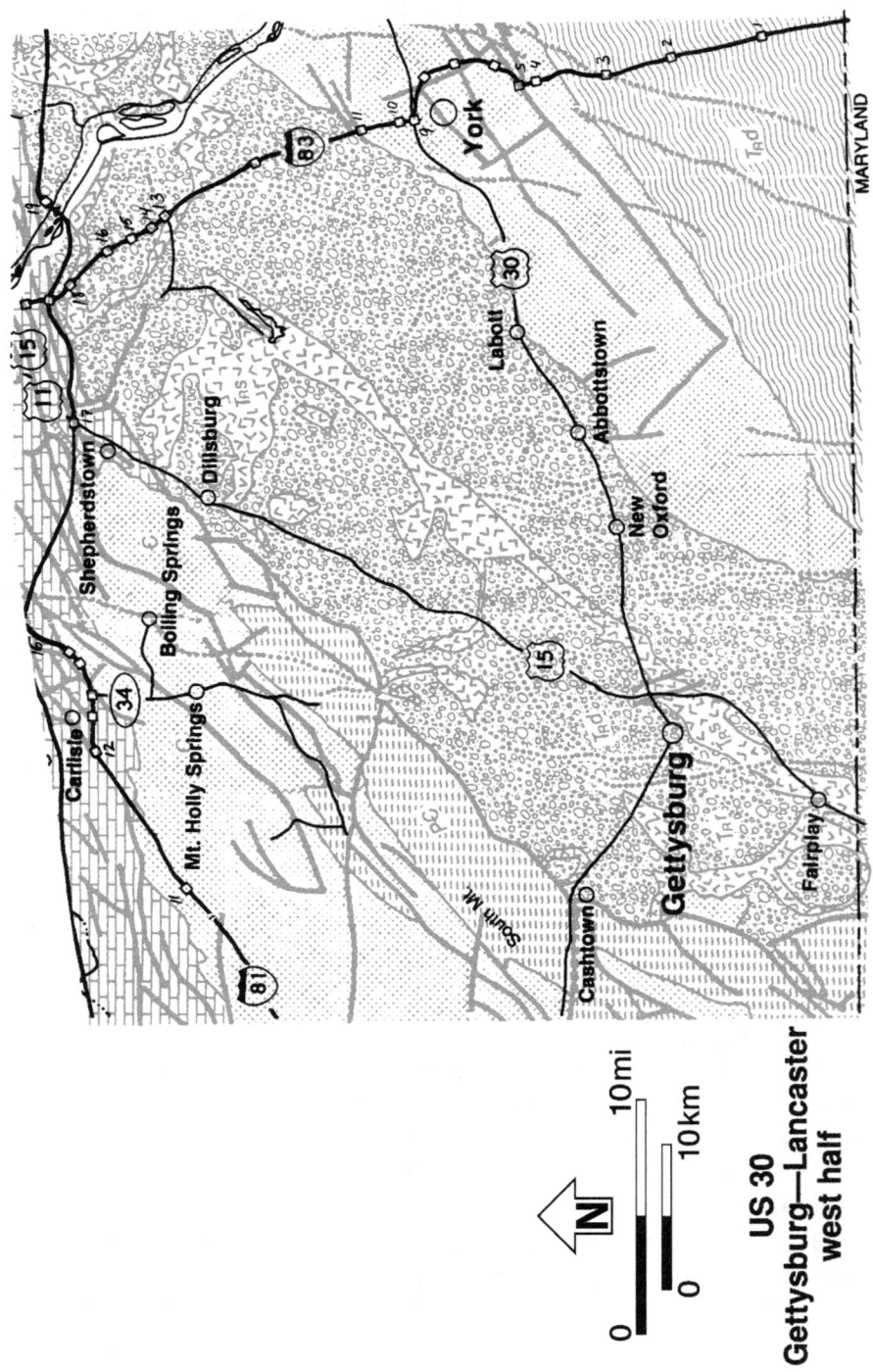

US 30
Gettysburg—Lancaster
west half

Stone house near York, made of gray, coarse-grained sandstone and conglomerate, with a slate roof.

Some excellent views appear from the slopes on either side of the river and from the bridge. This section of river is actually a reservoir called Lake Clarke.

The former Susquehanna and Tidewater Canal followed the river for 45 miles downstream from the bridge, operating from 1840 to 1894. You can see lock and canal masonry just south of Wrightsville.

Chickies Rock

Chickies Rock is a cliff about 200 feet high and 800 feet long on the eastern side of the Susquehanna River, by the railroad tracks, a little more than a mile north of US 30. It exposes a dramatic anticlinal fold in Cambrian Chickies quartzite, the backbone of Chickies Ridge that extends for several miles eastward. The ridge is faulted along both flanks. The highly resistant rock deflects the river to the west and produces the rapids in front of the cliff. The fold is probably Taconian.

Coal Dredging in Lake Clarke

During the period 1951-1973, Pennsylvania Power and Light Company ran a peculiar mining operation dredging bottom sediments from Lake Clarke to extract anthracite. The recovered coal, more than

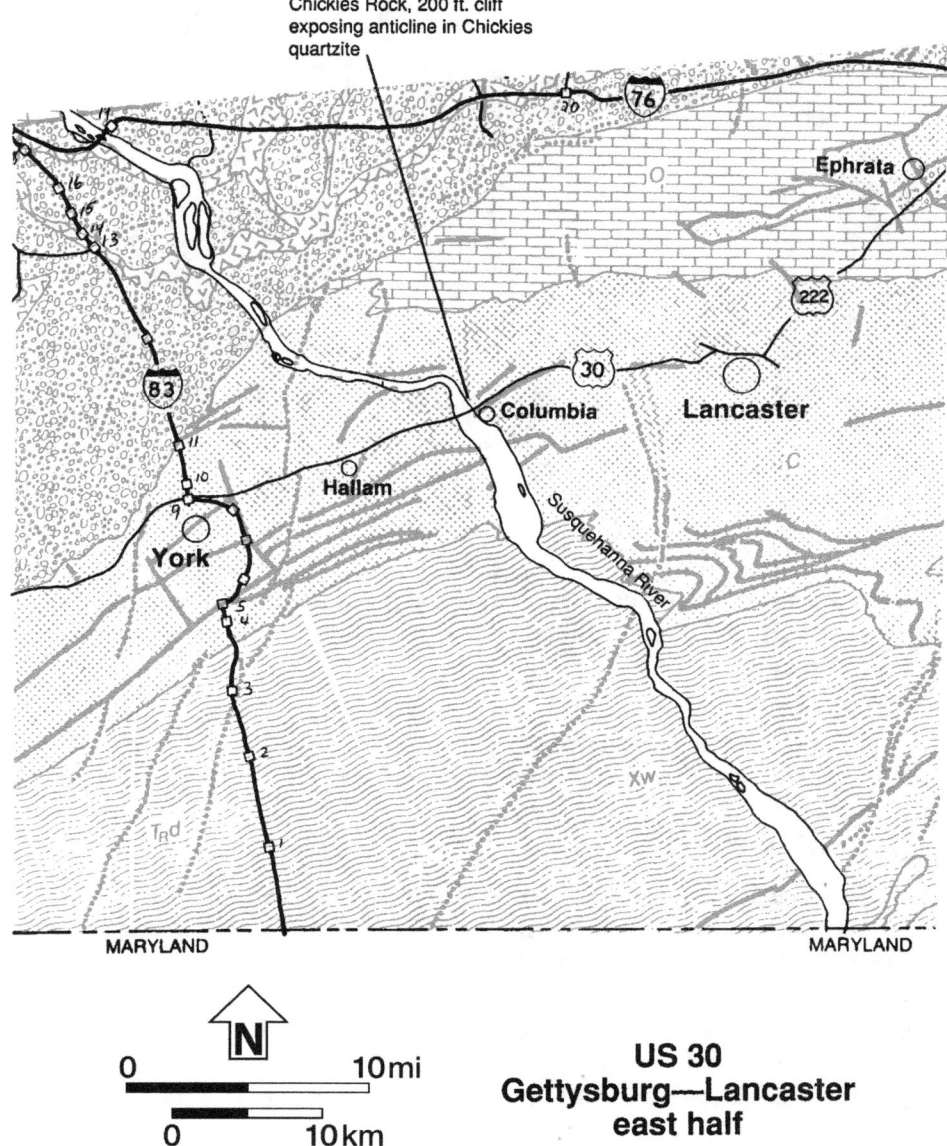

10 million tons of it, was used for fuel in the Holtwood Power Plant. Dredging greatly reduced the volume of lake sediment, which came from the coalfields of northeastern Pennsylvania during the 19th century when primitive practices allowed much coal waste to enter the streams. Today, modern processing methods and reclamation of old mine waste, called culm, has virtually cut off the supply of coal to the river.

Lake Clarke is the highest of three reservoirs in the Upland Gorge section of the Susquehanna Valley. The dam is at Safe Harbor about 11 miles downstream from US 30. Pennsylvania Power and Light Company previously dredged coal in Lake Aldred, the next reservoir downstream.

• You continue across Cambrian carbonate rocks in the 19 miles between Columbia and the Pennsylvania 283 junction at Lancaster. Chestnut Ridge, which rises immediately north of the highway over most of the way, is Cambrian Antietam quartzite on the southern limb of a large domical anticline.

Lancaster is in the heart of Lancaster County, famous for its Amish and Mennonite populations and its picturesque farms. The rollercoaster landscape is eroded almost entirely in Cambrian limestones.

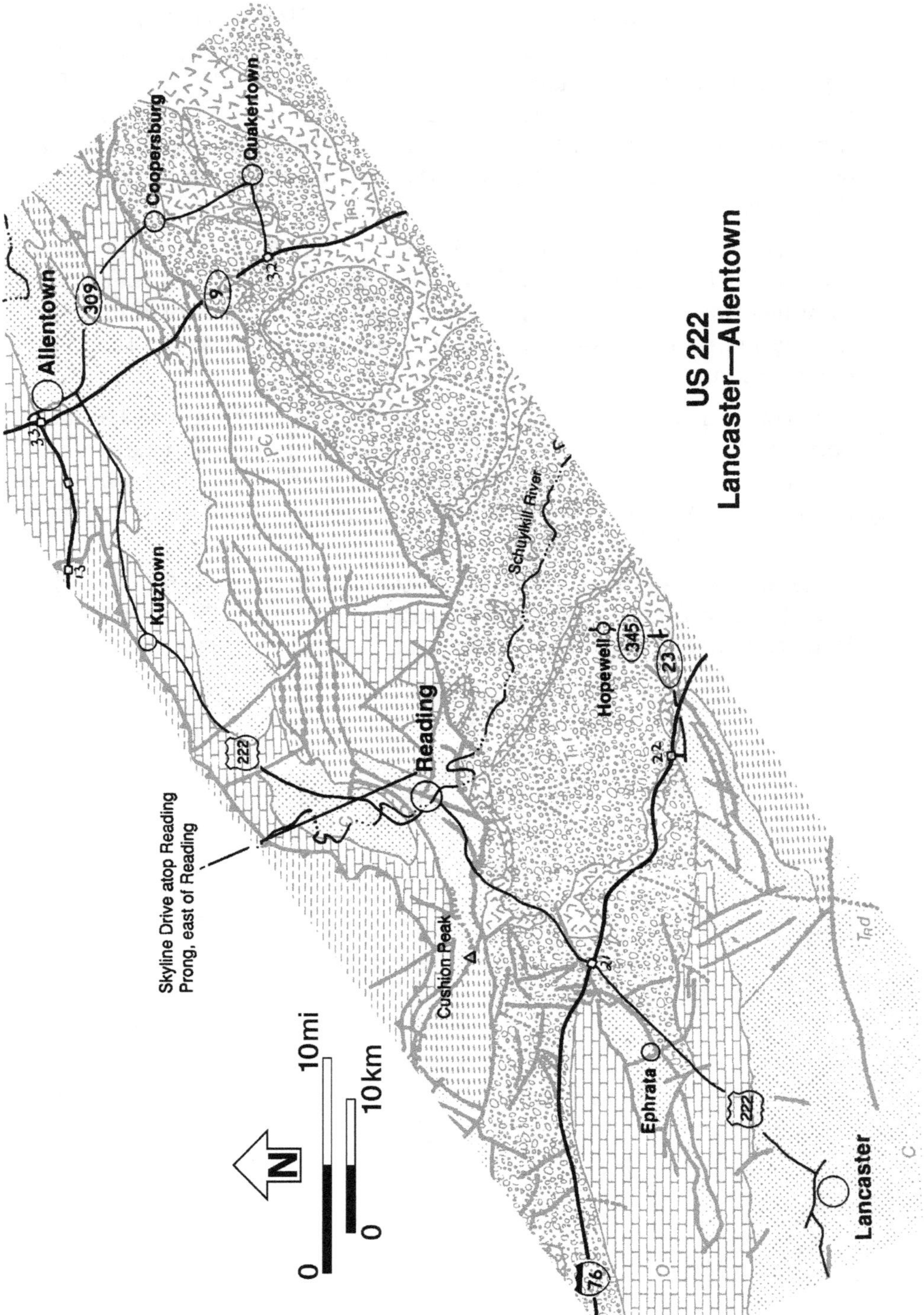

US 222
Lancaster—Allentown

Cambrian Buffalo Springs formation four miles northeast of Lancaster consists of interbedded limestone and dolostone with shale partings.

US 222
Lancaster—Allentown
58 mi./94 km.

Between Lancaster and Ephrata, ten miles, you cross rocks of the Piedmont province, almost entirely limestones and dolostones of Cambrian and Ordovician age, but outcrops are scarce. The attractive landscape is low and rolling with many farms. The carbonate rocks are generally gray, some with darker shaley partings and beds that contain irregular lenses and nodules of chert. The chert is very fine-grained quartz that tends to weather in relief. Some cherts formed from aggregates of silicious marine organisms, such as diatoms, radiolaria, or sponges, that lived in the sea at the same time that the host carbonate rock accumulated on the sea floor.

Ephrata lies just inside the southwestern margin of Newark Basin, where Triassic redbeds rest on an erosional unconformity over upper Ordovician rocks. The gap represents about 260 million years.

In the 13 miles between Ephrata and Mohnton, on the southwest side of Shillington, you cross the red plains of the Newark Basin. Near I-76 and again near the village of Sinking Spring, knobby hills rise 400 to 500 feet above the valley. Diabase of a large sill supports most

Limestone house and barn near Lancaster.

of them, but some of the lower hills are carved from resistant Triassic quartz-pebble conglomerates interbedded with the shales and sandstones.

Mohnton is near the northern edge of the Newark Basin, where a steep, south-dipping fault dropped the basin side several thousands of feet. Rocks north of Mohnton are practically all weak carbonates or shales of the Lebanon Valley section of the Great Valley. Cushion Peak is a piece of the Reading Prong near Fritztown, eight miles west of Reading, that moved on a fault then was isolated by erosion.

In the ten miles between Mohnton and Temple, you cross the Schuylkill River and pass through Reading. Mount Penn, which overlooks the city, is a raised fault block of the Reading Prong comprised of Precambrian gneiss partly covered with Cambrian dolostone and quartzite. The Reading Prong is on the southwestern end of a Precambrian massif that continues across New Jersey as the Ramapo Mountains and across southeastern New York as the Hudson Highlands. It also comes to the surface in the Berkshire Hills of Massachusetts, the Green Mountains of Vermont and to the south as South Mountain and the Blue Ridge in Virginia. The rocks were deformed and metamorphosed during the Grenville orogeny along the edge of ancestral North America 1100-1300 million years ago. Erosion has since removed miles of cover.

Between Temple and the Pennsylvania 309 junction at Allentown, 25 miles, you cross a rather featureless plain developed principally on Ordovician carbonates on the south side of the Lehigh Valley. Slightly more resistant shales, sandstones, and graywackes support some of the low hills that break the monotony of the valley floor. The higher ridge of Reading Prong rises only two or three miles to the south along the whole way.

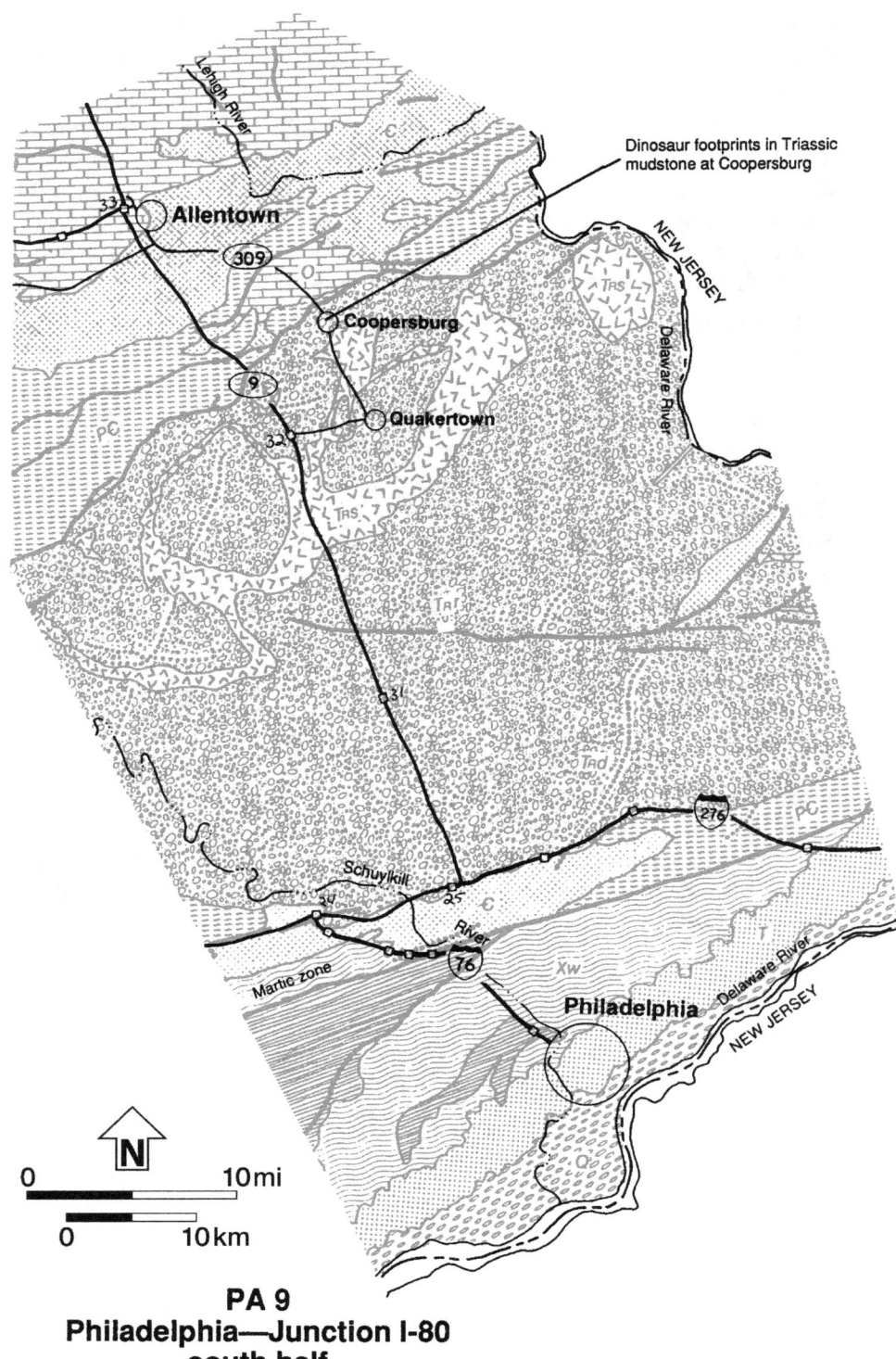

**PA 9
Philadelphia—Junction I-80
south half**

PA 9
Philadelphia—Junction I-80
75 mi./121 km.

The Pennyslvania Turnpike Northeast Extension

The southern end of this route at Exit 25 of the Pennsylvania Turnpike is on the southern boundary of the Triassic Newark Basin. The boundary is an unconformity representing a gap in the geologic record of about 300 million years; Triassic sedimentary beds rest on an erosional surface over Cambrian rocks of the Pennsylvania Piedmont. Between miles 0 and 28, the road spans the widest part of the basin. You see well-bedded shales, siltstones, and sandstones dipping about 30 degrees to the north. Dip of the beds towards the opposite side of the basin is the result of faulting that accompanied deposition of the sedimentary rocks. Basins — low areas — wherever they are or whatever their nature, tend to fill up with sediments to reconstruct a horizontal surface. Here, the strata thicken toward the faulted margin forming a wedge-shaped deposit, because that side was always deeper during sedimentation. As the fault movement continued for millions of years, the lower beds were tilted more and more in that direction. The upper beds now exposed near the fault margin are flat or nearly so.

Near the southern basin margin are buff-colored sandstones called arkose, that have a composition similar to granite; they derived by rapid erosion of Precambrian granitic gneisses and burial of the resulting sediments before they were deeply weathered. Chemical weathering would produce clay from the breakdown of the feldspars that are distinctive and abundant in arkose. Reddish sediment colors become dominant northward, with a few greenish-gray interbeds, and even some black shales that can easily be confused with basalt or diabase.

Why Triassic Redbeds?

Sedimentary rocks of the Triassic grabens of eastern North America, including Gettysburg and Newark basins, are dominantly red-colored, some vividly so. Redbeds are widespread in Triassic rocks of North and South America, Europe and Africa, commonly with dune

Triassic red sandstone and shale interbeds at mile 14 of Pennsylvania 9.

sandstones and evaporite deposits, such as rock salt and gypsum. The combination unquestionably indicates desert climatic conditions during deposition of the original sediments.

Why were deserts so prevalent during the Triassic period? By early Triassic time, the supercontinent of Pangaea was completely assembled. It was the culmination of continental drift that had been going on for about 200 million years, had clamped the Proto-Atlantic Ocean basin shut, and had produced the Appalachians and other great mountain chains. The result was a single vast landmass on one side of Earth, and a vast ocean covering the rest. The regions of the four modern continents mentioned above that were previously near the coast, were then deep within Pangaea, about as far from the sea as they could get. Gone, too, were the shallow inland seas that had covered large segments of the continents during much of the Paleozoic Era. Most of the supercontinent was in the southern hemisphere. Under these ideal conditions, an arid climate prevailed, Triassic deserts spread over the interior of the continent, and red sediments, dune sand, and evaporites accumulated.

The iron oxide mineral, hematite, colors the sediments red, and very little is required; a tenth of a percent iron pigment may make a rock intensely red. It forms by oxidation of iron in minerals like hornblende, pyroxene, biotite, or iron sulfides, which are fairly common in the regions bordering the Gettysburg Basin. Hematite is not evenly distributed in the redbeds, but generally forms a thin coating on grains, especially those supplying the iron.

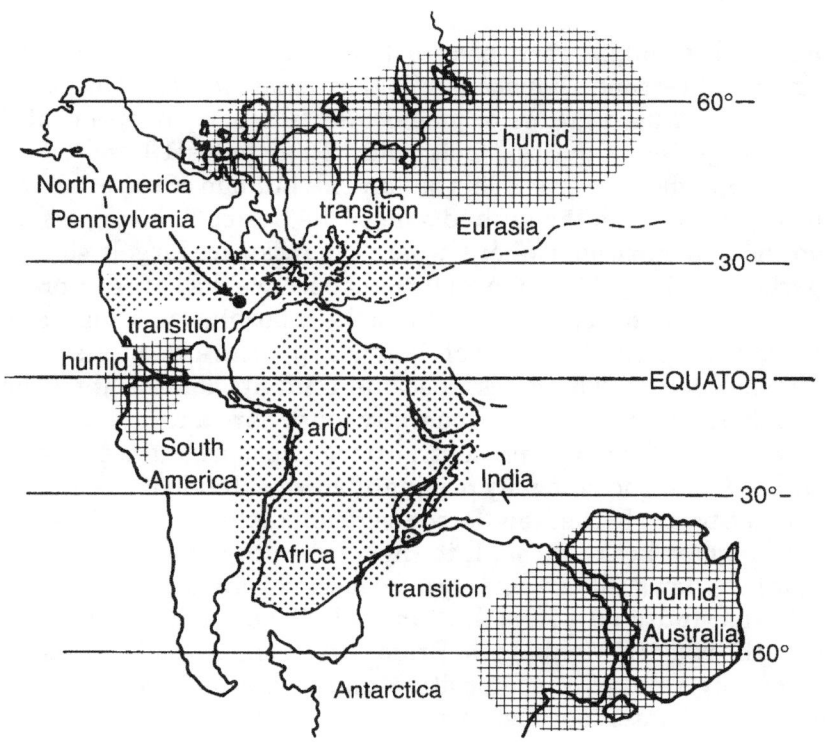

Paleoclimatic map showing approximate areas of arid, humid, and transitional climate during the Triassic period as deduced from the sedimentary rocks formed during that time. Note the position of Pennsylvania. —Adapted from Hubert, J.F., Reed, A.A., Dowdall, W.L., Gilchrist, J.M., 1978, Guide to the Mesozoic Red beds of Central Connecticut, State Geological and Natural History Survey of Connecticut Guidebook #4

The Gettysburg, Newark and other east coast Triassic grabens can be considered as failed ocean basins along the margin of a successful one, the Atlantic. When continental crust begins to pull apart, the stretching is accommodated by many down-dropped valleys—grabens. Eventually one or several take over, and fill with a long, narrow sea. The Triassic grabens with their desert regime and their red sediments preceded the narrow sea, which in time, was to become the modern Atlantic Ocean. Opening of the ocean basin changed the climate dramatically and, as a result, redbeds are less common in rocks younger than Triassic.

• Between mile 18 and 22, you cross the largest diabase sill of the basin, but it is poorly exposed. The diabase stands above the surrounding lowlands because it is more resistant to erosion. Because of their ruggedness, diabase terranes are less developed, and therefore more woodsy, making them easy to distinguish from the surrounding sedimentary terrane. Diabase at mile 18 is dense, black and porphyritic with small whitish grains of plagioclase feldspar. It also contains scattered grains of pyrite, "fools gold." Near miles 19 and 20, it is coarser and medium gray. This sill and all but one of the other large ones are crowded to the faulted side of the basin. The magma came from the upper asthenosphere, where the rocks are largely solid, but superheated. They do not normally melt because they are compressed by the weight of the lithosphere. When the lithosphere got thinner and cracked in response to rifting of Pangaea, pressure was relieved and the rocks melted. The cracks became conduits for the intrusions, which then worked their way generally southward and updip between the beds. However, the present surface exposures are erosionally truncated, and the sills may originally have extended much farther across the basin. In the New Jersey segment of the Newark Basin, the same magmas broke through to the surface to form the Watchung basalt lava flows. At the northern tip of the basin on the New Jersey side of the Hudson River, is the famous Palisades sill, best seen from across the river in Manhattan, Bronx, and Yonkers. All of these sills are roughly similar in composition and origin and were emplaced more or less contemporaneously.

Between miles 22 and 28, you see Triassic redbeds with some greenish-gray interbeds and several small diabase sills. Conglomerates, with whitish, rounded pebbles and cobbles set in a red sandy matrix are rather common here. Their original sediments eroded from Precambrian gneisses of the adjacent Reading Prong.

Dinosaur Footprints

In 1978, an amateur paleontologist found dinosaur footprints in Triassic red mudstones of the Newark Basin. The site is on Pennsylvania 309 by a tractor-trailer parking lot exactly four tenths of a mile south of Station Avenue in Coopersburg, which is about six miles southeast of Allentown. You can reach it from turnpike Exits 32 or 33. The prints, on the exposed bedding planes, range from one to five inches long and represent four different species: 1) three-toed *Anchiosauripus* 2) four-toed armored, crocodile-like *Rutiodon* 3) four-toed *Chirotherium* and; 4) a tiny lizard called *Rhyncosauroides Brunswicki*. The Triassic was truly a momentous period of Earth history, ushered in with Pangaea, the rise of the dinosaurs, and vast deserts in the continental interior, and terminated with the tearing apart of the supercontinent that initiated our modern geography.

Beautiful syncline-anticline pair in Ordovician limestone and dolostone exposed in a 150-foot quarry headwall north of Allentown. For scale, note the person standing left of the anticline core.

- You cross the Reading Prong with its highly folded and faulted rocks between miles 28 and 36. Strongly-foliated, light-colored gneisses near mile 30 are Precambrian. Between miles 31 and 36, you cross Cambrian and Ordovician limestones and dolostones of the northern flank of the Reading Prong bordering the Lehigh Valley. Thrust faults, possibly of Taconian or Acadian age have shoved the rocks upward to the north, placing old Precambrian rocks over younger Cambrian over still younger Ordovician, opposite to their original stratigraphic positions. The thrust slices are now truncated to the south by the steeply-inclined border faults of the Newark Basin. A limestone/dolostone quarry, is visible from mile 40, north of Allentown on the eastern side of the turnpike. Called Kerns Quarry, the 120-foot quarry face shows a textbook example of a chevron-like syncline-anticline pair in well-bedded lower-to-middle Ordovician Beekmantown carbonates.

Between mile 36 and the Lehigh Tunnel at mile 52, across Lehigh Valley, almost all bedrock is dark shale and slate of the upper Ordovician Martinsburg formation. The abundance of slate, the low-grade metamorphic equivalent of shale, in Lehigh and Northampton counties, makes this the heart of the Pennsylvania slate industry. In roadcuts, slate can often be distinguished from shale by very flat cleavage planes that cut obliquely across bedding. A good example is the cut near mile 46, but there are many others. Most of the slaty cleavage is of a particular type called axial plane cleavage, that is, parallel to imaginary planes that divide folds into two approximately equal halves. It develops contemporaneously with the folding and

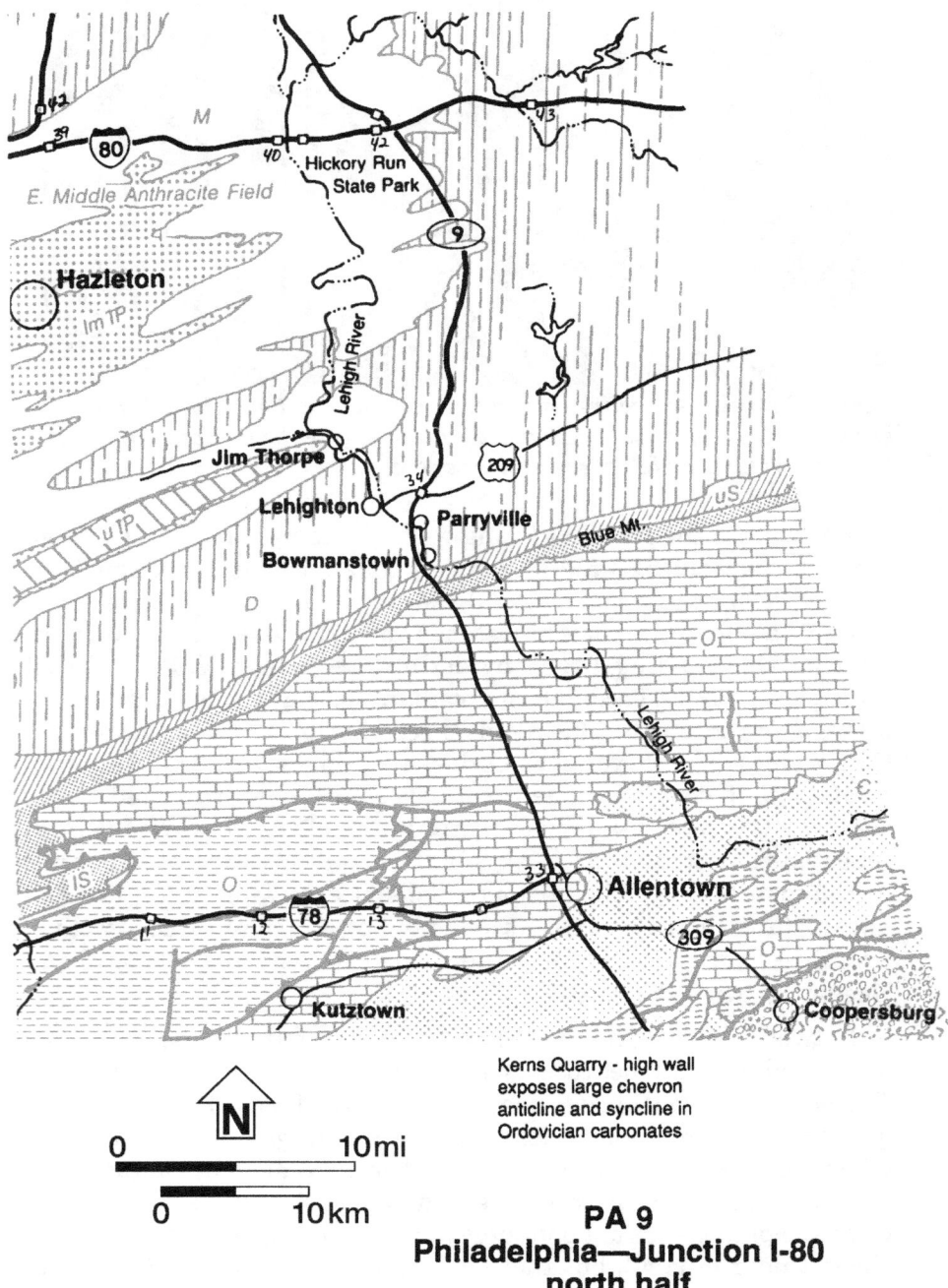

Kerns Quarry - high wall exposes large chevron anticline and syncline in Ordovician carbonates

**PA 9
Philadelphia—Junction I-80
north half**

Slaty cleavage in middle Devonian Mahantango formation dips 45° to right and cuts across nearly horizontal bedding. Site is at mile 46 of Pennsylvania 9.

metamorphism. Pennsylvania slate is used mainly for roofing, billard tables, blackboards, and electrical and structural applications. In recent years, many brush fires in western states have increased demand for roofing slates as a substitute for flammable asphalt tiles.

Views of flat-topped Blue Mountain and Lehigh Gap are excellent in the few miles south of the Lehigh Valley tunnel. This ridge marks the southeastern border of the Valley and Ridge province, held up, like many ridges, by Shawangunk conglomerate. The southern portal is near a thrust fault that places dark Martinsburg shales up and to the northwest against the Shawangunk beds. The shales crop out near the portal, dipping steeply southward.

The tunnel is about nine-tenths of a mile long. Between the northern portal and Exit 34, mile 56, views are nothing less than spectacular, especially near the Lehigh River; the roadcuts are enormous and colorful. The best is a mile-long cut in highly folded, faulted, and cleaved Devonian rocks visible across the river. The cut is on Pennsylvania 248 between Parryville and Bowmanstown and can be easily reached from Exit 34. It is described in the US 209 roadguide on page 230.

Between miles 56 and 58, you cross the Lehighton anticline through brown and black shales of the middle Devonian Mahantango and Marcellus formations in its core. At mile 58.5, you are in the middle of Sawmill Run Gap in Indian Hills ridge. The ridge, carved from red to greenish-gray sandstones of the Catskill formation exposed in the gap, borders the Allegheny Front at the edge of the Pocono Plateau. Between miles 59 and 64, the road crosses the 600-foot high scarp from bottom to top alongside Pine Run, with splendid views to the southeast encompassing Indian Hills and, in the distance, Blue Mountain. Red to greenish-gray sandstones, siltstones, and shales of the roadcuts are all in Catskill. Along the edge of the scarp, the more resistant Catskill beds form well-defined steps with gentle northwest-dipping treads. Mile 64, at about 1500 feet elevation on the Pocono Plateau, is near the northeast-pointing nose of the Southern Anthracite syncline. The mines are to the southwest principally in Pennsylvanian Llewellyn formation in the core of the fold.

Between miles 64 and I-80, you skirt the nose of the Eastern Middle Anthracite syncline, cutting through some of the Mississippian rocks that frame it. Look for brownish, gently folded, lower Mississippian sandstones between miles 71 and 73.

Vertical shale and sandstone beds near the north side of the Lehigh Tunnel on Pennsylvania 9.

Springfield Falls of Neshannock Creek cascade over resistant Pennsylvanian Pottsville sandstones. Location is 5 miles south of I-80 Exit 2 in State Game Lands No. 284.

Glossary

Alluvium: stream-laid, unconsolidated clay, silt, sand, gravel.

Anticline: arched rock layers.

Asthenosphere: the partially melted part of the Earth's mantle that supports the rigid lithosphere with its moving tectonic plates.

Basement rocks: normally very old rocks upon which younger sediments and volcanic rocks are deposited; for example, Precambrian rocks of the Piedmont.

Bed: a layer or stratum within stratified sediments or sedimentary rocks.

Bedding, graded: a gradation from coarse-to-fine grain size from bottom to top of a single sandstone bed.

Bedrock: the solid rock that underlies unconsolidated soils and sediments.

Block faulting: a type of faulting in which the crust is divided into more or less parallel blocks, generally as a result of crustal extension.

Caprock: as used here, a resistant rock layer, usually sandstone or conglomerate, that holds up steep slopes, despite weak supporting rock such as shale.

Catskill Delta: huge delta complex of upper Devonian sedimentary rocks that underlie the Allegheny Plateau of Pennsylvania and New York.

Chill zone: rapidly-cooled, fine-grained margin of an igneous intrusion such as a dike.

Cleavage, rock: property or tendency of a rock to split along parallel fractures imparted to it by deformation or metamorphism.

Composite stratigraphic column: graphical composite of two or more rock sections with rock strata stacked in their proper order from oldest at the bottom to youngest at the top.

Continental glacier: ice sheet of continental proportions.

Continental shelf: the gently-sloping, submerged edge of the continent that reaches a maximum depth of about 600 feet below sea level.

Continental slope: the part of the continental margin that slopes seaward from the outer edge of the continental shelf.

Correlation, stratigraphic: determination of time-equivalence of strata from place to place.

Cross-bedding: arrangement of strata inclined at an angle to main stratification.

Crust: the rigid outermost thin shell of Earth, the upper layer of the lithosphere that includes both the continents and ocean floors.

Crystal: geometric form of a mineral with plane faces that are the external expression of an internal atomic order.

Crystalline rock: igneous and metamorphic rock.

Cyclothem: a series of sedimentary beds and coal seams deposited during a single sedimentary cycle of the type that prevailed during the Pennsylvanian and Permian periods.

Decomposition, rock: chemical weathering of rock, as in the solution of limestone, or in the conversion of feldspars in granite to clay.

Deglaciation: uncovering of glaciated land by ice wastage.

Delta: generally fan-shaped body of sediment formed at the mouth of a river.

Dike: tabular intrusion of magma into fractures that cut across bedding.

Dip: angle of slope of rock layers, faults, or joints, as measured down from the horizontal.

Dipslope: slope roughly parallel to the dip of the underlying strata.

Discharge, river: rate of flow through a certain cross-section of a river over a specific time, often expressed as cubic feet per second (cfs).

Disintegration, rock: mechanical weathering of rock, as by root pry, frost wedging, or stream abrasion.

Drift, glacial: all sediments of glacial origin, including those reworked by meltwater streams.

Drumlin: till hill streamlined by overriding ice, with an ideal shape of an overturned spoon.

Epoch: geologic time unit next smaller in order to period.

Era: geologic time unit that includes several periods.

Erosion: sum of all geologic processes that tend to wear down the land.

Erratic, glacial: boulder transported by a glacier and then dropped when the ice melted; it is generally different from the bedrock underneath.

Esker: long, low, winding ridge of sand and gravel dropped by streams that flowed through tunnels in a wasting glacier.

Fault: surface or zone of rock fracture caused by bodily movement of one mass of rock against another.

Fault scarp: cliff or slope on the upthrown side of fault that moves vertically.

Flagstone: a hard sandstone that splits easily into thin, even layers.

Flour, glacial: pulverized rock formed by milling of rocks in and under glacial ice.

Flysch: sedimentary rock assemblage that often includes dark shales and turbidites and represents sedimentation in the deep waters of a basin.

Fold: rock deformed by real or apparent bending of rock layers.

Foliation: general term for thin layering in metamorphic rocks.

Formation: mappable rock unit that is more or less homogeneous.

Fossil: any remains, trace, or imprint of plant or animal naturally preserved in sediments or rocks.

Frost heave: expansion and upward movement of water-soaked soil when it freezes.

Gap, water: pass occupied by a river through a resistant ridge.

Gap, wind: similar to a water gap and presumably cut by a river that no longer passes through it.

Graben: in block faulting, an elongate, down-dropped crustal block, typically bounded by parallel, inward-dipping faults; a rift valley.

Group, stratigraphic: major rock-stratigraphic unit next higher in rank than formation, consisting wholly of two or more contiguous formations.

Ice sheet: thick, sheet-like glacier, generally covering a very large area; it spreads radially over the land under its own weight; continental glacier.

Interglacial: time period between major Pleistocene ice ages.

Joint: simple rock fracture along which rocks do not move sideways against each other.

Jointing, columnar: fracturing common in lava flows, sills and dikes that separates the rock into columns perpendicular to contacts with adjacent rocks. It forms by shrinkage during crystallization of the magma.

Kame: generally conical hill of sand and gravel deposited in a lake in or adjacent to a glacier by meltwater streams.

Karst: irregular, pitted topography characterized by caves, sinkholes, disappearing streams and springs, and caused by water solution of underlying limestone, dolostone, or marble.

Kettle: depression in glacial drift where outwash partly or completely buried a residual block of ice that later melted.

Kettle lake: a water-filled kettle.

Klippe: isolated erosional remnant of an overthrust fault.

Lava: magma that erupted, and the rock it formed.

Lithosphere: in plate tectonics, the rigid outer layer of Earth above the asthenosphere; includes the crust and the uppermost zone of the mantle.

Magma: rock melt.

Mantle: the zone of Earth below the crust and above the core.

Meta-(prefix): signifies that the rock has been metamorphosed.

Mineral: natural inorganic solid with limited chemical variability and distinctive internal crystalline structure.

Molasse: sedimentary sequence of conglomerates, sandstones, shales, and sometimes coals and carbonates generally deposited during and after an orogeny.

Moraine, ground: blanket of till with no marked relief, formed under a glacier.

Moraine, terminal: ridge of till left at the glacier front at a stand where ice remains in fixed position for a long time in response to steady climatic conditions.

Mudflat: low-lying strip of muddy ground by the shore, typically submerged at high tide; also tidal flat.

Mudstone: sedimentary rock made of consolidated mud.

Orogeny: processes by which fold-belt mountains are formed, including folding, faulting, metamorphism, volcanism, and igneous intrusion; produced by plate tectonic convergence.

Outcrop: surface exposure of in-place bedrock.

Outcrop, pinnacled: patchwork exposures of carbonate bedrock representing a soil-mantled solutional topography, or karst.

Outwash, glacial: glacial debris transported and deposited by meltwater streams.

Oxbow lake: crescent-shaped lake formed in an abandoned meander loop which has become separated by a change in the river course.

Parting, shale: thin shale beds that separate thicker sandstone, siltstone, or carbonate beds.

Peat: unconsolidated deposit of semi-coalified plant remains in a bog.

Period: geologic time unit longer than an epoch and a subdivision of an era.

Petroleum: a general term for all naturally-occurring hydrocarbons, including oil, gas, and tar.

Petroleum reservoir: a natural trap in which large quantities of oil and gas may accumulate.

Piracy, stream: capture of one stream by another with a lower streambed, leaving the lower reach of the captured stream dry.

Plate tectonics: a field of geology which recognizes that the entire lithosphere of Earth is made up of a small number of rigid plates that constantly move on the underlying asthenosphere.

Plucking, glacial: breaking away and removal by glacial ice of blocks from bedrock.

Pothole: rounded hole ground into streambed rock by sand and gravel in swirling eddies.

Rebound, glacial: buoyant rise of the land after the weight of an ice sheet is removed.

Recession, glacial: the melting of a glacier, in response to climatic warming.

Redbeds: sandstone, siltstone, shale, and conglomerate colored red by an iron oxide called hematite.

Reef: ridge or mound of coral and other shallow marine organisms.

Rhyolite: a volcanic rock with the composition of granite, generally recognized by its fine-grain size and light color.

River terraces: rather flat-surfaced, step-like remnants of former floodplain levels bordering a river that has cut downward.

Rock city: large, rectangular blocks of resistant caprock that have been separated along joint and bedding planes.

Salamanca re-entrant: area of southwestern New York and northwestern Pennsylvania that never was covered by glaciers.

Scarp: long, more or less continuous cliff or slope separating relatively flat land into two levels.

Scour, glacial: grinding, scraping, gouging, bulldozing action of a glacier.

Shear zone: intensely disrupted rock zone marked by numerous closely spaced, parallel shears; generally indicative of faulting at considerable depth.

Sill: tabular igneous intrusion between the layers of the intruded rock.

Sinkhole: pit caused by solution and collapse of a cave.

Stalactite: calcium carbonate deposit suspended from a cave ceiling at the site of dripping water.

Stalagmite: post-like mass of calcium carbonate rising from the floor of a cave where water drips from the ceiling; typically below a stalactite.

Strata: the layers of sedimentary and volcanic rocks.

Strip mining, areal: a type of strip mining practiced in flat-lying coal seams (or other mineral resource) with a uniformly thick overburden that is easily removed in all directions.

Strip mining, contour: a type of narrow, linear strip mining practiced in flat-lying coal seams (or other mineral resource) that crop out on the slopes of hilly country, which follows the topographic contours.

Stromatolites: calcareous algal structures that commonly resemble cabbage heads.

Subduction: in plate tectonics, the process of one lithospheric plate descending beneath another.

Superimposed stream: a river whose established course was maintained as it eroded downward through underlying rocks, cutting across geologic structures and rocks different from those on which it began to flow.

Syncline: downfolded rock layers.

Talus: fragmentary rock deposit at the base of a cliff.

Tectonic: large-scale deformation of Earth's crust.

Thrust fault: low angle push fault produced by plate tectonic convergence.

Till: unsorted and unstratified glacial drift deposited directly from the ice without subsequent reworking by streams; it typically contains material of a wide range of sizes from boulders or pebbles to dirt.

Travertine: limestone formed by chemical precipitation of calcium carbonate from water, common around springs and in caves.

Unconformity: a surface of erosion or non-deposition separating older from younger rocks and constituting a gap in the rock record.

Varves: clays deposited in lakes supplied by glacial meltwater streams in which the laminations form yearly, like tree rings.

Wave cut bench: level or nearly level bedrock surface cut by wave erosion.

Wave cut cliff: cliff produced and maintained by waves undercutting rock along the shore.

Weathering: the physical disintegration and chemical decomposition of rock.

Weathering, spheroidal: a form of weathering characterized by repeated spalling of concentric shells of decayed rock, leaving an increasingly spherical core of fresh rock.

Suggested Readings

Berg, T.M., Edmunds, W.E., Geyer, A.R., and others, compilers (1980), *Geologic map of Pennsylvania*, Pennsylvania Geological Survey, 4th series, Map 1, scale 1:250,000, 3 sheets

Bolles, William H. and Geyer, Alan R. (1976) *Pennsylvania Interstate 81 Geologic Guide*; Pennsylvania Geological Survey

Brown, Andrew (1962) *Geology and the Gettysburg Campaign*: Pennsylvania Geological Survey, 4th series, Education Series #5

Cleaves, Arthur B. and Stephenson, Robert C. (1949) *Guidebook to the Geology of the Pennsylvania Turnpike, Carlisle to Irwin*: Pennsylvania Geological Survey, 4th series, Bulletin G24

DeKok, David (1986) *Unseen Danger*, University of Pennsylvania Press, Philadelphia

Delano, Helen (1987), *Stop 6, Walnut Creek access area,* in Thomas, D.J., and others, *Pleistocene and Holocene geology of a dynamic coast,* Annual Field Conference of Pennsylvania Geologists, 52nd, Erie, Pa., Guidebook, p. 65-67

Edmunds, William E. and Koppe, Edwin F. (1968) *Coal in Pennsylvania*: Pennsylvania Geological Survey, 4th series, Educ. Series #7

Faill, R.T. (1981), *The Tipton block—an unusual structure in the Appalachians,* Pennsylvania Geology, v. 12, no. 2, p. 5-9

Frakes, L.A., Glaeser, J.D., Wagner, W.R., and Wietrzychowski, J.F. (1963), *Stratigraphy and structure of Upper and Middle Devonian rocks in northeastern Pennsylvania,* Annual Field Conference of Pennsylvania Geologists, 28th, Stroudsburg, Pa., Guidebook, 44 p.

Geyer, Alan R. and Bolles, William H. (1979) *Outstanding Scenic Geological Features of Pennsylvania*: Pennsylvania Geological Survey, Environmental Geology Report #7, Vol.1 and 2

Geyer, Alan R., Smith II, Robert C., Barnes, John H. (1976) *Mineral Collecting in Pennsylvania*: Pennsylvania Geological Survey, General Geology Report #33

Giddens, Paul (1964) *Early Days of Oil*: Peter Smith, Cloucester, Mass.

Guidebooks of Field Conferences of Pennsylvania Geologists, especially: 28th, Stroudsburg, 1963; 32nd, East Stroudsburg, 1967; 34th, Hazelton, 1969; 40th, Bartonsville, 1975; 41st, Titusville, 1976; 46th, Wellsboro, 1981;48th, Danville, 1983; 49th, Reading, 1984; 52nd, Erie, 1987

Hoskins, Donald M., Inners, John D. and Harper, John A. (1964) *Fossil Collecting in Pennsylvania*: Pennsylvanisa Geological Survey, 4th series, General Geology #40

Inners, Jon D. (1987), *Upper Paleozoic Stratigraphy along the Allegheny Topographic Front at the Horseshoe Curve, west-central Pennsylvania:* in Centennial Field Guide, Vol. 5, Northeastern Section of the Geological Society of America, ed. by David C. Roy (1987)

MacLachlan, D.B., Buckwalter, T.V., and McLaughlin, D.B. (1975), *Geology and mineral resources of the Sinking Spring quadrangle, Berks and Lancaster Counties, Pennsylvania*, Pennsylvania Geological Survey, 4th series, Atlas 177d, 228 p.

Pennsylvania Geological Survey (1982), *Geologic map of Pennsylvania*, Pennsylvania Geological Survey, 4th series, Map 7

Pohn, H.A., and Purdy, T.L. (1981), *A major (?) thrust fault of Towanda, Pennsylvania: an example of faulting with some speculation on the structure of the Allegheny Plateau*, in Berg, T.M., and others, *Geology of Tioga and Bradford Counties,* Annual Field Conference of Pennsylvania Geologists, 46th, Wellsboro, Pa., Guidebook, p. 45-55

Schooler, Elizabeth E. (1974) *Pleistocene Beach Ridges of Northwestern Pennsylvania*: Pennsylvania Geological Survey, General Geology Report #64

Shepps, Vincent C. (1962) *Pennsylvania and the Ice Age*: Pennsylvania Geological Survey, Educ. Series #6

Swanson, Peter L. (1974), *Valley Forge State Park, History of the Rocks*, Pennsylvania Geological Survey Park Guide #8

Swartz, Frank M. (1965) *Guide to the Horse Shoe Curve Section Between Altoona and Gallitzin, Central Pennsylvania*: Pennsylvania Geological Survey, 4th Series, Bulletin G50

Theisen, J.P. (1983), *Is there a fault in our gap?*, Pennsylvania Geology, v. 14, no. 3, p. 5-11

Van Diver, Bradford (1985) *Roadside Geology of New York*, Mountain Press Publishing Co., Missoula, Montana

Wagner, W.R. and Lytle, W.S. (1968) *Geology of Pennsylvania's Oil and Gas*: Pennsylvania Geological Survey, Educ. Series #8

Wagner, Walter et al (1970) *Geology of the Pittsburgh Area*: Pennsylvania Geological Survey, 4th series, General Geology Report #59

Way, John H. (1986) *Your Guide to the Geology of the Kings Gap Area, Cumberland County, Pennsylvania*: Pennsylvania Geological Survey, 4th Series, Environmental Geology Report #8

Willard, Bradford (1962) *Pennsylvania Geology Summarized*: Pennsylvania Geological Survey, Educ. series #4

Wilshusen, J. Peter (1983) *Geology of the Appalachian Trail in Pennsylvania*: Pennsylvania Geological Survey, General Geology Report #74

A list of other publications of interest that may be ordered through the Pennsylvania Geological Survey, Harrisburg, PA, either free or for a small fee can be found in a booklet entitled, *Pennsylvania Geological Publications*

Index

Acadian foredeep, 127, 131, 193, 227, 303
Acadian Mountains, 31, 33, 92, 96, 115, 136, 139, 152, 163, 172, 243
Acadian orogeny, 28, 31, 33, 92, 95, 96, 165, 171, 243, 311
Acadian uplift, 86
Adirondack Mountains, 27, 295
African plate, 24
Aliquippa Gap, 172
Allegany State Park, 117, 121, 123
Alleghenian deformation, 49, 255
Alleghenian folding, 176
Alleghenian mountain building, 115, 247, 262
Alleghenian Mountains, 167
Alleghenian orogeny, 33, 34, 45, 159, 161, 169, 311
Alleghenian uplift, 43, 73, 84, 86, 163
Allegheny beds, 111, 142, 143, 146, 167
Allegheny County, 64
Allegheny crest, 127
Allegheny formation, 58, 142, 167
Allegheny Front, 13, 33, 36, 45, 47, 51, 57, 84, 86, 95, 105, 125, 127, 145, 153, 155, 157, 161, 162, 167, 168, 190, 201, 213, 220, 221, 223, 224, 229, 233, 235, 238, 239, 242, 243, 245, 247, 252, 328
Allegheny group, 51, 53, 55, 57, 69, 77, 79, 81, 82, 84, 145, 242
Allegheny Mountains, 55. 59, 117, 167
Allegheny Plateau, 11, 12, 13, 16, 33, 36, 45, 47, 48, 49, 51, 53, 54, 55, 57, 59, 75, 79, 81, 83, 86, 92, 95, 96, 105, 127, 140, 152, 153, 159, 161, 167, 201, 221, 235, 239, 241, 243, 255, 256
Allegheny Portage, **117**
Allegheny Reservoir, 121, 122
Allegheny River, 51, 55, 67, 79, 83, 117, 122, 125, 131, 133, 134
Allegheny tunnel, 167
Allegheny Valley, 119, 122, 131
Allentown, 276, 319, 325
Allenwood, 219
Altoona Valley, 242
Altoona, 242, 243
Ames shaley limestone, 58
Amity Hall, 215, 247
ancestral Taconic Mountains, 30
Ansonia, 113, 115
Anthracite Museum, 258
anthracite synclinorium, 201, 202
Antietam quartzite, 272, 281, 285, 315
Appalachian field, 41
Appalachian mountain-building, 24
Appalachian Mountains, 13, 23, 33, 35, 36, 159, 170
Appalachian Trail, 276 282
Appalachian Uplands, 12
Appalachian water gaps, 195
Appalachians, 322
Arch Rock, 248

Archbald Pothole, **105**, 106
Armstrong County, 64
Ashland, 258
Athens, 151
Atlantic Coastal plain, 12, 152, 167
Atlantic Ocean basin, 291
Atlantic Ocean, 24, 34, 35, 67, 83, 123
Avalonia, 33
Axemann formation, 4
Azilum, 107, 108

Balanced Rock at Boulder Point, 294
Bald Eagle conglomerate, 86, 181
Bald Eagle Creek, 179, 238, 239, 238
Bald Eagle formation, 175, 176, 200, 219, 249, 251
Bald Eagle Mountain anticline, 157
Bald Eagle Mountain, 86, 87, 127, 129, 145, 179, 220, 233, 235, 236, 241, 219, 232
Bald Eagle ridge, 238
Bald Eagle sandstone, 172, 251, 252
Bald Eagle Valley, 252
Barclay syncline, 111, 152
Bardwell, 106
basalt, 8
Battle of Gettysburg, 295, 306, 308, 310, 311
Bear Mountain, 232
Beartown Rocks, 78, **79**
Beaver Creek, 293
Beaver Falls, 53
Beaver River, 54, 67
Bedford Narrows, 171, 172
Bedford syncline, 171
Bedford Valley, 245
Bedford, 171, 245
 Beech Glen, 153
Beekmantown carbonates, 325
Beekmantown group, 4, 276
Bellefonte dolostone, 87, 172, 181, 236, 252

Bellefonte formation, 4
Bellefonte limestone, 175
Bellefonte, 87, 88
Beltzville Lake, 229
Beltzville, 229
Benner formation, 87
Berkshires, 15
Berlin coal basin, 167
Berry Mountain, 215
Berwick anticline, 91, 208
Berwick, 183
Big Bend, 119, 122
Big Cove Creek Valley, 304
Big Cut, 179, 181
Big Injun, 58
Big Mountain, 181, 253, 258
Big Pocono Ski Area, 190, 213
Big Pocono State Park, 190
Bilger Run, 143
Bilgers Rock, **143**
Biling Springs dike, 287
Birch Mountain #5 coal seam, 204
Birmingham window, 241
Blacklog Hill, 174
Blackwell, 115
Bloomsburg beds, 221
Bloomsburg formation, 176, 183, 200, 221
Bloomsburg redbeds, 181, 219
Blossburg syncline, 111, 130
Blossburg, 129, 130
Blue Mountain Gap, 299
Blue Mountain Ridge, 190
Blue Mountain Tunnel, 176, 177
Blue Mountain, 176, 191, 195, 199, 201, 202, 249, 261, 276, 280, 289, 297, 299, 300, 327, 328
Blue Ridge Mountains, 15, 27
Blue Ridge, 280, 308, 318
Blue Rocks, 188
Boalsburg, 251
Boiling Springs, **285**, 287
Boot Jack Summit, 140
Bowmanstown, 229, 327
Bradford, 139
Brallier formation, 86, 127, 242, 303

341

Brandy Camp, 140, 142, 143
Breezewood, 172, 303
Breezy Point, 303
Bridge Street thrust fault zone, 151, 152
Bridgeville, 64
Broad Mountain, 203
Broad Top syncline, 174, 303, 304
Brockport, 142
Brodheadsville, 191, 227, 229
Brooks Model Mine, 91
Brookville, 81
Brookville-Clarion coal, 142
Brusce Lake, 97
Brush Creek Gap, 163
Brush Creek oilfield, 55
Brush Mountain, 241, 242
Buffalo Mountain, 170, 215
Buffalo Springs formation, 317
Burgoon formation, 83, 86, 115, 239, 243
Burgoon sandstone, 48, 58, 84, 114, 127, 129, 167, 239
Burnham Freedom Forge, 249
Bushkill Falls, 223, 224
Bushkill, 223
Butler Mountain, 208
Buttermilk Falls formation, 193
Byrnesville, 258

Caledonia iron ore furnace, 307
Caledonia syncline, 82
Camelback Mountain Overlook, **190**
Camelback Mountain, 191, 213, 229
Campaign, 308
Canadaway group, 99
Canadian Shield, 27
Carlisle, 167, 177, 261, 280, 281, 287, 308
Carrolltown, 145
Carter Camp, 115
Cashtown fault, 306, 307
Cashtown Gap, 306, 307, 310
Cashtown, 308
Catawissa Mountain, 183
Catskill beds, 108, 146

Catskill conglomerate, 191
Catskill Delta, 28, 31, 32, 48, 84, 92, 95, 96, 119, 136, **139**, 152, 172, 243
Catskill formation, 33, 48, 86, 95, 96, 106, 110, 114, 115, 117, 129, 130, 139, 152, 153, 155, 163, 168, 174, 184, 189, 190, 193, 202, 213, 215, 215, 217, 232, 233, 239, 243, 304, 328
Catskill Mountains, 13, 29
Catskill redbeds, 163, 165, 168, 172, 247
Catskill sandstone, 156, 157, 212
Catskill, 11
Cattaraugus Creek, 123
caves, **236**
Cemetery Hill, 310
Cemetery Ridge, 310
Centerfield coral reef, 192,**193**
Centerfield fossil zone, 193
Centerville, 245
Central Lowlands, 16
Centralia Mine Fire, **256**
Centralia syncline, 216, 256
Centralia Valley, **255**, 256, 257
Centralia, 257
Chadakoin formation, 75, 119, 137, 139
Chambersburg, 280, 306, 308, 310
Chemung River, 93, 108, 113
Cherry Spring tower, 115
Chesapeake Bay, 83
Chestnut Ridge anticline, 83, 143, 146
Chestnut Ridge, 57, 315
Chickies quartzite, 313
Chickies Ridge, 313
Chickies Rock, 313
Civil War, 285
Clarion River, 79, 140, 142
Clarks Run, 55
Clarks Summit, 105
Claysburg, 245
Clearfield syncline, 83
Clearfield, 84
Cleeland Rock, 70
Clinton group redbeds, 170, 174,

176, 183, 219, 242, 248, 249, 249
Coal Hill, 143
Coal Knob ridge, 119, 123
Coal Measures, 39
coal dredging in Lake Clarke, 313
coal, 7, **39**
Coastal Plain, 15
Colebrook, 267
colors of the Catskill formation, **96**
Colton Point State Park, 113
Columbia, 311, 315
columnar jointing, **271**
Conemaugh beds, 51, 143, 146
Conemaugh formation, 58, 146
Conemaugh group, 49, 55, 57, 64, 69, 81, 82, 142, 145, 167
Conemaugh River, 146, 243
Conewago Falls potholes, **263**
Conewago Falls, 264, 265, 266
Conewago Mountains, 293
Conewango Creek, 123
Conewango River, 125
conglomerate, 5
Conneaut group, 99
Conneaut Lake village, 74
Conneaut Lake, **73**
Conneaut Marsh, 73
Conneaut Outlet, 73
Connoquenessing Creek, 67, 70
Conococheague Creek Valley, 175
Conodoquinet Creek, 288, 297
Conyngham Valley, 184, 208
Coppersburg, 324
Coral Caverns, 169
Cornerplanter Bridge, 119
Cornwall iron ore banks and furnace, 266
Coudersport Ice Mine, **115**
Coudersport, 115, 117
Council Cup Scenic Overlook, **185**
Cove Mountain, 299, 300, 305
Coxes Valley, 249
Cranberry Creek, 191
Crescent Lake, 213
Crooked Creek, 113

cross-bedding, **212**
cross-fold joints, **239**
Culpepper Basin, 16, 295
Culps Hill, 310
Cumberland Valley, 14, 176, 177, 197, 261, 262, 279, 280, 285, 287, 297, 305, 306, 308
Curtin Gap, 86, 238
Cushion Peak, **268**, 318
Cuyahoga formation, 133, 136
Cuyahoga group, 74
cyclothems, **81**

Deep Leap Falls, 223
Deep River Basin, 16
Deer Park anticline, 168
Delaware River bridge, 195
Delaware River, 95, 221, 223, 224, 226,
Delaware Water Gap, 95, 191, 195, 196, 213, 224, 225, 226, 229
Deneen Gap, 165
Denton Hill, 115
Devil's Elbow, 249
Devil's Potato Patch, 188
Devil's Race Course, 188
Devil's Turnip Patch, 188, 219
Devils Den, 310
dikes, 8
Dingman Falls, 221, 223
Dingmans Ferry, 223
Dinosaur Footprints, 324
Dinosaur Rock, **267**, 270
dolostone, 7
Dorrance, 201
drainage basins, 13
Drake discovery, 38
Drake Well, 134
Driftwood, 117
DuBois, 82, 143, 146
Duncan Island, 215, 247
Duncannon, 300
Dundore, 216
Dunkard group, 59, 161
Dunning Mountain, 245
Dushore, 153, 155

East Freedom, 245
Eastern Middle Anthracite Field, 185, 189, 207
Eastern Middle Anthracite syncline, 328
Ebensburg, 145, 146
Echo Lake lowlands, 224
Edinboro Lake, 75
Egypt Lake, 97
Elbrook formation, 51, 287
Ellwood City, 67
Emigsville, 291, 293
Emporium Junction, 117
Enola, 297
Ephrata, 317
Erie County, 99
Erie Extension Canal, **79**
Erie lowlands, 12, 48
Erie Plain, 75, 99
Erie Railroad, 92, 118
Erie scarp, 16, 48, 75, 99
Erie, 103
erosion and landsliding on the Erie Shore, **103**
Everhart Museum, 91
Evitts Mountain, 171, 172, 245

Factoryville, 106
Fall Line, 15
Falmouth, 263
Fayetteville, 306
Ferguson Valley, 249
Finger Lakes, 93, 152
Fishing Creek Gap, 235
Fishing Creek, 179
Flagstaff Mountain Overlook, **232**
folded Appalachians, 15
Foreknobs formation, 174
Forest City, 89
Fort Augusta, **253**
Fort Davis, 306
Fort Franklin, 131
Fort Littleton, 174
Fort Loudon, 304, 306
Fort Machault, 131
Fort Marshall, 306
Fort McCord, 306
Fort McDowell, 306

Fort Venango, 131
Fort Waddel, 306
Fort Wyoming, 108
Frackville, 258
Franklin, 131, 133, 134, 136
Frankstown branch of the Juniata River, 245
Frankstown Valley, 243
Frankstown, 241
Freeport coals, 142
French Creek, 131, 270
Friends Cove anticline, 171, 172
Friends Cove thrust fault, 172
Fritztown, 268, 318
Front Mountain, 250
Fuller Lake, 285
Fullmer, 223

gabbro, 8
Galeton, 115
gas, **35**
Gatesburg formation, 179
Genesee River, 115
Geologic Time Scale, 2
Geologic Time-Travel Map, 197, 200, 279
Gettysburg Basin, 15, 34, 261, 262, 266, 267, 291, 293, 295, 297, 307, 308, 311, 322,
Gettysburg Campaign, 295, 311
Gettysburg conglomerate, 293
Gettysburg diabase sill, 295, 310, 311
Gettysburg, 295, 308, 310, 311, 321
Gifford Pinchot State Park, 293, 294
Girard shale, 75
Glacial map, 17
Glen Mawr, 156
gneiss, 10
Golden Triangle, 51, 66, 67
Gondwana, 34, 43
Gowanda moraine, 123
grabens, **262**
Grampian, 143
Grand View, 153
granite, 8

344

graywacke, **275**
Great Bend of the Susquehanna River, 92
Great Lakes, 93, 151
Great Valley, 12, 14, 159, 176, 261, 268, 275, 276, 308, 318
Green Mountains of Vermont, 318
Green Mountains, 15
Greene formation, 49, 59
Greenville, 79
Grenville continent, 27, 276
Grenville orogeny, 27, 318
Grenville province, 27
Grenville supercontinent, 287-288, 307
Groins, 100
Gulf of Mexico, 123
Gunter Valley, 175, 176

Hagerstown Valley, 14
Hamburg klippe, 276, 287, 288, 297
Hamburg sequence, 275, 276, 287
Hamburg, 202, 276
Hamilton group, 157, 201, 238, 245, 247
Hammon Reservoir, 130
Hammond Dam, 130
Hardyston quartzite, 269
Harlem coal seam, 58
Harmonsburg, 74
Harpers formation, 282
Harpers quartzite, 272
Harrell formation, 86, 127, 242, 303
Harrisburg, 195, 215, 233, 247, 264, 288, 289, 297, 300, 306, 308, 311
Hartford Basin, 15
Haven formation, 239
Hebron anticline, 14C
Hegins Valley, 203
Helderberg group, 168
Helen iron furnace, 78
Hell Run, 71
Hepburnville, 127
Hess Mountain, 185
Hickory Ridge, 247

Hickory Run boulder field, 186, 187
Hickory Run State Park, 186
Hickory Run Valley, 186
Hickory Run, 188
high-calcium limestone at Bellefonte, **87**
hogbacks, **195**
Hollidaysburg, 235, 241, 242, 243, 245
Holtwood Power Plant, 315
Homewood Falls, 55
Honey Brook Upland, 270
Honey Creek, 249
Hopewell Furnace National Historic Site, **270**
Horseshoe Curve, 242, 243
Horseshoe Falls, 196
Houtzdale-Snow Shoe syncline, 84
Hudson Highlands, 15 318
Hudson Valley, 151
Hughesville, 155, 157
Hunter, 134, 136
Huntington Mountain, 183
Huntley Mountain formation, 48, 114, 115, 127, 130, 153, 233
Huntley Mountain, 155
Hyndman Peak, 245

Ice Age, 16
igneous rock, 7
Indian Hills ridge, 328
Indian Hills, 328
Irish Valley, 215
Irvine, 125

Jacks Mountain, 249
Jacksonburg formation, 276
Jake's Rocks, 121, 123, 125
Jersey Shore Pike, 115
Jersey Shore, 114, 117
Jervis trough, 221
Jetties, **100**
Jim Thorpe, 231, 232
Johnstown Expressway, 146
Johnstown Flood, 146
Johnstown, 146
Joliet anticline, 203

Juniata Canal, 247, 248
Juniata formation, 172, 175, 176, 181, 200, 249, 249, 250, 252, 299
Juniata iron, 249
Juniata River, 215, 247, 300
Juniata sandstone, 29
Juniata shale redbeds, 171

Keating, 117
Keefer sandstone, 248
Kerns Quarry, 325
Keyser and Tonoloway limestone, 165
Keyser formation, 243, 245, 248
Keyser limestone, 169, 181, 249
Kingston, New York, 195
Kinzua Bridge State Park, 118
Kinzua Creek, 118, 121, 140
Kinzua Dam, **119,** 122
Kinzua Pass, 121, 122
Kinzua Viaduct, **118**
Kishacoquilllas Creek, 249
Kitchen Creek, 153
Kittanning coal, 142
Kittanning State Forest, 79
Kittanning, 167
Kittatinny Mountain, 95, 175, 195, 196, 221, 225, 226, 229
Kittatinny ridge, 223
Knob Mountain, 175, 176
Kregeville, 229
Kulpmont, 256

Labott, 311
Lackawanna Iron Company, 91
Lackawanna syncline, 92, 105, 202, 209, 211
Lackawanna Valley, 89, 105, 208
Lake Aldred, 315
Lake Arthur, 69
Lake Carll, 121, 122
Lake Clarke, 313, 315
Lake Edmond, 69
Lake Erie basin, 74
Lake Erie plain, 16
Lake Erie, 21, 29, 75, 99, 100, 102, 103, 122, 131

Lake Redmond, 69
Lake Sciota, 196
Lake Warren, 16, 21, 75, 99, 100, 103
Lake Watts, 69
Lake Whittlesey, 16, 21, 75. 99. 100
Lancaster, 315, 317, 318
Lantz Corners, 117, 119, 139, 140
Laporte, 155
Larke formation, 4
Laurel Creek Reservoir, 249, 250
Laurel Forge Pond, 282, 284
Laurel Hill anticline, 57, 84, 243
Laurel Hill tunnel bypass, 57, 8
Laurel Hill, 58, 146
Laurentia, 28, 33, 43
Lawrence Furnace, 71
Lawrenceville, 129, 130
Lebanon Valley, 14, 318
Ledger dolostone, 272
Lee-Penobscot Mountain, 185
Lehigh Canal, 231, 232
Lehigh Gap, 229, 327
Lehigh River, 229, 230, 232, 327
Lehigh Valley Tunnel, 325, 327, 328
Lehigh Valley, 14, 201, 202, 232, 319, 325
Lehighton anticline, 227, 229, 231, 327
Lehighton Bridge, 231
Lehighton, 227
Lewis Run, 139
Lewisburg, 217, 219
Lewistown Narrows, 249
Lewistown, 249
Liberty, 129
Lickdale, 199
Licking Creek Valley, 303
Limestone Ridge, 165
Lincoln Highway, 303
Little Conemaugh River, 146
Little Juniata River, 241, 242, 300
Little Mountain, 253
Little Pine Creek, 113
Little Poutz Valley, 248
Little Round Top, 310

Little Toby Creek, 142
Little Tonoloway Creek, 165
Liverpool, 215
Llewellyn beds, 207
Llewellyn coals, 202
Llewellyn formation, 89, 91, 203, 204, 256, 328
Llewellyn sandstone, 106, 203, 258
Lock Haven formation, 110, 111, 127, 130, 152, 233, 243
Lock Haven, 145, 179, 181, 233, 235, 236
Lock Mountain, 241
Loganville basin, 291
Long Point, 103
Long Run Gap, 181
Longhouse, 119
Longshore Drift, 100
Loop Mountain, 241
Lost Creek Ridge, 248
Lower Lake, 97
lower Freeport coals, 167
Loyalhann formation, 242-243
Loyalhanna sandy limestone, 58
Loyalsock Creek, 155
Luthersburg, 143
Lymansville, 115

Macedonia, 249
Mahanoy Mountain, 258
Mahanoy Railroad, 216
Mahantango Creek, 216
Mahantango Falls, 216
Mahantango formation, 127, 169, 174, 193, 201, 221, 224, 247, 327
Mahantango group, 95
Mahantango Mountain, 203, 215
Mahantango shale, 86, 171, 181, 229, 231
Mammoth Spring, 249
Manchester Beach, 101
Mann's Choice, 169
Mansfield, 111, 113, 130
marble, 10
Marcellus formation, 227, 327
Marcellus shale, 86, 193, 201, 229

Markes, 306
Marsh Creek, 113
Marsh-Crooked Creek valley, 114
Martha Furnace, 252
martic zone, 271
Martinsburg formation, 175, 199, 280, 299, 306, 325,
Martinsburg shale, 176, 177, 195, 200, 225, 276, 297, 327
Martinsburg siltstones, 200
Marvin Creek, 117
Marys syncline, 142
Matamoras, 95, 221
Mauch Chunk formation, 33, 49, 83, 86, 92, 129, 167, 205, 207, 208, 211, 242-243, 243, 258
Mauch Chunk group, 48, 57, 58
Mauch Chunk redbeds, 163, 174, 184, 185, 209, 304
Mauch Chunk Ridge, 231, 232, 229
Mauch Chunk rocks, 91
Mauch Chunk, 232
McConnells Mill State Park, 69, 70
McConnellsburg, 303 304, 305
McGees Mills, 143, 145
McKee Gap, 245
McKees Half Falls, 216
Meadow Ground Mountain, 303, 304
Meadow Ground syncline, 303, 404
Meadow Grounds Lake, 304
Meadow Lake, 97
Meadville, 125
Mercersburg, 306
Meshoppen, 212
metamorphic rocks, 8
mid-Atlantic ridge, 24
mid-Atlantic rift, 35
Middle Anthracite Field, 202, 203-204
Miesery Bay, 102
Mifflintown formation, 249, 183
Mifflintown redbeds, 219
Mifflintown, 248, 249
Mifflinville, 183

Milan, 151
Milesburg, 86
Milford Factory, 223
Milford, 95
Millersburg, 261
Millerstown, 247, 248
Mississippi River, 83, 123
modern Atlantic Ocean, 323
Mohawk Valley, 93, 151
Mohnton, 317, 318
Monongahela group, 49, 57, 64, 142
Monongahela River, 49, 51, 67, 125
Monroetown, 152
Montalto quartzite, 282
Montandon sand dunes, 217
Montandon, 217
Montebello sandstone, 247, 248
Montgomery Ferry, 215
Montgomery, 219
Montour Ridge, 183
Moosic Mountain, 105, 211, 212, 213
Moraine State Park, 69, 72
Mount Penn, 318
Mountain Creek valley, 282, 285
Mountour Ridge, 217
Mt. Davis, 167
Mt. Holly Springs, 281, 285
Mt. Holly, 281
Mt. Jewett, 118
Mt. Joy, 272
Mt. Minsi, 195, 225
Mt. Misery, 272
Mt. Pisgah, 232
Mt. Pleasant, 249
Mt. Tammany, 195, 225
Muncy Valley, 153, 155
Muncy, 233, 235
Murrysville antilcine, 61

Nanticoke, 91, 217, 253
Negro Mountain anticline, 167
Negro Mountain. 146
Nescopeck Creek, 184
Nescopeck Landslide, 183
Nescopeck Mountain, 183, 208,

Nescopeck, 183
Nesquehoning Mountain, 232
New Albany, 152, 153
New Baltimore landslide, 168
New Baltimore, 167, 168
New Boston syncline, 204
New Smithville, 276
Newark Basin, 15, 35, 269, 270, 272, 317, 318, 321, 324, 325
Newberrytown, 293, 294
Nittany anticline, 233, 235, 241
Nittany Arch, 127, 219
Nittany formation, 4
Nittany Mountain, 87, 88, 251, 252
Nittany Valley, 86, 87, 88, 179, 235, 236, 252
North American plate, 24
North Bend, 145
North East, 103
Northern Anthracite Field, 105, 209
Northumberland sycline, 253
Northumberland, 145, 217, 253

Ohio River, 51, 66, 67, 75, 83, 123, 125, 131, 142, 143
Oil City, 134
Oil Creek Valley, 38
Oil Creek, 37, 134
oil, 35
Old Port formation, 235, 236
Old Toboggan Run, 294
Onondaga limestone, 201
Onondaga limey shale, 181
Onondaga-Helderberg escarpment, 92, 108
Oriskany formation, 245, 130
Oswayo beds, 140
outcrop patterns of northwestern Pennsylvania, 140

Pacific Ocean, 26
Pacific plate, 26
Paleozoic strata, 14
Paleozoic, 159, 161
Palisades sill, 324
Pangaea, 16, 59, 322, 324

Park Forest Village, 251, 252
Parryville syncline cross-section, 230
Parryville syncline, 227
Parryville, 229, 231, 327
Patchinville, 145
Paxinos, 253
pegmatite, 8
Pennfield, 82
Penns Cave, **88**
Penns Valley, 251
Pennsylvania Canal, 151 216, 247, **253**
Pennsylvania landscape, 11
Pennsylvania slate industry, 325
Penobscot Mountain, 208
Peters Mountain, 215, 247, 299, 300
Pfoutz Valley, 248
Phillipsburg, 84
phyllite, 10
physiographic provinces, 11
Piedmont rocks, 29
Piedmont, 12, 15, 27, 272, 262 270, 291, 293, 311, 317, 321
Pigeon Cove, 165
Pike Township, 143
Pinchot Falls, 223
Pinchot Lake, 293
Pine Creek Gorge, 66, 113, 114
Pine Creek, 115
Pine Grove Furnace State Park, **285**
Pine Run, 328
Pittsburgh coal seam, 40, 41, 49, 55, 64
Pittsburgh formation, 64
Pittsburgh Plateau, 49, 59, 75
Pittsburgh, 61, 125, 134
Pittston, 89, 107
plate tectonics, 23
Pleistocene, 16
Pocono Creek, 190
Pocono formation, 83, 91, 92, 153, 155, 163, 165, 184, 189, 202, 208, 211, 242-243, 303
Pocono Mountains, 125
Pocono Plateau, 96, 189, 190, 191, 213, 328
Pocono sandstone, 48, 174, 186, 203, 215, 216 248, 253, 299, 304
Pocono State Park, 213
Pocono, 96
Pohopoco Creek, 229
Pole Steeple, 282, 284
Port Allegany, 117
Port Jervis trough, 95
Port Matilda, 238, 239
Port Treverton Basin, 216
Port Treverton, 216
Portage Creek, 117
Portersville, 70
Portland, 224
Potato Creek, 117
Potters Mills, 249, 251
Pottsville beds, 105, 111, 114, 130, 143, 146
Pottsville beds, 232
Pottsville coals, 202, 203
Pottsville conglomerate, 78, 201, 204, 205, 207, 253, 258
Pottsville formation, 129, 133, 208
Pottsville group, 57, 77, 79, 84, 86, 89, 123, 143, 204, 209, 242, 256
Pottsville sandstone, 55, 58, 91, 125, 139, 140, 142, 167, 201, 204, 207, 253, 258
Pottsville siltstone, 203
Powys, 127, 129
President, 136
Presque Isle Bay, 102
Presque Isle, 101, 102, 103
Primrose coal, 203
Princeton Station, 70
Promised Land State Park, 96, 97
Proto-Atlantic Ocean, 27, 28, 33, 139, 172, 276, 288, 295, 305, 307, 311, 322
Punxsutawney-Driftwood Gas Field, 83

quartzite, 10
Queenston Delta, 29, 30, 31, 170

Ramapo Mountains, 15
Rays Hill Tunnel, 174

349

Rays Hill, 163
Raystown branch of the Juniata River, 168, 171
Reading Prong, 12, 15 261 276
Red Rock Mountain, 153
Reedsville formation, 172
Reedsville shale, 251
Reedsville, 249
Revolutionary War, 253, 266, 270, 272
rhyolite, 8
Ricketts Glen State Park, 153, 155
Ringdale, 155
River Hill, 183
Roaring Brook Gap, 209
Roaring Brook, 91
Roaring Run Gap, 249
Roaring Spring, 241
Rock City, 123
rock salt, 7
Rockwell formation, 83, 84, 86, 163, 239, 243,
Rockwell sandstone, 165
Rose Hill shale, 249, 250
Rouseville, 134 Rummerfeld, 110

Sabinsville anticline, 81, 130
Salamanca Re-entrant, 117
San Andreas fault, 26
sandstone, 6
Sawkill Falls, 223
Scherr formation, 174
schist, 9
Schuylkill River, 271r, 272, 276
Scranton Antracite Museums, 91
Scranton, 89, 92, 93, 105, 106, 107, 209
Scranton Iron Furnaces, 91
Second Mountain, 201, 202
sedimentary rocks, 4
Seneca Trail, 115
serpentinite, 10
Seven Mountains, 249
Shade Mountain, 249
shale, 6
Shamokin Creek, 253
Shamokin Dam, 216, 217

Shamokin, 253, 255, 256, 257
Sharpsville, 79
Shawangunk conglomerate, 195, 225
Shawangunk formation, 221
Shawangunk Mountains, 95, 195, 223
Shawangunk quartzite, 224, 226
Shawangunk sandstone, 225
Shawmut syncline, 81
Shawmut, 142
Shenandoah Valley, 14
Shenango beds, 140
Shenango formation, 73, 133, 139
Sherman Creek, 253
Shickshinny Mountain, 91
Shickshinny, 89, 91
Shikellamy State Park, 217, 253
Shinnemahoning Creek, 115, 117
Short Mountain Valley, 241
Short Mountain, 245
Sideling Hill Tunnel, 174
Sideling Hill, 163, 165
sills, 8
siltstone, 6
Silverthread Falls, 223, 223
slate, 10
Slippery Rock Creek, 69, 70, 73
Slippery Rock Gorge, 66, 70, 71
Smethport, 117
Snow Shoe, 84, 86
Snydersville, 227
Somerset, 58, 146, 167
South American plate, 24
South Mountain, 12, 15, 177, 261
South White Deer Ridge, 219
Southern Anthracite Field, 201, 202, 203, 232, 229, 247
Spechty Kopf formation, 92, 189, 211
spheroidal weathering, 185, 186
Spring Mountain, 205
Sproul, 245
Spruce Mountain, 250
St. Clairsville, 245
St. Lawrence River, 93
St. Paul limestone, 177
State College Bypass, 251, 252

State College, 247
Steamburg, 123
Stonehenge formation, 4
Stony Ridge diabase dike, 261
strip mining, **53**
stromatolites, **169**, 170, 171
Stroudsburg, 95, 227
Suedberg Fossil Site, 201
Suedberg, 200
Sugar Notch, 208
Sugar Valley Narrows, 181
Sugarloaf Mountain, 208
Sullivan's March, 107, 108
Sunbury, 91, 252, 255
Susquehanna Canal, 215, 217
Susquehanna Railroad, 216
Susquehanna River, 84, 89, 92, 93, 105, 107, 108, 110, 114, 115, 117, 129, 145, 151, 183, 185, 203, 215, 216, 217, 233, 247, 248, 253, 264
Susquehanna River, west branch, 83
Susquehanna system, 142
Susquehanna Valley, 93, 108, 157, 184, 220, 233
Susquehannock State Forest, 115
Swatara Creek, 199, 200, 201
Swatara Gap Fossil locality, 199
Swatara Gap, 199, 200
Sweden Valley, 115
Switchback Railroad, 232

Taconian deformation, 177
Taconian mountain building, 176, 262, 269
Taconian Mountains, 31, 165, 170, 177
Taconian orogeny, 28, 29, 31, 171, 175, 243, 249, 276
Taconic unconformity, 200
Tannersville Peat Bog, 191
Tannersville, 190
Taylor Run, 73
Tethy Sea, 43
Thompsontown, 248
Threemile Island, 263

Tiadaghton, 113, 114
Ticklish Rock, 155, **156**, 157
Tidioute, 136, 137
Timmons Mountain, 175, 176
Tioga Dam, 130
Tioga Junction, 130
Tioga Point, 108
Tioga Reservior, 130
Tioga Valley, 113
Tioga, 113
Tionesta Reservoir, 136
Tionesta, 134, 136
Titusville, 37, 71, 131, 134
Tivoli, 155
Tonoloway formation, 245, 248, 249
Tonoloway limestone, 171
Tonoloway Ridge, 165
Tonoloway, 243
Twanda anticline, 110
Towanda Creek, 152
Towanda formation, 110
Towanda, 108, 111, 151, 152
Town Hill, 163
Tremont syncline, 203
Treverton Railroad, 216
Triassic and Jurassic dikes, 261
Triassic basin, 269, 263
Triassic diabase sill, 266
Triassic lowlands, 12, 15, 16, 268
Triassic redbeds, 263
Trimmers Rock beds, 153
Trimmers Rock formation, 92, 95, 155, 169, 247
Trimmers Rock sandstone, 168, 183
Trimmers Rock shales and silstones, 168, 181, 183-184, 229
tropical storm Agnes, 184
Tully limestone, 86, 169
Tunkhannock Creek, 106
Tunkhannock formation, 110
Tunkhannock, 105, 106, 107, 108
Tunungwant Creek, 139
turbidite, 6
Tuscarora conglomerates, 201
Tuscarora formation, 4

Tuscarora Mountain, 175, 248
Tuscarora quartzite, 87, 95, 127, 170, 171, 172, 176, 181, 183, 195, 200, 201, 215, 217, 219, 220, 235, 238, 241, 245, 248, 249, 250, 252, 276
Tuscarora sandstone, 29, 31
Tuscarora talus, 86
Tuscarora Tunnel, 174
Tuscarora Valley, 248
Tussey Mountain, 172, 251
Tyrone Gap, 241
Tyrone, 239, 242

unglaciated landscape, 117

Valley and Ridge province, 11, 13, 14, 33, 45, 47, 51, 57, 86, 107, 127, 159, 161, 162, 168, 175, 176, 181, 188, 197, 201, 213, 215, 219, 233, 238, 239, 240, 242, 243, 247, 261, 262
Valley Creek, 272
Valley Forge National Historic Park, **271**
Valley Forge, 272
Vanport limestone, 71
Venango formation, 75, 119, 136, 137

Wallenpaupack Creek, 97
Wallpack Bend, 221, 223
Wambaugh Run, 168
Wapwallopen, 185
Ward's Pit, 41
Warfordsburg, 165
Warren, 117, 119, 122, 125
Warrendale, 55
Warrior limestone, 172

Washington formation, 49, 59
Waterville, 113
Waves, **100**
Waynesboro formation, 51, 49, 59
Wellsboro anticline, 111, 130
Wellsboro, 113
West Liberty Esker, 69, **72**
West Sunbury, 69
west branch of the Susquehanna River, 179
Western Anthracite Field, 202, 203-204
Western Middle Anthracite Field, 216, 253, 255
Westmoreland County, 55, 64
White Deer Creek Valley, 181
Wilkes-Barre Mountain, 208
Wilkes-Barre, 89, 107, 208, 217, 253
Williamsport, 117, 145, 157, 215, 217, 219, 220, 233, 235, 238, 253
Wills Creek formation, 51, 171, 183, 195, 248
Wills Creek shale, 161, 165, 167, 175, 181
Wills Mountain anticline, 170
Wills Mountain, 245
Wilmot anticline, 153
Winfield, 217
Winona Falls, 223
Winter of Despair, 271
Woodland, 84
Wurtenburg, 70
Wyalusing Rocks, 108
Wyalusing, 110

York Haven Dam, 263
Youghiogheny River, 49, 51

We hope you enjoy this title from Echo Point Books & Media

Visit our website to see our full catalog and take 10% off your order total at checkout!

Arctic Adventure

Peter Freuchen

Freuchen relates tales of polar bear hunts, of meeting Eskimos who had resorted to cannibalism during a severe famine, and of the thrill of seeing the sun after three months of winter darkness. In this memoir, he writes about the Inuit with genuine respect and affection.

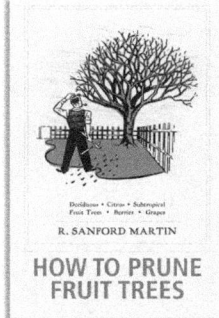

How to Prune Fruit Trees

R. Sanford Martin

While the act of pruning is simple enough, knowing where and when to prune can confound even experienced gardeners. For more than half a century, Robert Sanford Martin's *How to Prune Fruit Trees* has been the go-to guide for pruners of all levels of expertise.

Freshwater Aquaculture

William McLarney

William McLarney, scientist and pioneer in the field, describes every aspect of aquaculture, from the underlying scientific concepts to step-by-step instructions for each type, size, and phase of culture in this definitive guide.

Fruits of Warm Climates

Julia F. Morton

In one definitive volume, Morton explores the world of tropical and subtropical fruit, providing information on the history of the plants, cultivation techniques, food and alternative uses, nutrition, varieties, and much more.

Vermont Nature Guide

Sheri Amsel

With descriptions and full-color illustrations of more than 260 species of birds, mammals, insects, reptiles, amphibians, wildflowers, and trees, this little book is the perfect Vermont companion.

The Spotted Hyena

Hans Kruuk

Kruuk redefines the image of the spotted hyena, not as a common scavenger, but as a complex matriarchal predator with links to human evolution. This was the first study to capture the true behavior and ecology of these formidable predators.

Buy direct and save 10% at www.EchoPointBooks.com

DISCOUNT CODE: PENNSYLVANIA

www.ingramcontent.com/pod-product-compliance
Lightning Source LLC
Chambersburg PA
CBHW082144230426
43672CB00015B/2838